Practical estuarine chemistry
A handbook

Estuarine and brackish-water sciences association handbooks

Practical estuarine chemistry

A handbook

Edited by

P. C. HEAD

North West Water Authority, Warrington, England

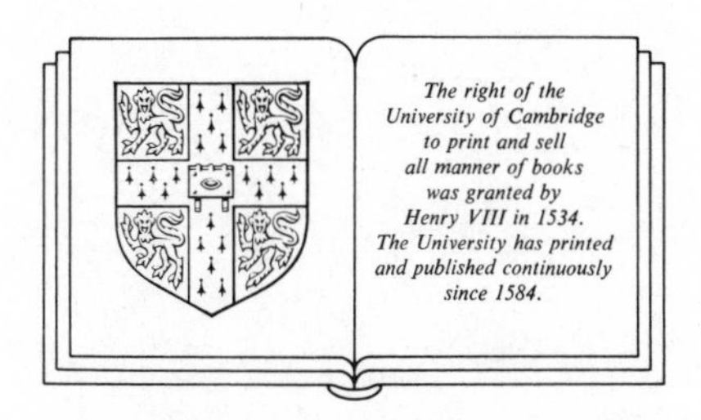

CAMBRIDGE UNIVERSITY PRESS

Cambridge

London New York New Rochelle

Melbourne Sydney

Published by the Press Syndicate of the University of Cambridge
The Pitt Building, Trumpington Street, Cambridge CB2 1RP
32 East 57th Street, New York, NY 10022, USA
10 Stamford Road, Oakleigh, Melbourne 3166, Australia

First published 1985

Printed in Great Britain by the University Press, Cambridge

Library of Congress catalogue card number: 84–21368

British Library Cataloguing in Publication Data

Practical estuarine chemistry: a handbook.
1. Estuarine oceanography 2. Chemical oceanography
I. Head, P. C.
551.46′09 GC 97

ISBN 0 521 30165 3

UP

Contents

Contributors

S. R. Aston, International Laboratory of Marine Radioactivity, Oceanographic Museum, Monaco

P. C. Head, North West Water Authority, Warrington, UK

T. M. Leatherland, Forth River Purification Board, Edinburgh, Scotland

A. W. Morris, Institute for Marine Environmental Research, Plymouth, UK

M. Whitfield, Marine Biological Association of the United Kingdom, Plymouth, UK

P. J. leB. Williams, Department of Marine Microbiology, University of Gothenburg, Gothenburg, Sweden

Preface

This book is concerned with the 'how' and 'why' of estuarine chemistry. The authors have drawn on their considerable experience of working in estuaries to provide guides to various aspects of estuarine chemistry which are designed to explain how and why particular types of information are collected. Although there is no substitute for experience and we must all, to a certain extent, learn from our mistakes, it is hoped that the information presented will increase the reader's understanding of estuarine processes to an extent which will greatly reduce the chances of the more obvious mistakes being repeated. From the information provided the reader should be in a position to evaluate the techniques available for collecting and interpreting the data relevant to a particular problem but will need to refer to the original publications for many of the details of particular methods. Although concentrating on the practicalities of data collection and interpretation the book should also provide an introduction to estuarine chemistry suitable for the advanced undergraduate and postgraduate teaching courses.

This is the second of a series of handbooks sponsored by the Estuarine and Brackish-Water Sciences Association. The first handbook dealt with aspects of estuarine hydrography and sedimentation and a further volume dealing with estuarine biology is in preparation. Although originally planned at the same time as the handbook on hydrography and sedimentation this volume was subject to a number of delays, including the withdrawl of the original editor from the project, which meant that the first set of manuscripts needed to be completely revised before publication was possible. Most of the revision was carried out during 1982 and 1983 but one of the chapters has only received minor modifications since it was completed in 1980. I am convinced, however, that these unfortunate delays have not decreased the relevance of the book and that it will fulfil the need identified when it was planned. Indeed, the revisions have allowed recent developments to be assessed and the fruits of even more experience incorporated.

I should like to thank the authors for the time and effort they have put in to making, what must have seemed to have been a lost cause, the useful volume I believe this to be.

Finally I should like to thank my wife for allowing me to become involved in this apparently never ending saga and also for typing and retyping various sections and preparing the index.

Warrington, May 1984 P. C. HEAD

1

Estuarine chemistry and general survey strategy

A. W. MORRIS

This chapter is intended firstly to provide a description of the important characteristics of estuarine systems, a knowledge of which is an essential prerequisite to attempting chemical investigations. Secondly, the application of such information to the planning and execution of estuarine chemical studies and to the recognition of the practical limitations involved, together with their implications with respect to the results and conclusions which may be obtained, are discussed.

It is impossible in such a short account as this to include comprehensively and in detail all the individual aspects of estuaries which may be pertinent to the study of any particular process occurring in any individual estuary. This treatment is therefore limited, by and large, to general considerations but it is hoped that sufficient reference material has been provided to enable the reader to consolidate and expand this framework as necessary. To this end, a short bibliography of informative texts covering theoretical and practical aspects of the chemistry, physics and biology of estuaries has been included at the end of this chapter.

There are no chemical (or physical and biological) processes which are unique to estuaries. The individuality of estuarine chemistry lies in the way in which the functioning and extent of natural chemical processes are regulated and modified within this particular environment. Certainly, the results of chemical determinations in estuaries cannot be fully interpreted without a knowledge of the physical, geological and biological processes occurring at the position and time of sampling, for the complexity of estuaries is such that any instantaneous product of these processes and their interactions, for example, the synoptic distribution of a particular chemical constituent, is unlikely to be exactly reproduced again. Classification systems which describe idealized estuarine systems are useful for clarifying the basic characteristics which exert an overall control of the transports, distributions and reactions of chemical constituents within estuaries, but it must always be remembered that idealized descriptions of estuaries provide only an approximation to any real situation.

Until the last few years, practical studies of estuaries have necessarily been analysed under the assumption of a steady-state condition, or as a series of steady-state conditions, whereas the true picture is one of continuous variability in response to the highly dynamic nature of the physical system. Bowden (1967) considered that the development of adequate methods of dealing with the variable state is one of the outstanding requirements for studying problems of estuarine circulation. Logically, one can extend this requirement to the study of all other estuarine properties, including chemical state and reactivity, which are influenced by the continuous movement and mixing of waters. However, recent developments in computer technology and corresponding mathematical techniques now allow the construction of realistic simulations of complex estuaries incorporating the effects of all known (or assumed) processes and giving almost unlimited resolution in both time and space. Advances in this direction have been so rapid that the availability of high quality field data together with quantitative data on the mechanisms and rates of estuarine processes commensurate with the capabilities of such powerful tools, are now emerging as essential and immediate requirements of estuarine chemical research. Such needs will not readily be satisfied, for they demand considerable improvements in methodology and investigational techniques applied both to field studies and to laboratory investigations and simulations of chemical processes of importance.

One of the fascinating problems in studying these complex and highly dynamic systems is that their nature necessarily demands the use of specialized and sophisticated investigational techniques yet, at the same time, severely limits the methodologies and practical approaches which can usefully be employed. Considerable attention is paid in the following discussions to the continuous variability, in both time and space, of the chemistry of estuarine systems, for it is this property, above all, which determines the strategy to be applied during field investigations.

Basic characteristics of estuaries

Definitions

Pritchard (1967) has reasoned that an estuary can best be defined as follows:

> *An estuary is a semi-enclosed coastal body of water which has a free connection with the open sea and within which sea water is measurably diluted with fresh water derived from land drainage.*

This wording was carefully designed to include only those systems which have a basic similarity in their distributions of salinity and density and, furthermore, to indicate that these distributions are the consequence of characteristic circulation patterns and mixing processes which are strongly influenced by lateral boundaries. It is also implicit that the connection with the open sea must be sufficiently free to allow continuous transmission of both tidal energy and sea salt.

The definition of an estuary has been discussed in detail by Fairbridge (1980). He argues that the tidally affected fresh-water region should be considered an integral part of any estuary. Accordingly he has proposed this definition:

> *An estuary is an inlet of the sea reaching into a river valley as far as the upper limit of tidal rise, normally being divisible into three sectors:* a. *a marine, or lower estuary, in free connection with the open sea;* b. *a middle estuary, subject to strong salt- and fresh-water mixing; and* c. *an upper or fluvial estuary, characterized by fresh water but subject to daily tidal action.*

Not surprisingly, any definition of an estuary must remain something of a compromise; for example, there is no sharp discontinuity within the continuum of descriptions ranging from what is clearly an estuary to what is clearly a lagoon. These two definitions, however, go a long way towards identifying those areas which may be grouped together for scientific purposes. They embrace the majority of brackish-water regions, covering coastal plain estuaries (drowned river valleys and bar-built estuaries), fjords, certain gulfs, sounds and inlets and embayments formed behind offshore bars provided they have a salinity significantly lower than the open sea (Bowden, 1967). All these are often collected under the general term positive estuaries. Negative estuaries, in which evaporation exceeds the combined effects of river run-off and precipitation producing hypersaline conditions, and the rarely encountered intermediate system, where fresh-water inputs are balanced by evaporation, are excluded although these may behave intermittently as positive estuaries.

An important aspect of estuaries which is not highlighted by the definitions is the extent to which estuaries influence, and are influenced by, events occurring at considerable distances both seaward and landward from the defined area. For example, tidal oscillations in the estuary of the Columbia River have been found to reverse the river flow as far as 53 miles upstream and to produce measurable fluctuations in water level at 140 miles even though the maximum intrusion length of saline water is usually less than 23 miles (Neal, 1972). The seaward limit of an estuary, defined by

Pritchard in practical terms of measurability, is not precisely locatable, for in this region only gradual transitions in properties are encountered.

Although the definitive criteria draw together coastal sea-water regions with many common characteristics, the essence of positive estuaries undoubtedly lies in the continuous variability in the balance of physical factors which determine individual estuarine hydrodynamics. River flow, tidally induced movements, precipitation and evaporation, wind stress and the distribution of sediments are all fluctuating continuously. Human activities such as dredging, land reclamation, bridge and sea-wall construction and flood control measures impose further instability. In these circumstances, it is illogical to consider estuaries as steady-state systems whatever the time scale used. Generalized interpretations of estuarine behaviour in time-averaged terms should therefore necessarily include an indication of the degree of variability involved, and, if possible, the relevant values of the more important controlling parameters.

Tidal motion

For practical purposes in planning field investigations and in the interpretation and presentation of the results obtained, a knowledge of tides in the estuary under consideration is essential. Fortunately, the importance of tidal motion and elevation to navigation within coastal waters has resulted in the accumulation of sufficient tidal information from which the tidal characteristics of most estuaries are accurately predictable.

The period and amplitude of the tidal wave within an estuary are directly governed by the characteristics of the tide in the adjacent open sea. The volume of water entering and draining at each cycle (the tidal prism) and the corresponding strength of the tidal currents vary in response to the tidal amplitude at the mouth, giving rise to spring and neap conditions varying continuously but cyclically throughout the year. In the absence of complicating hydraulic factors, tidal motion within estuaries is confined by the lateral boundaries to axial flow. This flow at any point in an estuary is often expressed as the tidal excursion, which can be defined as the distance a particle of water moves along the axis of the estuary during the time between high and low water. Generally, tidal amplitudes and velocities decrease landwards as the tidal energy is dissipated by friction. This pattern may however be perturbed by sharp changes in the dimensions of the estuary; rapid shoaling or narrowing can markedly increase tidal amplitudes and velocities and, consequently, the length of the tidal excursion. As the head of the estuary is approached, the strength and duration of ebb flow increasingly predominate over flood flow through river flow augmentation, with corresponding changes in the timing of high and low water.

Frictional and turbulent dissipation is often insufficient to degenerate completely the progressive tidal wave within the length of an estuary. Reflections of the wave at the upstream end can occur, producing standing waves. Most estuaries have a tidal response which exhibits both standing and progressive contributions. The relative importance of these contributions influences the timing of high and low water relative to tidal

Fig. 1.1. The tidal response of estuaries. From Dyer (1973) reproduced by permission of John Wiley & Sons Ltd.

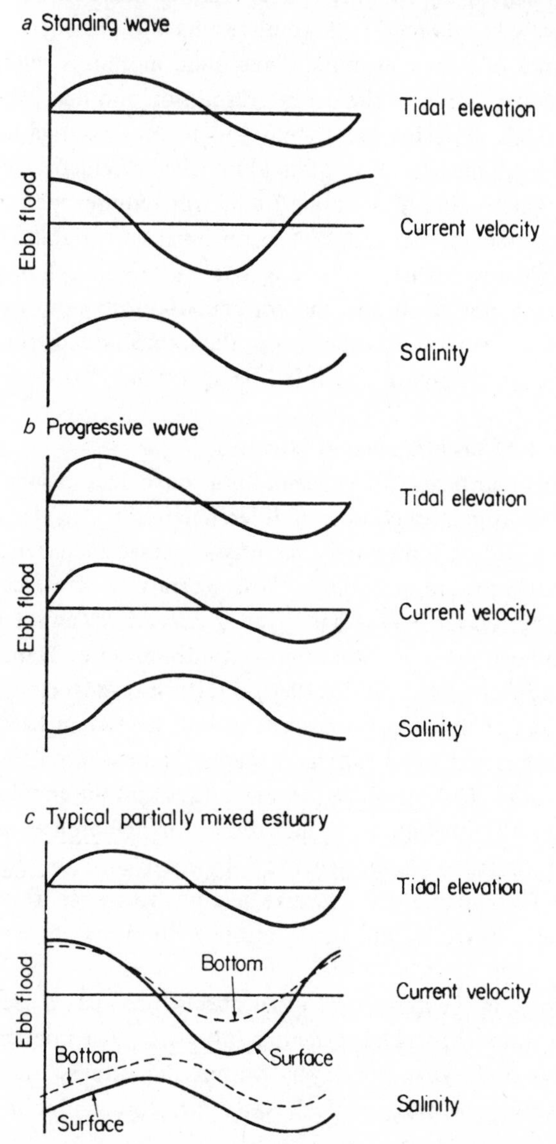

conditions at the mouth and also the strength and timing of the ebb and flood currents. The tidal responses in an estuary characteristic of standing or progressive wave contributions and their combination result in a typical partially mixed estuary as shown in Fig. 1.1.

The oscillatory tidal advection of water does not directly influence the dispersion of water and the corresponding distribution of dissolved estuarine components. For the purpose of many estuarine studies, particularly those concerned with processes confined to the bulk aqueous phase and of effective duration appreciably longer than a tidal period, tidal motion can effectively be ignored in recording results, but not of course in terms of the logistics of their collection. Thus, tidal motion is relatively unimportant to the derivation of the overall dispersion and flux of water and materials through the estuarine system and to the investigation of longer term cyclic phenomena such as annual biological production. Other studies however, will be tidally orientated and will require operational investigations which include tidal motion as an intrinsic variable. These include, for example, investigation of changes in bottom morphology in response to sediment movement and the determination of water quality characteristics at fixed geographical positions, the immediate distribution of material around the vicinity of an outfall being a typical example.

Mixing and the classification of estuaries

For practical purposes, it has been found convenient to subdivide positive estuaries according to a number of different criteria. Useful classifications have been based on topography, salinity structure and circulation pattern, a stratification parameter and, most recently, a stratification/circulation parameter. Dyer (1973) has given a concise review of these schemes. The stratification/circulation treatment intruduced by Hansen & Rattray (1966) and developed by Smith (1980) and Rattray & Uncles (1983) is probably the most comprehensive treatment, but the system based on non-tidal circulation pattern and salinity structure is most useful for the purpose of this treatise. This classification recognizes that the circulation, and hence the salinity distribution, in any estuary is determined by the extent to which turbulent mixing, generated by tidal currents, acts on the river inflow. This mixing gives rise to a diffusive transport which controls the overall dispersion and distribution of fluid materials entering an estuary from any source.

River inflow alone, in the absence of other effects, produces a seaward transport of fresh water superficial to denser salt water. The interactions between the fresh- and salt-water layers within the estuary result in characteristic mixing processes, vertical water structure and non-tidal circulation patterns which are driven by the density difference between them. Density is predominantly controlled in estuarine waters by the salt content; temperature differences are relatively ineffective in comparison with salinity differences in their influence on estuarine dynamics and are usually ignored.

There are two quite distinct processes involved in the vertical transport and mixing of fresh and salt waters across the interface between them; namely entrainment and turbulent or eddy diffusion (see Dyer, 1973; Bowden, 1980). Mixing in estuaries is produced by a combination of both. The relative degree of mixing induced by each process depends on the degree of turbulence in the two layers and, to a first approximation, is a function of tidal amplitude. Generally, where river flow predominates, entrainment is the predominant process; the greater the tidal amplitude, the more eddy diffusion will be effective.

The variable mixing conditions across the interface and the consequent circulation and vertical salinity profile which are induced, give rise to a recognizable sequence of positive estuarine types on which a classification system can be based. This sequence is outlined in Table 1.1, using the nomenclature introduced by Pritchard (1955) and illustrated schematically in Fig. 1.2. Type B, the partially mixed estuary, is by far the most commonly encountered. The important physical parameters which determine the position of any estuary within this classification are the volume and rate of fresh-water inflow, the strength of the tidal currents and depth. The way in which changes in each of these properties tend to influence the position of any estuary within the sequence is noted in Table 1.1. This combination of features can be represented numerically in a simple way by the ratio of the volume of water flowing up the estuary through a given section during the

Fig. 1.2. Estuarine circulation patterns, isohaline structure and typical vertical profiles of salinity and residual velocity in mid-estuary.

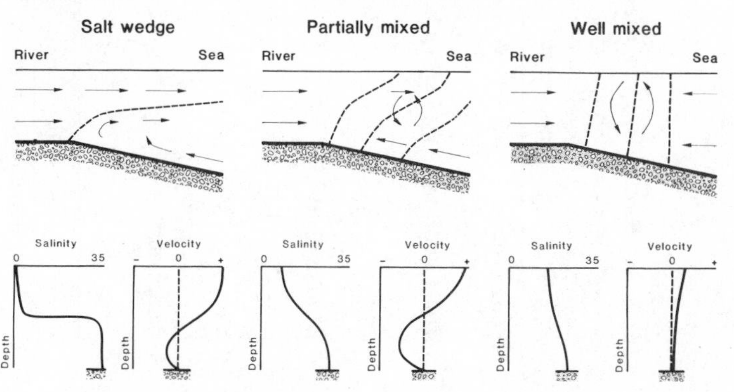

Table 1.1. *Sequence of estuaries according to mixing and circulation pattern*

Estuarine type description	Highly stratified salt wedge	Fjord	Partially mixed	Vertically homogeneous	
Mixing process	River flow dominant	River flow with entrainment	River flow with tidal mixing	Tidal mixing dominant	
Vertical structure	Very sharp halocline	Sharp halocline; deeper layering	No marked interface; increasing salinity from surface to bottom	Zero or very small vertical differences in salinity	
Nomenclature (Pritchard, 1955)	A		B	D Without lateral salinity gradient	C With lateral salinity gradient
Ratio[a]	~ 1		10–100	~ 1000	
Effect of physical parameters			decreasing river flow → decreasing depth → increasing width → increasing tidal velocity →		

[a] $\text{Ratio} = \dfrac{\text{volume of water flowing up the estuary through a given section during the flood tide}}{\text{volume of fresh water flowing into the estuary above the section during a complete tidal cycle}}$

flood tide (the tidal prism) to the volume of fresh water flowing into the estuary above the section during a complete tidal cycle (Pritchard, 1967). This ratio ranges from approximate unity for a salt-wedge type A estuary to values of the order of 10^3 in vertically homogeneous type C or D estuaries.

The essential difference between types C and D is a function of width. Type D is relatively narrow and deep so that appreciable lateral stratification is not attained. The direction of water movement is symmetrical about the longitudinal axis, and the net flow is seaward at all depths, with a diffusive flux of salt towards the head of the estuary maintaining the salt balance. In the type C, vertically homogeneous system, with sufficiently large width to depth ratio, Coriolis Force induces lateral inhomogeneity such that, in the northern hemisphere, water on the right-hand side of an estuary, looking seaward, is of lower salinity. There is a net seaward flow of water on this side which is compensated by a flow of higher salinity to the left-hand side. A horizontal cyclonic circulation system is produced.

Conformity by any estuary to a single type description may well be limited both in areal extent and in time. Irregularities in shape, variability in bottom roughness, and subsidiary fresh-water inflows all produce localized distortions to regular circulation patterns and salinity distributions. Tidal periodicity and variability in climatic conditions may also exert temporal disruptions to conformity to a single classification. The Columbia River Estuary well illustrates this point (Neal, 1972). Near the mouth, the estuary is of type B, but may become type D during periods of low river discharge and spring tides. At a point further upstream, type B is the predominant form but periods of high river flow induce type A behaviour, or even riverine, fresh-water conditions for very high flows. Horizontal salinity gradients also occur in the estuary, but are not sufficiently pronounced to justify a type C classification.

Knowledge of the basic type of estuary under consideration, and the extent of its variability in mixing and circulation patterns in response to extremes of environmental control is obviously an essential requirement for the planning of *in situ* estuarine investigations. Estuaries which approximate to vertical homogeneity in their salinity structure can often be adequately characterized with respect to many constituents by sampling at a single depth; other estuarine types and certain properties of all estuaries (e.g. suspended particulate load) will require sampling throughout the depth profile.

The recognition of conformity to a particular type, even over restricted time intervals, does not imply complete regularity in gradients of water properties either with depth or along the length of the estuary according to that type. Such simplification is representative only of time-averaged conditions; in the short-term, small-scale inhomogeneity about mean conditions is a fundamental property of the turbulent mixing process involving eddies which is properly described only in probability terms. Lewis & Stephenson (1975) have discussed this point in relation to pollution problems in estuaries. On a larger scale, irregular boundaries, subsidiary stream inflows and effluent discharges can produce discrete patches of

Fig. 1.3. *a* Tidally dominated salinity variations at an anchor station in the Severn Estuary (51° 20.2′ N, 3° 8.0′ W). *b* Axial profile of salinity in the Tamar Estuary at neap tide. In each case, irregularities about the regular trend are highly significant. These are probably generated by temporary isolation of water parcels from the main tidally oscillating stream through topographical effects, e.g. in bays, tributary junctions and subsidiary channelling.

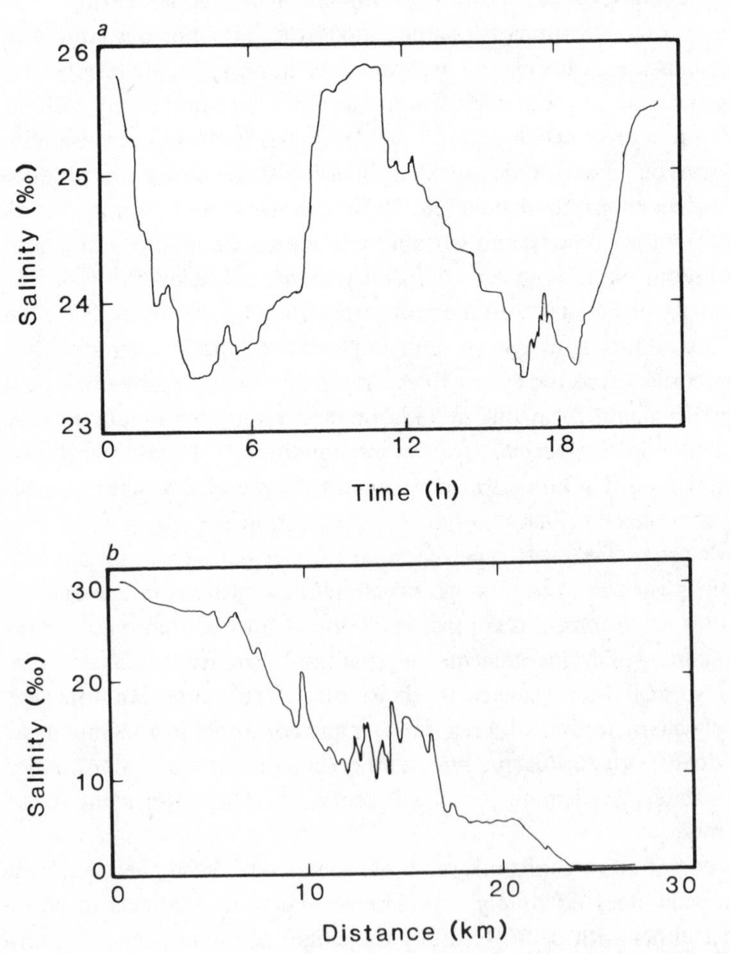

water which maintain singular characteristics for appreciable time intervals (see Fig. 1.3; Garvine & Monk, 1974; Perkins, 1974; Morris *et al.*, 1981).

Quantification of the mixing processes and mathematical descriptions of estuarine circulation and related phenomena are based on applications of the principles of salt balance and dynamic balance. Equations expressing the continuity of volume of water and mass of salt and the balance of forces acting to produce motion, can be derived for any element of water within the system. Detailed reviews of this topic have been given by Bowden (1967; 1980) and Dyer (1973). Festa & Hansen (1976) and Hess (1976) have developed numerical models of estuaries in two- and three-dimensional forms, respectively. Such treatments are strictly applicable only to total salt, the conservative behaviour of which is a basic tenet (see pages 24 and 312).

Dissolved constituents and the finer, colloidal material are transported in a similar manner to the total salt and may also behave conservatively, so that their estuarine distributions can be deduced directly from a knowledge of the salinity or fresh-water distributions, together with the constituent inputs. Any *in situ* process which alters the concentration of a constituent relative to the total salt content imparts non-conservative behaviour to that constituent; components subject to such processes are sometimes referred to as reactive or interactive. Modelling the mixing and transports of reactive components requires the addition of appropriate reaction kinetic terms to the continuity equations. There has been little progress in the development of realistic simulations of this type due to a lack of reliable kinetic data and, especially, of data which adequately cover the temporal and spatial variability in the rates of estuarine chemical reactions in response to changing environmental conditions.

Suspended material and sediments

The suspended particulate and superficial sedimentary materials present in an estuary take part in a considerable variety of processes which affect the chemistry of the system. Numerous examples of complex, non-biological and often poorly characterized interactions between the dissolved and particulate components of estuaries have been described. Abiological transfers of dissolved silicate (Liss & Spencer, 1970) and boron (Liss & Pointon, 1973) by sorption onto suspended particulate material; dissolved phosphate exchange with bottom sediments (Pomeroy *et al.*, 1965); the desorption of trace elements from river-borne particles on entering estuaries (Kharkhar *et al.*, 1968); the complex behaviour of artificial radionuclides released to the estuarine system (Forster, 1972);

salinity-controlled humic acid solubility (Hair & Bassett, 1973) and coagulation of riverborne microcolloids (Sholkovitz, 1976) represent just a few examples.

Origins and nature

Estuarine particulate materials can be classified under three main categories indicative of their origin; lithogenous, hydrogenous and biogenous. Lithogenous particles are inorganic and are derived from erosive weathering of crustal material. They occur as discrete mineral grains usually of a crystalline nature and are transported into the estuary by the influent river and also from the adjacent coastal region; there may also be an appreciable atmospheric contribution. Lithogenous material is often subdivided into an essentially unreactive detrital phase, comprising the bulk crystalline structure, and a reactive phase, comprising associated sorbed constituents which are therefore subject to exchange and transfer reactions with their ambient environment. Hydrogenous particulate material is generated within the aquatic phase and may occur as a coating on lithogenous particles or as a discrete phase. Iron, and other hydrous metal, oxide precipitates and humic aggregates are typical *in situ* products.

Biogenic particles are generated by biological processes occurring within the estuary, together with transported contributions from the adjacent land, the fresh-water sources and the adjacent coastal marine environment. Typical biogenic constituents include living biota and their dead and decaying remains, faecal pellets and terrestrial plant debris. Besides these natural components, anthropogenic products of industrial and domestic waste disposal such as coal dust, sewage solids and plastic particles are not uncommon in some estuarine regions.

The suspended and sedimented particle population at any point in an estuary comprises a heterogeneous mixture of these components. Moreover, individual mineral grains are often partially or totally encrusted with amorphous, inorganic phases together with inanimate organic material and living organisms. Loose aggregates, which consist of an amorphous matrix in which are embedded a wide size range of lithogenic particles, often abound (Johnson, 1974; Pierce & Siegel, 1979). The flocculation processes which bring about the formation of these aggregates are not well understood (Hahn & Stumm, 1970; Kranck, 1973; Edzwald *et al.*, 1974).

A more discriminative chemical characterization of estuarine particulate materials than that provided by bulk analysis is not readily obtained. A number of chemical leaching schemes have been developed in attempts to quantify the various chemical phases and/or physico–chemical states comprising a bulk sediment or suspended particle population. Inevitably, such methods are only partially specific, to an extent that the operational basis of such separations should always be stressed (Luoma & Bryan, 1981). Nevertheless, the results are often considerably more useful than bulk analyses for assessing transport pathways and reactivity of chemical species within estuaries.

Population properties and reactivity

Whatever their nature, suspended particulate materials are present in estuaries in sizes which range from the point of solution, through the colloidal state, to discrete visible entities. It is customary, and satisfactory for many purposes, to use an arbitrary distinction between the particulate and the dissolved state which is determined by the pore-size of the filter which is used for separation (usually 0.45 μm). However, the majority of interactive chemical processes occurring in estuaries are essentially surface phenomena, taking place at the solid–aqueous interface. The particle size distribution, through its control of total available surface areas, is a fundamental controlling factor in such reactions. The microcolloidal fraction ($< 1\ \mu$m) contributes a practically negligible fraction of the total weight of suspended material in estuaries, but it is numerically in abundance and, importantly, it constitutes a significant proportion of the total particulate surface area. At present, little is known of the chemical and physical characteristics of this component although recent observations (Sigleo & Helz, 1981) have emphasized its distinctive nature.

A predominance of evidence (Bayne & Lawrence, 1972; Neihof & Loeb, 1972, 1974; Myers *et al.*, 1975; Hunter & Liss, 1979, 1982; Hunter, 1980) indicates that all estuarine suspended particles, regardless of composition, carry a net negative charge, the magnitude of which is responsive to the ionic composition of the ambient medium. Moreover, at any point in an estuary, the electrophoretic mobilities of particles fall within a very narrow range of values. It is probable that this remarkable electrokinetic uniformity is due to ubiquitous adsorption of organic material. It follows that electrostatic repulsions between particles are also independent of their basic composition. Hunter & Liss (1982) point out that this can account for the lack of evidence for differential flocculation in estuaries which was remarked by Meade (1972). A corresponding unification of the chemical reactivity of particle surfaces is not a necessary corollary. However, experiments (Meyers & Quinn, 1974; Hunter, 1980; Lion *et al.*, 1982) have shown that adsorbed organic coatings can strongly influence adsorption and exchange processes at particle surfaces. In contrast, Sayles & Mangelsdorf (1979) found that the cation exchange properties of suspended particles in the Amazon River catchment and estuary were compatible with their constituent clay mineral suites, with no evidence that the organic content had any significant effect. The extent to which chemical interactions between particles and the aquatic phase are regulated by surface attached organic material requires further investigation.

Presently, we must accept that each of the constituent phases of particulate material has characteristic mechanisms of material exchange with the soluble phase, the rates of which are modified according to their physico–chemical state and the degree of mechanical entrapment involved. Appreciation of the reactions and processes involving particulate material requires characterization of the material with respect to the physico–chemical state and stability of each of its constituent phases. A bulk particulate material should not be considered as a single uniform entity.

A further problem in studies of the reactivity of particulate materials arises from the variability in both time and space of the total particulate loading and of its constituent chemistry and mineralogy. Determinations of individual chemical constituents of the bulk particulate load are therefore subject to a degree of both sampling and analytical variability which together often severely limit their significance. Hence, interelement ratios may be more informative than total element concentrations. More specifically, the constituents of suspended particulates, and of sediments, should ideally be determined for separate particle size fractions; a major proportion of the variability in bulk sediment chemistry of a region is often attributable to differences in particle size spectrum alone.

Distributions and transports

Particulate material within estuaries is transported both as bed load and in suspension, the suspended material being subject to the transient states of turbulent diffusion and vertical settling. The settling velocity of any particle will depend on its size, morphology and specific gravity relative to the specific gravity of the surrounding water. Conomos & Gross (1972) have collated information on the settling rates of various particulate types and the extent to which the rate is influenced by salinity or temperature.

The erosion, suspension and depositional cycles in the transport of both suspension and bed load alternate with tidal motion, and the resultant net transport is governed by the time-averaged velocities within the tidal cycle (Ippen, 1966). There is some inertia in mobilization of settled sediment. Bed load will be moved initially only when a minimum velocity is exceeded, but the particles may then be transported at lower velocities. In general, an individual particle may be carried back and forth under the influence of both tidal and residual estuarine motion and deposited and eroded many times before it settles permanently or for a long period. During transport and after deposition, sedimentary material may undergo changes as a result of chemical and biological processes. Even when permanently settled, sedimentary material should not be considered as unreactive with respect to the water column. Recent work has shown that appreciable fluxes of dissolved nutrients (Rowe *et al.*, 1975) and trace metals (Elderfield & Hepworth, 1975) may traverse the sediment/water interface as a consequence of diagenetic alterations within the sedimentary column of nearshore sediments. Indeed, Rowe *et al.* (1975) have contended that coastal sediments are a tightly coupled component of the nearshore ecosystem, and that the function of the shallow sea floor in nutrient regeneration has been underestimated, and may represent a highly significant proportion of the nutrient supply for inshore primary production.

High non-detrital levels of a particular constituent in the sediments of an estuary do not necessarily indicate that the sediments are acting as a trap or sink for that constituent. The potential extent of a sink is dependent on whether or not total sediment is accumulating and also on the relative rates of sorption and desorption reactions involving the constituent in exchange between the aqueous phase and the bulk sediment. The key to understanding chemical retention phenomena within estuaries is not through a description of standing levels, but from a knowledge of particulate sedimentary and transport processes and the rates of attainment of, or movement towards, the corresponding positions of heterogeneous chemical equilibria.

Fig. 1.4. Suspended sediment variations in the outer Severn Estuary (51° 20.2′ N, 3° 8.0′ W). Resuspension of bottom material at higher tidal velocities is evident.

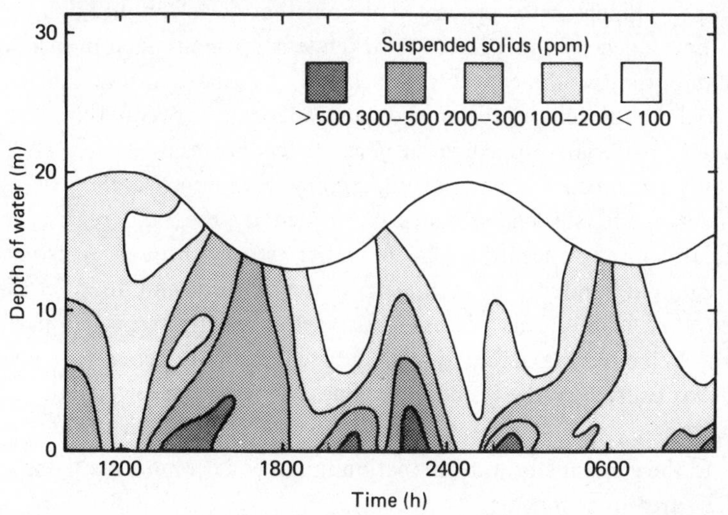

The continuous periodic resuspension and deposition of particulate materials, especially in estuarine regions where tidal currents are strong, can produce extremely high suspended sediment loads (see Fig. 1.4). Average suspended loads are normally greater than can be maintained by the river-borne supply and there is a correspondingly low residual seaward transport. Instantaneous values of suspended load are closely coupled to tidal-stream velocity and therefore show characteristic semi-diurnal and spring–neap variability. In high energy systems, the sediment–water interface may at times become indistinguishable, with only a gradual transition in bulk density between the two phases (the fluid mud phenomenon).

The estuarine turbidity maximum

Residual, non-tidal, estuarine circulation can alone produce a turbidity maximum within the low salinity region of salt-wedge and partially mixed estuaries. Particles introduced into the estuary by river water, or by any other means, will, on average, be transported seaward. Sinking, however, tends to transfer them to the lower, landward-moving more saline water which also carries material of marine origin upstream, but is counteracted by upward entrainment and eddy diffusion. An individual suspended particle can be carried back and forth along the estuary before escaping this residual circulation. The net result is a sediment trap wherein particles of either fresh or marine origin may be circulated many times, thus producing a turbidity maximum. This process is selective with respect to particle settling velocity. Typical examples of this phenomenon have been described by Postma (1967), Schubel (1968), D'Anglejan & Smith (1973) and Gallene (1974). Festa & Hansen (1976) and Officer (1980) have developed models of the process. The contribution to the turbidity maximum developed by this process is dependent on many factors including most importantly, the concentration levels of suspended material in the river and the adjacent sea region and the strength of the residual circulation. However, material trapped in suspension by the residual circulation is normally swamped by material undergoing continuous cycling between sedimented and suspended states controlled by the tidal regime. This material is often concentrated in the upper estuary through up-estuarine 'tidal pumping' induced by asymmetry in the tidal ebb and flood velocities (Allen *et al.*, 1980). This process is also selective with respect to particle settling velocity. Maximum tidally induced turbidity tends to coincide with that caused by the residual circulation within the low salinity zone owing to:

1. the general landward reduction in aqueous volume overlying unit area of sediment;

2. local geomorphological enhancement of current velocities;
3. additional frictional stresses arising from the fresh-water flow.

The estuarine turbidity maximum thus comprises a temporally variable mixture of two principal components. Although resuspendable material is normally in abundance, its dependence on tidal currents means that its contribution is least at, or just after, slack water during neap tides. The extent to which aggregation and disaggregation processes influence the settling velocities and transports of specific particle types and contribute to the dynamics of the turbidity maximum phenomenon remains a matter of controversy (Schubel *et al.*, 1978; Krone, 1978).

Upstream from the turbidity maximum, the amount of suspended sediment decreases, but enhanced turbidities often extend into the fresh-water region. This can result from both diffusion of suspended material and tidal resuspension effects. Whatever the cause, it is important to take this phenomenon into account when characterizing the quality of river waters entering an estuary. Obviously, the optimum sampling position is as near to the saline intrusion as possible, but this must be upstream of the region of turbidity enhancement.

There have been few systematic studies of the composition and physical properties of estuarine suspended particle populations or of their constituent phases. However, it is apparent from available data (Martin *et al.*, 1978; Sholkovitz, 1979; Duinker *et al.*, 1980; Gobeil *et al.*, 1981; Loring *et al.*, 1983) that physical transport processes dominate in controlling the temporal and spatial variability of particle population properties. *In situ* production processes are also occasionally significant, especially in larger estuaries (Sholkovitz & Price, 1980). Basically, spatial variations in population composition are compatible with mixing between riverborne particles and particles of marine origin, modified locally by resuspension of sediment or by subsidiary inputs. If tidally controlled sediment resuspension is pronounced, resuspended material can predominate over most of the estuary. Riverine inputs occasionally dominate during short periods of storm run-off which both carries an enhanced suspended load and represses tidal resuspension.

These processes in combination lead, directly or indirectly, to appreciable zonation of suspended particle types. Resuspended material is distinctly different in composition to influxing materials due to selectivity during hydrodynamic transports and to diagenetic alteration during residence in the sediment. Particles of low to negligible settling velocity which become trapped within the turbidity maximum zone can also be distinguished compositionally (Duinker *et al.*, 1980; Morris *et al.*, 1982*c*).

It is apparent that predicting the fate of particulate materials entering an estuary, quite apart from considerations of changes in loading and composition brought about by chemical and biological activity, is by no means simple. General principles have been described, but these are of limited use in any individual real estuary and are not likely to be developed to such an extent that field work will become unnecessary. It is clear, however, that an appreciable proportion of solid material introduced into an estuary, whatever the source, may well be retained within the system for a considerable period of time. This material will be dispersed throughout the whole length of the estuarine system if maintained in the water column, and tend to sediment out in regions of reduced tidal scour. Material of appropriate hydraulic characteristics can become concentrated within the turbidity maximum. Studies of the dispersion and effects of particulate effluents within estuaries must not be confined solely to the immediate vicinity of the outfall or dumping region.

McCave (1979) has provided a detailed introduction to methods for the sampling, analysis and characterization of suspended particle populations in estuaries. Methods for the sampling and physical examination of sediments have been described by Buller & McManus (1979).

Chemical characteristics of inputs

The saline input

The saline coastal water component which is diluted by fresh water within an estuary is, in general, more consistent in chemical composition than the fresh-water component. In estuarine studies therefore, it is often reasonable to assume a constant composition for the coastal water, particularly over time intervals of weeks or less. Such an assumption is, of course, most effective when the concentration of any component under investigation within the estuary is much greater in the sea water than in the river water, as is the case for the major constituents which effectively comprise the salinity of sea waters. It is also applicable, however, to some of the minor components including, at times, the essential nutrients.

Coastal sea water, even away from the immediate influence of a single river, is usually modified chemically by the presence of land-derived waters. Within a few tens of kilometres of the shoreline, fairly regular gradients in the distribution of salinity and of the dissolved and particulate chemical species are apparent. Within this region, individual river effluents may be recognizable as plumes of distinctive chemical and physical characteristics. The extent of the plume is predominantly determined by the rate of input of volume of fresh water from the estuarine source, but it is modified in shape and geographical position in response to local hydrographical and climatological conditions which also control its rate of dispersal (Duxbury, 1972). Foster & Morris (1974) have shown that the ultraviolet absorption of coastal waters can be used as a characteristic property associated with individual fresh-water contributions and many other individual chemical species, e.g. dissolved silicate (Liss, 1969) may be useful at times. A multi-component approach is, however, much more effective as illustrated by the investigation of the intrusion of River Mersey water into the Irish Sea by Abdullah & Royle (1973).

The transport and mixing processes which diffuse individual plumes and limit the extent to which they are recognizable, give rise to a mixed coastal water such that the chemistry of any estuary is affected to a degree by other fresh-water sources within the surrounding region. This is readily shown by studies of the distribution of radionuclides in coastal waters, for their source can usually be recognized unequivocally; the distribution of caesium-137 within the coastal waters of the British Isles (Jefferies *et al.*, 1973; Wilson, 1974) provides a good example. The chemistry of coastal waters is, of course, subject to temporal variability in response to *in situ* physical, chemical and biological processes. Descriptions of the seasonal variations in the abundance in coastal waters of dissolved trace metals, (Knauer & Martin, 1973; Morris, 1974) and nutrients (Ewins & Spencer, 1967; Armstrong & Butler, 1959) are typical and indicate the major sources of variability and the extent to which the effects of *in situ* coastal water processes are recognizable over transport-dominated fluctuations.

The fresh-water input

The natural chemical constituents of fresh water, whatever their physico-chemical state, are derived ultimately from the processes of rock weathering. The term weathering embraces a variety of contributory mechanisms, in almost all of which water is intimately involved both as the primary solvent and as the transport medium for introducing attacking chemicals such as CO_2 and organic acids and removing particulate and dissolved weathered products. One might expect therefore that the composition of a stream water reflects the lithological environment of its drainage area. However, studies of the geochemistry of many types of fresh waters have shown a lack of direct relationships, particularly with respect to the dissolved constituents. Turekian *et al.* (1967) were unable to correlate the chemistry of the dissolved constituents along the length of the Neuse River, N. Carolina with the local lithology, although this was varied and well-defined. They proposed that the stream chemistry was being modified continuously by the incorporation of ground waters and/or by *in situ* biological and chemical processes.

Gibbs (1970) has concluded that there are three major mechanisms controlling the general pattern of surface water chemistry; atmospheric precipitation, rock dominance and evaporation/crystallization processes, the relative importance of each process being determined by the prevailing geological and climatological conditions. In addition to the natural products of these processes, domestic and industrial waste-disposal products must be taken into account. Empirical studies of individual rivers or comparisons of rivers have yielded a number of examples where the general nature and response of the chemistry of rivers to prevailing environmental conditions and changes respectively, have been elucidated. For example, Beck *et al.* (1974) have drawn attention to the similar and distinctive nature of several coastal plain rivers of the south-eastern United States with respect to their organic and inorganic geochemistry. Investigations of individual chemical constituents have yielded more detailed, and in some circumstances, semi-quantitative descriptions. Dillon & Kirchner (1975) have demonstrated that the annual average exports of phosphorus from drainage basins have characteristic values determined by the local geology and land use. Differences in phosphorus export amongst the various combinations of forest or pasture and igneous or sedimentary domain are highly significant. Moreover, for the forested igneous drainage basin, Kirchner (1975) has demonstrated a direct dependence of annual phosphorus export on the drainage density (the length of the stream per unit drainage area). Agricultural utilization and urbanization both produce a vastly increased phosphorus export, giving rise to elevated river concentrations which reflect the intensity of local land use.

The dissolved metal contents of a number of Welsh rivers which receive very little domestic and contemporary industrial waste have been investigated by Abdullah & Royle (1972). Different metals occurred locally at high concentrations in individual rivers and these were found to be largely due to the presence of mineralized zones and/or to localized remains of past mining activities. Temporal variations were quite considerable, but the differences between rivers were maintained on a comparative basis. A more complex situation, however, is usually apparent in the incidence, variability and controlling factors for trace elements in fresh waters. Andelman (1973) has concluded that only by distinguishing between the various physical and chemical states, such as the size of particles and the incidence and nature of complexes and chelates, can one accurately characterize the behaviour of trace elements in stream waters. The physical and geochemical characteristics of river-transported suspended solids have been examined by Ongley *et al.* (1982). They found marked systematic seasonal changes in particle size distributions, organic content and chemical composition of transported particles. Flow related changes in principal source provenance, from surface soil erosion at higher flows to river bed/bank mobilization at low flows coupled with seasonal changes in biological utilization, accounted for much of the variability. The mineralogy of the suspended material was relatively constant. A quantitative assessment of the proportion of total metal transported by each of five separate mechanisms in the Amazon and Yukon Rivers has been attempted by Gibbs (1973); his results are summarized in Table 1.2.

The rates of input of soluble material to an estuary is not normally directly proportional to the volume flow of fresh water (Whitfield & Schreier, 1981). By far the major proportion of the input of particulate materials may be supplied in the short periods during and following high rainfall when the river is in spate and the high volume flow also carries a disproportionately high particulate load. Under these circumstances, the local rainwater composition becomes more predominant in determining the river solute composition as the mobilization of land-derived dissolved materials is reduced by the increase in superficially transported water relative to that which has percolated through soil.

Edwards & Liss (1973) have noted that in comparison with other major dissolved constituents of rivers, dissolved silicon tends to be remarkably uniform on a worldwide basis and shows little response to changes in discharge. They suggest that an abiological buffering mechanism involving sorption reactions on solid phases may be operative. Such a mechanism

Table 1.2. *Proportion of metals in identified physico–chemical forms transported by the Amazon and Yukon Rivers*

		Percentage of total metal					
Transported form	River	Fe	Ni	Co	Cr	Cu	Mn
In solution (passing 0.45 μm filter)	Amazon	0.7	2.7	1.6	10.4	6.9	17.3
	Yukon	0.05	2.2	1.7	12.6	3.3	10.1
Desorbable from solids	Amazon	0.02	2.7	8.0	3.5	4.9	0.7
	Yukon	0.01	3.1	4.7	2.3	2.3	0.5
In precipitates attached to solids	Amazon	47.2	44.1	27.3	2.9	8.1	50.0
	Yukon	40.6	47.8	29.2	7.2	3.8	45.7
In organic solids	Amazon	6.5	12.7	19.3	7.6	5.8	4.7
	Yukon	11.0	16.0	12.9	13.2	3.3	6.6
Detrital	Amazon	45.5	37.7	43.9	75.6	74.3	27.2
	Yukon	48.2	31.0	51.4	64.5	87.3	37.1

Adapted from Gibbs (1973).

does not however, fully obscure discharge-controlled variability. An intensive study by Kennedy (1971) of the dissolved silica in the Mattole River in northern California has shown a consistent pattern of variability which is determined by the relative proportions of source waters entering the stream. Silica is more concentrated in water which seeps through the soil in surface water or in ground water. Hence, during a spate, silica initially decreases in the stream as surface flow predominates, and then subsequently increases as soil seepage once again increases in dominance. With decreasing discharge, ground water supplies an increasing proportion of the stream flow and silica slowly decreases to the level controlled by the sorption exchange reactions in soil.

The few examples quoted above illustrate the complexity of natural factors controlling the composition of a river water at any one time, and the consequent temporal variability in the input of dissolved and particulate materials to an estuary. Present knowledge of the processes involved allow the responses of riverine chemical composition to changes in basic river properties such as discharge to be described only in semi-quantitative terms. Furthermore, the extent to which quantitative and qualitative descriptions of those rivers for which adequate data have been obtained can be utilized analogically for general predictive purposes appears to be quite limited. Under these circumstances, and in view of the paramount importance of the immediate fresh-water contribution in determining the instantaneous chemical state of any estuary, synoptic fresh-water data collection will often be a necessary and integral part of estuarine field studies. Empirical observations of compositional variations of an individual river in response to environmental changes, regular, cyclic, or intermittent, will provide major benefits to the understanding and characterization of their influence on estuarine behaviour.

Estuarine chemical reactivity

The dominant feature controlling the distribution, speciation and reactivity of chemical components within estuaries is the mixing of fresh and saline waters. The two previous sections have outlined the essential nature of, and variability associated with, the off-shore sea water and influent fresh water. Both of these impose a corresponding nature and variability on any parcel of water within an estuary, to an extent determined by their relative contributions to the admixture, the intensity of *in situ* reactivity, and the turnover time of waters. A more detailed discussion of the factors determining the reactivity and distribution of various chemical species in estuaries is given by Mantoura & Morris (1983).

The mixing environment

Differences in the nature of the fresh and saline mixing components produce gradients and transitions of physico–chemical properties within an estuary in response to the circulation and mixing pattern. Changes in the following are of considerable importance: ionic strength, the relative proportions of individual ions, molecules and complexing components, temperature, pH, Eh and the availability of surfaces. These physico–chemical properties, moreover, are not directly proportioned by the degree of mixing. For example, the pH of estuarine waters is responsive to changes in carbonate equilibria. Sharp increases in the first and second apparent dissociation constants of carbonic acid with increasing salinity result in an estuarine pH distribution which has a minimum at very low salinities when the pH of the river water is less than that of the sea water; for higher pH fresh waters, a sharp decrease in pH at low salinities is obtained (Mook & Koene, 1975). However, this equilibrium thermodynamic treatment is not always applicable because *in situ* production and respiration processes can significantly perturb the carbonate equilibria (Morris, 1978; Morris *et al.*, 1978, 1982*a*). Field observations of the distributions of chemical constituents in estuaries, aimed at elucidating reaction mechanisms, should wherever possible be supported by simultaneous observations of important physico–chemical variables.

Estuaries are classical examples of complex thermodynamically open systems, subject to constantly changing input and output fluxes and to continuous internal chemical reactions. Open systems can, of course, reach a dynamic steady-state or pseudo-equilibrium condition if the fluxes of reactants and products between the system and its surroundings and the kinetics of internal reactions involved are invariant. However, the fluctuations in supply of individual chemical species, *in situ* biological activities and the continuously changing state variables act together to inhibit the attainment of a steady state, so that chemical reactions within estuaries involve continuous movement towards a thermodynamically stable condition which is itself transient. In view of this, it is not surprising that almost all estuarine chemical reactions which have been demonstrated empirically by field studies or by mass-balance calculations are not well understood with respect both to the mechanisms and the kinetics of the reactions involved. Furthermore, processes reported as occurring in one estuary have not been detected in another or even, at different times, in the same estuary. Despite the fundamental importance of kinetic factors, in conjunction with hydrodynamic behaviour, in determining the extent to which chemical reactions proceed within any estuary, there remains a general lack of knowledge of the mechanisms and rates of estuarine chemical processes and of the factors which control them.

Conservative and non-conservative behaviour

In spite of the potential reactivity of the estuarine mixing environment outlined in the previous section many chemical constituents pass through estuaries without any apparent involvement in reactions and thus may be said to behave conservatively. Even the biologically essential primary nutrients, dissolved silicate, nitrate and phosphate, under conditions where their utilization during primary production is readily demonstrable, may be pseudo-conservative in their estuarine occurrence when their *in situ* rates of transformation are slow relative to their rate of transport through the estuary. Conditions in the Astoria Estuary provide a good example (Park *et al.*, 1972). Figure 1.5 shows the distributions of nutrients as a function of salinity at approximately monthly intervals for the years 1966 and 1967. The range of salinities within the estuary is quite variable in response to changes in the fresh-water input. Whatever the season and salinity range, however, the nutrients approximated to conservative

Fig. 1.5. Annual variations in the distributions of nutrients in the Astoria Estuary, plotted against salinity. Reproduced, with permission, from Park *et al.* (1972).

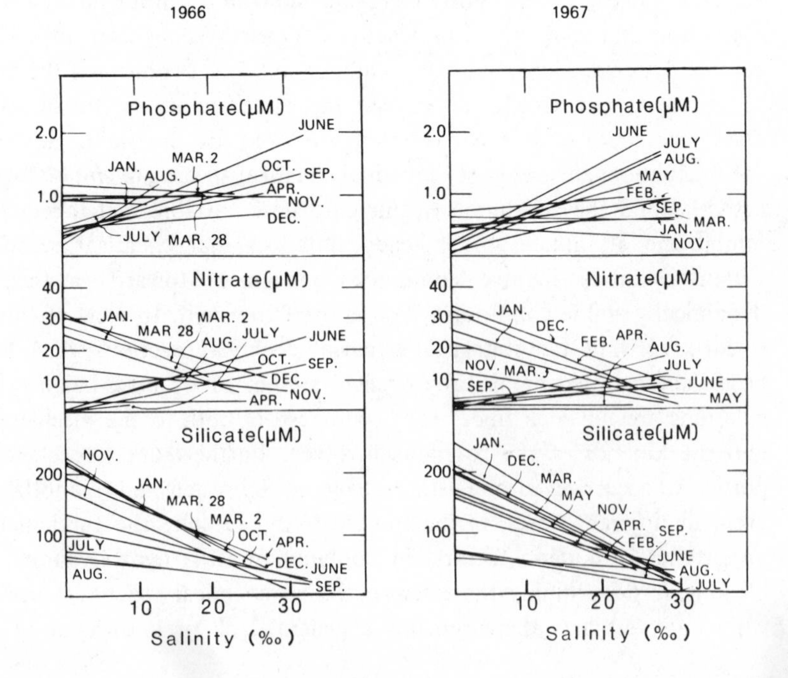

distribution within the estuary and there were apparently no significant mid-estuarine sources or sinks. Concentrations of these nutrients relative to salinity are therefore controlled solely by their concentrations in the contributory sources. Both sources exhibit seasonal variability and, noticeably, the fresh-water concentrations of nitrate and phosphate fall below the sea-water concentrations during the summer when the fresh-water silicate contribution is also substantially reduced. In other estuaries, significant nutrient interactions are found. The dissolved nutrients in San Francisco Bay show seasonal variability in their distribution as a function of salinity which clearly indicates a major (biological) sink for silicate in the estuary during the summer months; a complexity of sources and sinks apparently controls the nitrate distribution (Peterson *et al.*, 1975). Non-biological processes have also been shown to be important in controlling nutrient distributions, especially silicate and phosphate (Liss, 1976; Morris *et al.*, 1981), in some estuaries.

The extent to which either conservative or non-conservative chemical behaviour of any individual dissolved constituent within an estuary is detectable will depend on the following:

1. the thermodynamic distance from equilibrium and the corresponding rate of approach to equilibrium;
2. the rates of supply from all sources and their fluctuations;
3. advective and diffusive transport through the system and the dynamics of the mixing processes;
4. the rate and extent of contemporaneous biological perturbations;
5. the concentration level in relation to the accuracy and precision of analysis.

All these potential contributory factors must be taken into account when considering the implications of the analytical results obtained in the determination of a reactive constituent of estuarine waters. It is clear that the determination of a single chemical component in estuarine water sampled at a single point in time and space is of limited value, whatever degree of precision is attained. Any characteristic of the water is fully interpretable only when the nature of the sources, the degree of mixing, and the interim effects of *in situ* processes are known.

The salinity or total salt content is for many purposes ideal for indicating the degree of mixing, acting as a true internal standard. The large difference in total dissolved solid content between offshore sea waters (approximately 35 g l^{-1}) and fresh water (average, 0.12 g l^{-1}) allows the use of the simple formula:

$$\text{percentage fresh water} = 1 - S/S_0,$$

where S is the salinity of sample and S_0 is the salinity of sea water away from noticeable estuarine influence, with an accuracy compatible with most estuarine chemical determinations. Difficulties arise when the proportion of salt water is low (< 10 per cent) so that the total dissolved contents of both the sea- and fresh-water contributions are comparable. Under these circumstances the various indices of salinity diverge widely, although chloride content is apparently the most useful. The practical utility of chloride ion-specific electrodes for indicating the degree of mixing in the low salinity region of estuaries has been demonstrated by Morris *et al.* (1982*a*).

Characterization of estuarine samples with respect to salinity is a standard procedure for chemical investigations in estuaries and is known

Fig. 1.6. Model dissolved constituent–salinity relationships in an estuary under steady-state conditions. C_{FW} and C_{SW} are the concentrations of constituent C in the fresh-water and sea-water mixing component, respectively. Line *a* defines the theoretical dilution line for a non-interactive constituent. Curves *b* and *c* indicate relatively widespread estuarine input and removal of C, respectively. Curve *d* is typical of removal occurring only in the upper estuary. Curve *e* is generated when the rate of removal of C in mid-estuary exceeds the riverine input. Curve *f* indicates net input of C to the upper estuary coupled with net removal further seaward.

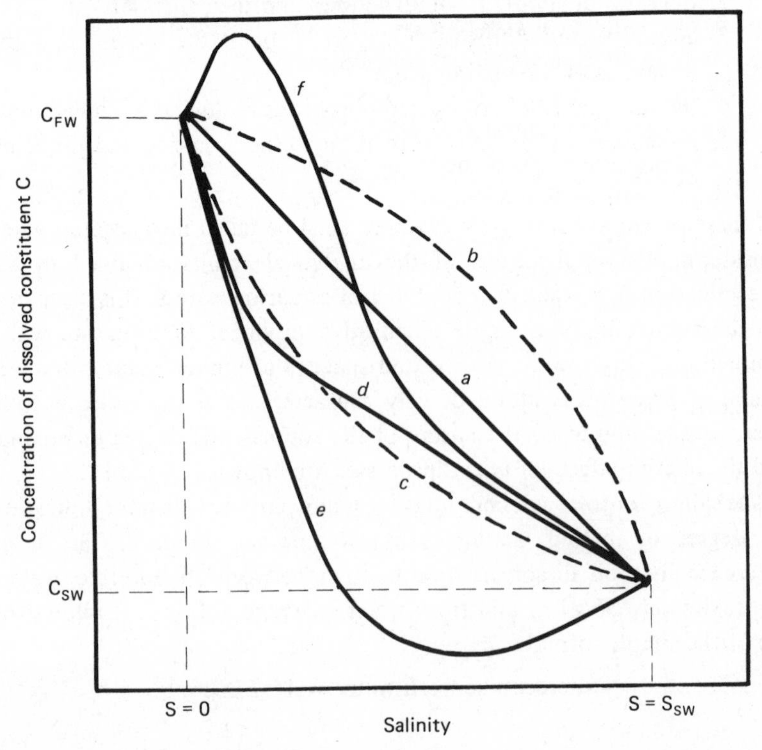

as the 'reactant method'. Correlation of the concentration of a dissolved chemical species with salinity for samples collected throughout the salinity range allows an assessment of gain, loss or conservation of the constituent and an indication of the relative contributions of the species from the separate marine and fresh-water sources (see Fig. 1.6). One can also deduce the salinity-related (and geographical) location of reactivity and the extent to which it has progressed. If hydrodynamic information is available, an estimate of the rate of the reaction can be obtained. Additionally, these indications of reactivity together provide circumstantial evidence of the reaction mechanism involved.

Where more than a single significant fresh-water source is apparent, of course, salinity is not a true internal standard with respect to the individual constituents of fresh water. A comparison of the chemical composition of each fresh-water source may indicate a constituent which can be used as indicative of the relative proportions of the fresh-water sources, but this is not often found in practice.

Theoretically, the sole limitation to the quality of these assessments is the precision of analysis of any constituent relative to the extent of non-conservative behaviour. In practice, however, there are three quite restrictive limitations to the applicability of this technique. Firstly, a steady-state condition is assumed so that, ideally, the contributory waters must have remained of constant composition over the time-span during which marine or fresh water was replaced within the estuary prior to the investigation. This is not usually the case, and they are better described as having a range of concentrations rather than a single value. Consequently, non-conservativeness can be deduced only when concentrations are outside a range, rather than a single value, relative to salinity (see Fig. 1.7). Secondly, a single fresh-water source is implicit although estuaries often have a number of significant water sources, entering at points through the length. Except in the unlikely event that all significant inputs contributing to the salinity value of an estuarine sample are uniform in concentration of the constituent under investigation, the resultant mixing curve will not be linear, notwithstanding conservative behaviour of that constituent. If conservative behaviour is prevalent, the resultant mixing curve will be composed of two or more line segments, each showing linear mixing. Such a curve often closely resembles, and may be mistaken for, the outcome of non-conservative behaviour and a single fresh-water source (see Fig. 1.6). Thirdly, quantitative and qualitative interpretations of mixing curves are severely restricted if the reactive constituent is subject to more than a single unidirectional reaction. This possibility is often disregarded.

These practical limitations, combined with imprecision arising from

Fig. 1.7. Model constituent-salinity relationships in an estuary of flushing time T, for a non-interactive constituent subject to variations in fresh-water input concentration. In each case, the saline end-member (salinity 32‰) is maintained at unit concentration and the time-averaged concentration in the fresh water is 2.5 units. *a* shows the theoretical dilution line for invariant fresh-water concentration. The family of curves in *b* defines the envelope covering the constituent–salinity relationships generated when the fresh-water concentration oscillates sinusoidally with an amplitude of 0.5 units and a period of $T/10$. the curves in *c* and *d* are generated when an additional long-term oscillation of amplitude 1.5 units and period 5 T is imposed. These cover the changes in constituent–salinity relationships through a period T, during increasing and decreasing phases, respectively, of the longer-term oscillation. Reproduced from Loder & Reichard (1981) with permission of the Estuarine Research Federation.

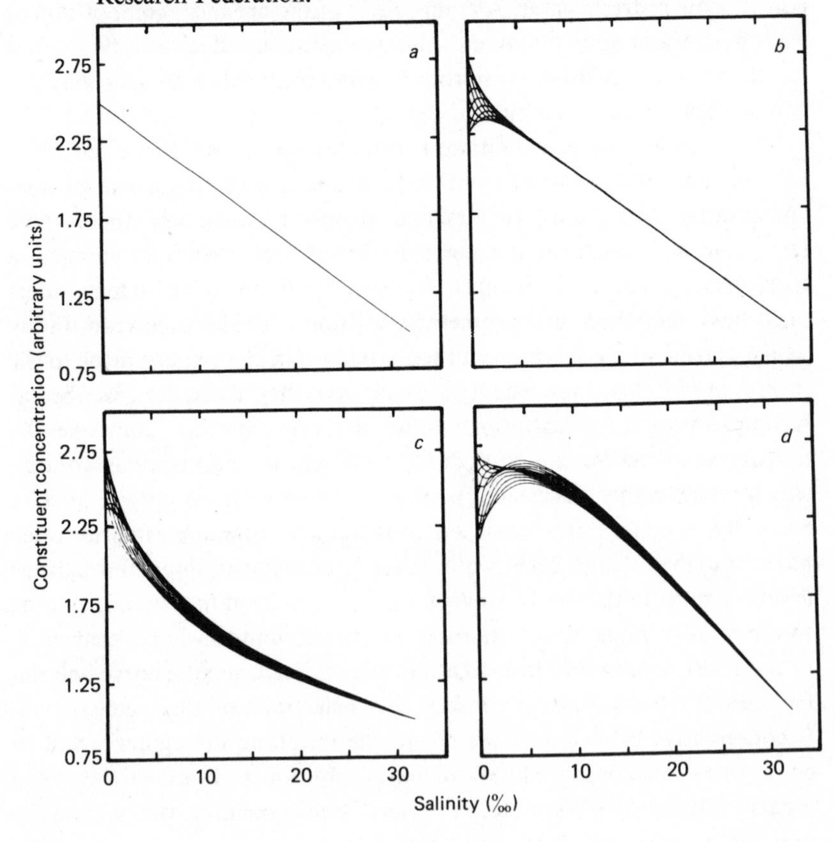

sampling and analytical errors and compounded by inherent estuarine variability, make the simple application of the reactant method quite insensitive. Deviations from conservativeness of less than 10–20 per cent will often be unresolved, or poorly characterized, unless they are very localized.

Model descriptions of the reactant method have been developed by Boyle *et al.* (1974), Li & Chan (1979), Officer (1979) and Rattray & Officer (1979, 1981). Applications of the method have been reviewed by Liss (1976), Aston (1981) and Bewers & Yates (1981). From these, it appears that indications of estuarine chemical transformations deduced from simple applications of the reactant method are numerous but that interpretations of the mechanisms and rates involved are limited and often equivocal. Generally, it is observed that reactions proceed most rapidly and extensively in the low salinity region which is characterized by the sharpest gradients in composition and physico–chemical properties and is often a region of greatly enhanced suspended load.

The interpretative potential of the reactant method can be enhanced in a number of ways. The use of continuously recording autoanalytical methods for field observations provides data of a spatial resolution and coherency which are far superior to those obtained using discrete sampling methods and considerably reduces the possibility of sample contamination and obviates deterioration during storage. The need to characterize compositional variability in the fresh-water inflow has often been stressed. Some investigators have attempted serial observations of fresh-water composition over a pre-investigational period equivalent to the fresh-water replacement time in the estuary. However, this is sometimes impracticable, especially for only minor deviations from conservativeness and for reactions concentrated in the low salinity region. An alternative solution is to apply the reactant method repetitively. This improves the sensitivity and provides a more accurate assessment of the time-averaged extent or rate of reactivity. It also enables non-random time variances in the process under investigation to be characterized.

As noted above, the field observations should be supported by simultaneous recordings of important rate and mechanism controlling variables. Comprehensive data of this kind are helpful for elucidating reaction mechanisms and rate controls, especially if comparative studies are made under differing environmental conditions in one estuary and the results can be compared with similarly comprehensive data from other estuaries. Nevertheless, it appears that field observations alone, however comprehensive or replicated, will rarely be totally conclusive regarding reaction mechanisms and consistently be unsatisfactory for precise evaluations of kinetic factors.

The recognition of conservative or pseudo-conservative behaviour of a dissolved chemical constituent allows the derivation of the dispersion, distribution and flux of that constituent in the estuary from a knowledge of the hydrodynamics of the system without recourse to practical or theoretical studies of that constituent, except for determination of the inputs to the estuary. When non-conservative behaviour is encountered, mathematical functions representing the rate of transfer of any constituent between physico–chemical states can be incorporated into the equations of state which define the hydrodynamic characteristics of estuaries. The utility of this approach, however, is at present severely limited by shortcomings in our fundamental knowledge of the mechanism and rates of estuarine chemical processes so that formulations are usually of empirical derivation and must therefore be based on reliable field data. Moreover, soundly based thermodynamic descriptions and predictions of the behaviour of chemical species within estuaries will require advances in our ability to quantify specific forms of occurrence of elements. The prevailing techniques of water analysis are predominantly applicable to the determination of total elemental concentrations, although some operationally defined species discriminations can be achieved using appropriate combinations of separation and analytical methods (Florence & Batley, 1976; Batley & Gardner, 1978).

The theoretical basis for the formulation and computerized solution of equilibrium speciation models of chemical constituents in natural waters is well established (Westall, 1980). The solution model for trace metals in estuarine waters which incorporates humic complexation developed by Mantoura *et al.* (1978) and expanded by Mouvet & Bourg (1983) to include adsorption onto particles provides a pertinent example. However, the use of such models is presently hampered by a shortage of information on the natures and concentrations of complexing organic molecules and of particle surfaces in natural waters, and also by a lack of reliable data on the stability constants for metal–ligand and metal–surface interactions (Vuceta & Morgan, 1978). Until these needs are overcome, the practical utility of equilibrium speciation models, which is dependent also on the severity of kinetic restrictions to the attainment of thermodynamic stability throughout the estuarine mixing region, must remain in question.

The practical recognition of pseudo-conservative behaviour of a dissolved constituent should not immediately be interpreted as indicating that the constituent is not accumulating within the estuarine system. Reactions involving undeterminably small mass transfers over time intervals of the order of the estuary turnover period can result in significant accumulations of some materials in the permanent or semi-permanent estuarine particulate and sedimentary material over extended periods. On the other hand, abnormally high sediment concentrations are not always indicative of *in situ* accumulatory reactions. Chester & Stoner (1975) showed that the < 61 μm fraction of lower Severn Estuary sediments have concentrations of lead, tin and zinc which are considerably higher than those of the average nearshore sediment. Although the lead and zinc anomalies were apparently mainly hydrogenous, the elevated tin concentrations were probably detrital in origin being compatible with the mineralogy of the local drainage area.

Models of estuarine systems

The numerous important physical, chemical and biological parameters and their interrelationships and fluctuations in time and space within estuaries create complex problems for those who need to describe the overall behaviour of any estuary and to predict the consequences of any changes in the regime brought about by natural or anthropogenic modifications. Yet such descriptions and predictions are often the main requirements for appropriate management decisions concerning these resources. The use of models, whether concerned with a single process or with attempting to integrate many processes and their interrelationships, often provides the only practical method of realistically simulating estuarine conditions and behaviour, and interpolating and extrapolating the results of field observations.

Hydraulic models

Hydraulic scale models have provided an accurate engineering and scientific tool for the interpretation of many of the physical phenomena of individual estuaries. Many complex hydrodynamical problems concerned with shoaling, channel dredging and spoil dumping, the diffusion, dispersion and flushing of wastes and the effects of surges and unusual tidal conditions have proved amenable to at least partial solution using this technique. A number of examples have been discussed by Ippen (1966) and Gameson (1973) provides further examples.

Simple mathematical models

Relatively simple mathematical techniques directed towards prediction of the flushing of pollutant materials introduced into an estuary can be formulated. These require data on the dimensions of the estuary, together with river flow and tidal information, and are mostly based on the tidal-prism concept. In its simplest form, the volume of water entering an estuary on the flood, the tidal prism, is assumed to be fully mixed within the estuary. On the ebb, the same volume of sea water retreats, but it is diluted by a fresh-water content equal to the river input, removing a proportion of the total fresh water within the estuary. The flushing time (T in tidal cycles) of the estuary is given by the ratio of $V+P$ to P, where V is the low tide volume and P the tidal prism (or inter-tidal volume). This simple technique gives an exaggerated estimate of the rate of flushing because the assumption of complete mixing over one tidal cycle is unrealistic. Ketchum (1951) has improved the technique by considering that the elements of mixing volume are constrained by the length of the tidal excursion. Length segments of the estuary are calculated such that the volume of water contained therein at high tide is equal to that contained in the next seaward segment at low tide; each segment is then treated as for the whole estuary in the simplest procedure. If the salinity distribution is known, this 'fraction of fresh water' method can be used to calculate the flushing time. The volume of fresh water in the estuary or in any segment, determined from the salinity values, is equated with the rate of fresh-water influx. The distribution of fresh water or salinity within the estuary can also be used to indicate the distribution of a conservative, dissolved pollutant introduced at any point

Fig. 1.8. Steady-state distribution of a conservative pollutant *P*, (——), introduced into mid-estuary, in relation to the salinity distribution (- - -). Numerical values are calculated for low run-off conditions in the Severn Estuary, with an input of 1 unit of *P* per day, at a distance 25 km below the limit of saline intrusion (R. J. Uncles, personal communication).

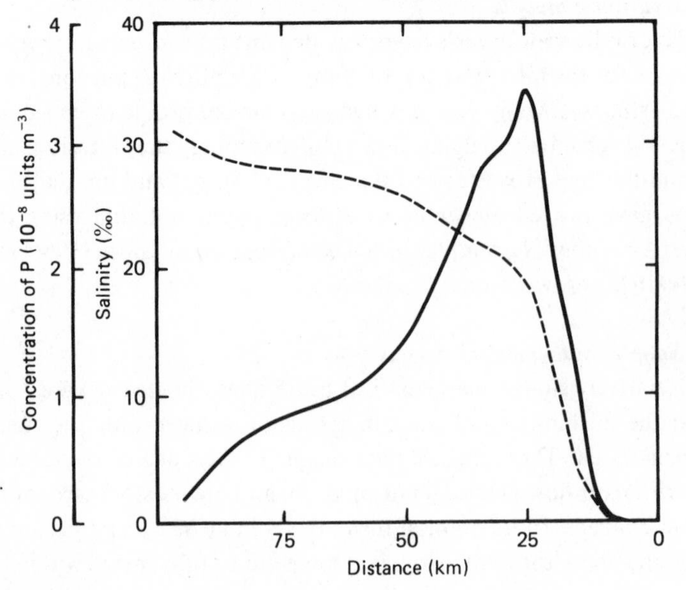

along the length of the estuary, if the rate of input is known (Ketchum, 1952). Whatever the point of input of a continuous discharge, the constituents will be distributed both upstream and downstream, with maximum concentrations in the vicinity of the discharge point. Downstream of the outfall, the distribution of dissolved unreactive constituents will have the same form as the salinity; upstream, it will be inversely proportional to the salt water fraction (see Fig. 1.8). Other treatments of estuarine flushing and pollutant dispersal have been described by Arons & Stommel (1951), Stommel (1953), Kent (1960), Pearson & Pearson (1965) and Pritchard (1969); see also the reviews by Dyer (1973, 1981). Neal (1972) has compared the results obtained using a number of these techniques when applied to the Columbia River estuary. An improved segmented prism model for well-mixed estuaries has been developed by Dyer & Taylor (1973). O'Kane (1980) has detailed the development of a one-dimensional, quasi steady-state model, with cost effective management of water quality as a primary requirement for its use.

More complex models

In order to provide descriptions and predictions of chemical and biological processes, their interrelationships and the resultant distributions of properties, the vastly increased flexibility of modern mathematical modelling techniques is distinctly advantageous. Many mathematical models of estuaries of all degrees of complexity have been, or are being, produced. These models range from fundamental to totally empirical, but optimum reality is usually obtained from a combination. The simplest models relate to steady-state conditions with gradients in one dimension, the length of the estuary, or two dimensions when vertical stratification must be taken into account. Although models such as these provide useful management tools (see, for example, Gameson, 1973; Mackay & Gilligan, 1972 and Chapter 7), they are limited in applicability, being restricted to a description of average conditions or a series of average conditions. Realistic descriptions and subsequent reliable predictions are best obtained by using models which incorporate continuous temporal variability. For example, the initial dispersion of material entering an estuary from an outfall is strongly influenced by stream velocity and therefore subject to continuous cyclic fluctuations dominated by the local tidal movements; a build-up of effluent around the outfall at times of slack water is obtained.

More particularly, the occurrence of *in situ* chemical and biological processes which induce non-conservative behaviour of individual constituents can only be realistically simulated by including mathematical expressions which account both for the actual processes and the temporal and spatial changes in the rates at which they are proceeding. Undoubtedly, present day computer technology has advanced to an extent that realistic, multi-dimensional, time varying models of estuaries and the processes occurring within them with almost unlimited resolution in both time and space, are quite feasible. Simplifying assumptions are not necessarily demanded, although they may be required for practicality and expediency, Whatever the degree of complexity of a model, however, it must always be remembered that the reliability of any descriptive or predictive output is controlled by the assumptions that are made, the accuracy of the mathematical formulations of the processes incorporated and the quality of the original data input. For example, there remains an appreciable degree of uncertainty with respect to the values of diffusion coefficients and their spatial and temporal changes within any estuary, although a knowledge of these coefficients is fundamental to the accurate mathematical representation of the mixing processes which control the distribution of water and chemical species. Also, many of the estuarine processes which control chemical speciation are not well understood. Useful empirical mathematical descriptions of estuarine chemical processes may be obtained from the results of field observations, but there remains, at present, an outstanding requirement for much more fundamental information of the mechanisms and rates, and their regulation by ambient conditions, of chemical reactions within estuarine systems.

Because of such uncertainties and also because there is a practical limit to the comprehensiveness of the basic information which can be incorporated, the resolution of any model will have a finite limit. Any abstracted information will inherently be subject to a degree of uncertainty. This uncertainty is best characterized and quantified by field observations during model testing, and, if found necessary, subsequent modification of the model must be carried out. Field validations must therefore be included as an integral part of model construction. An adequate reference frame both in space, controlled by the number and spacing of sampling stations within the estuary, and in time, ideally covering the extremes of the ranges of conditions produced by time-variant phenomena, is required but not always attainable. The extent to which model descriptions of estuaries are validated by experimentation is limited by practical, expediency and economic considerations (Ulanowicz & Flemer, 1978). Interpolation and extrapolation in space and time, both of which are readily performed using model techniques are valid procedures, but must be carried out with caution.

The present developments, limitations and probable future developments in mathematical modelling techniques applied to marine systems have been comprehensively treated in Nihoul (1975).

The planning and execution of estuarine chemical investigations

General considerations

Estuaries are regions of fundamental importance with respect to geochemical processes occurring on a global scale, for they represent the major route whereby weathered lithospheric material is transported to the oceanic sedimentary domain. Although many of the processes involved in determining the natural chemical state of estuaries are not well understood, the direct acquisition of basic knowledge in this context represents a minor proportion of present estuarine research effort. Rather, estuarine studies are predominantly initiated in order that these important natural resources, subject as they are to a wide spectrum of exploitations including transport, mineral extraction, food extraction, waste disposal and recreational and aquacultural usages, may be sufficiently comprehended to allow balanced and controlled management of their utilization. Basically, the scientist is required to predict the probable effects of any facet of this utilization, particularly with respect to changing exploitation. Typical questions concern the consequences of the addition of any type of pollutant, or of the abstraction of minerals, or of changes in run-off by impoundment and abstraction of river water (see Neu, 1982*a*, *b*).

It is important to recognize the accent on consequences or effects. Proper answers to such questions require two main categories of information; firstly, how does the system function now, and secondly, how will it respond to a perturbation. Furthermore the chemical nature of estuaries both influences and is influenced by a complexity of physical, chemical and biological processes. Thus, chemical investigations will only be fully useful in management terms if they are carried out with full recognition of all the estuarine processes and properties which are responsive to, or subject to disturbance by, chemical agencies. Similarly, studies of chemical properties and processes in isolation, without consideration of their physical and biological connotations, are in many circumstances unlikely to yield a complete solution to the problem in hand.

It has become customary to commission so-called 'base-line' studies of estuaries, aimed at providing a description of their present-day performance which is of use in forecasting or detecting the effects of future perturbations, accidental or planned. However, it should not be assumed that conditions will remain stable if the estuary is not subject to further anthropogenic disturbance. Estuaries are, in geological terms, metastable phenomena, being continuously subject to internal evolutionary processes which may, at times, produce relatively rapid irreversible changes, for example in morphology, quite distinct from response to external control by factors such as climatic trends. Furthermore, few estuaries are completely free from anthropogenic influences. Land use within catchment areas is continuously subject to change with consequent disturbance of river drainage characteristics and chemical mobilization processes. Within estuaries, industrial and urbanization pressures are increasing, but these are being counter-balanced by improving pollution control measures. Thus, an estuary may, at any one time, be responding to a variety of historical usages. Immediate past and present-day technology is likely to accelerate changes and introduce new factors which must be taken into account. These considerations argue that 'base-line' is an inappropriate term to apply to estuaries, and that initial descriptions should be based on data from a considerable period. The order of decades is not too long if slow but consistent trends are to be distinguished from the short and medium-term variability.

Many of the problems concerning the chemistry of estuaries, particularly in relation to pollutant discharge, will have a direct biological connotation, for it is a forceful and often applied argument that unless any utilization produces a noticeable detrimental effect, the estuary is operating below capacity in its ability to deal with the applied perturbations, and interference with the functioning of indigenous biota represents, probably, the most sensitive indicator of damage. Thus, a typical research programme may be directed towards elucidating the lethal or sub-lethal effects of a particular effluent on any of the various forms of marine life and determining whether injurious concentrations will be attained within the region. Although such problems are often expressed as single cause and effect phenomena, the natural consequences of interfering with any single biological process may often be far more complex and far reaching. There is a need for greater understanding of estuarine systems from a broad-based ecological viewpoint.

For a complete description of the instantaneous chemical state of any estuary, even with the simplifying assumption of steady-state conditions, the following information would be required:

1. the rates of supply and chemical forms of all constituents supplied by the sources, fresh water and offshore saline water and the net balance of material transport at the air/water and water/sediment interfaces;
2. the resultant distribution of these constituents as determined by the hydrodynamic characteristics which control the circulation and mixing of water and particulate materials;
3. the reactants, mechanisms, rates and products of all *in situ* chemical and biological processes occurring within the system;
4. quantitative knowledge of the physico–chemical rate-determining state variables and their distribution throughout the estuary.

The considerable increase in complexity produced when temporal and spatial fluctuations are included is readily apparent. The problems associated with satisfying these requirements are not all amenable to solution using a direct practical field approach, and the complete description of any chemical system will probably require a combination of results from both field and laboratory studies. Field studies yield information concerning the instantaneous distributions and temporal variability of the chemical constituents of estuarine waters, with limits of resolution and precision in space and time which are determined by the procedures adopted. Unless the constituents behave conservatively, complete interpretation of field results will require laboratory studies and simulations using simplified systems, for the effects of *in situ* processes, and their temporal and spatial variability in real situations will usually be too complex for investigation by direct sampling means alone.

The extent to which the results of any field investigation are interpretable will be determined by the tactics adopted during their collection, which must be considered with knowledge of the behaviour of estuaries in general, together with information pertinent to the particular estuary under consideration. The important general characteristics of estuaries introduced in earlier sections may be summarized as follows:

1. Variability is an inherent and essential property of estuarine systems. The distribution of properties and the occurrence of *in situ* processes undergo short-term, seasonal and longer term fluctuations, although many of the fluctuations have a cyclic nature. Studies of estuarine properties, be they physical, chemical or biological, entail investigations of a dynamic system within which steady-state conditions are unlikely to be attained. However, many estuarine functions approximate to a pattern of continuous adjustment about recognizable and, within finite limits, reproducible average conditions.
2. Whereas the hydrodynamic behaviour of any estuary cannot be fully described using readily available information such as physical dimensions, tidal range and river flow, the general influence of these important characteristics is sufficiently well documented that collation and utilization of available information should be a necessary prerequisite to the planning and execution of field investigations.
3. The rate of introduction of chemical constituents into an estuary by the saline and fresh-water contributions is not a direct function of the volume of supply, for each is subject to cyclic and erratic fluctuations in composition. These are most marked in the fresh-water

contribution for which the rate of mass input is also strongly influenced by fluctuations in volume discharge. Proper characterization of estuarine chemistry necessarily demands a knowledge of constituent inputs and their variability so that investigation of the chemistry of the source waters, particularly the fresh water, may well be a necessary part of any estuarine investigation.

4. True chemical equilibrium of any chemical constituent within an estuary with respect particularly to heterogeneous reactions is unlikely, even for transient periods. However, the balance of effects such as rates of reactions, turnover of water within the estuary and overall concentration levels often results in a quasi-equilibrium state whereby many dissolved chemical species apparently behave conservatively. The estuarine dispersion, distribution and flux of these components can be determined from hydrodynamic considerations alone, provided that the inputs are known.
5. However introduced, settlable particles exhibit distinctive transport and distributional patterns within estuaries and cannot be considered conservative in the hydrodynamic sense. They are likely to be dispersed throughout the estuary and may be maintained within the system for periods much longer than required for turnover of water. Tidally generated cyclic resuspension may maintain them in close contact with the bulk aqueous phase for extended periods. Even if permanently or semi-permanently settled, they must still be considered as an intimate part of the estuarine chemical system.
6. Present knowledge of the mechanisms and rates of *in situ* estuarine chemical reactions, and of their controls, is mostly insufficient to allow more than cursory, and mainly empirical predictions of the behaviour of reactive chemical constituents, even where non-conservative behaviour is pronounced.

These generalized attributes provide a framework of information pertinent to the design of any estuarine chemical study. However, practical investigations of estuarine behaviour will involve a choice from a wide range of types of investigation and methods of implementation in any individual circumstances.

Programme objectives and the 'desk study'

Whatever the reasons for the implementation of an estuarine investigational programme, the precise objective must first be clearly and unambiguously defined in order that the optimum procedure may be derived. Subsequently, it must be ascertained whether practical field and/or laboratory investigations are necessarily required, and if so, whether, and to what extent, they are feasible when the defined objective is considered in relation to the general nature of estuarine behaviour, the facilities available for investigational purposes, and the individuality of the particular estuary under investigation. It should always be considered whether or not the required information can be obtained, with sufficient accuracy, from information concerning estuarine chemical processes in general in conjunction with already available descriptions of the particular estuary under consideration. Even if such a 'desk study' approach does not fully suffice to answer the requirements, the collation and utilization of all available information should always be considered as an essential preliminary adjunct to any practical investigation.

A complete evaluation of the mechanism of an estuarine chemical process and, especially, the acquirement of quantitative kinetic data requires systematic laboratory investigations in addition to field observations. Two basic experimental approaches have been utilized. The first uses natural samples, abstracted from the mixing region, as experimental material. The responses of these samples to controlled chemical and/or biological manipulations are examined. For example, incubations of a sample, with and without modification of individual physico-chemical properties, provides time-series data on changes in reactant concentration which are amenable to kinetic interpretation (reaction mechanism, order, rate constants, stoichiometry). Strictly speaking, the results are relevant only to conditions at the time and place of sampling and generalized conclusions must be founded on experiments repeated through time and space scales. The design of these experiments should be based on the results of previous field observations, and they should, ideally, be carried out on samples collected whilst field observations are being made. See Morris *et al.* (1982*b*) for an example of combined field and laboratory studies utilizing these techniques.

The second approach, which endeavours to induce estuarine chemical reactivity by a simulation of estuarine mixing, can be divided into two categories, batch mixing and dynamic mixing. In the batch procedure, two estuarine samples (one is usually a river water), which may be real, artificial or pre-manipulated as required, are mixed in proportions which yield a range of samples of intermediate degrees of mixing. For example, Sholkovitz (1976) mixed filtered river water with sea water in an investigation of the production of particulate material during estuarine mixing (accordingly, he termed this the 'product method'). Other applications of the batch mixing method include studies of particle generation (Eisma *et al.*, 1980), trace metal sorption on particles (Duinker & Nolting, 1978) and humic-clay interactions (Preston & Riley, 1982). By sequential sub-sampling following mixing, kinetic aspects of reactivity may be evaluated, but little use has been made of this.

Dynamic simulation of mixing overcomes one major objection to the batch procedure, that it does not realistically simulate the natural situation where mixing occurs between waters of closely similar properties whose compositions have been modified throughout their previous mixing histories. Mixing is achieved in a series of well-stirred tank reactors which provide an adaptable method of simulating a wide range of conditions (Bale & Morris, 1981). As presently applied, this dynamic method is suitable for studies of dissolved and colloidal phases but it does not cope adequately with settlable particles.

A major problem in the study of environmental aquatic processes using natural samples is the separation of microbiological from chemical contributions to reactivity (Brock, 1978). Many published estuarine chemical studies have ignored this problem. The usual approach has been to add a biocide to reactant mixtures, but the choice of biocide is not straightforward, since it must be totally effective without interfering with non-biotic reactivity or with the chemical assays used. The demonstrations of an optimum reaction temperature is probably the most effective test of microbiological participation.

Field studies – the sampling and analytical programme

In general terms, the approach which will be adopted in field operations will be included in one of three categories:

1. surveying;
2. monitoring;
3. specific investigations.

Most problems will be properly and necessarily tackled using these techniques separately or in combination but some information concerning estuarine conditions may also be obtained by indirect means. For example, the unregulated nature of the accumulation of many trace elements by brown algae allows the use of algal analyses for determining mean ambient dissolved metal levels which may provide sufficient information for metal pollution studies (Bryan & Hummerstone, 1973; Morris & Bale, 1975).

Whilst considerations of both time and space are always important, surveying has a predominantly spatial orientation whereas monitoring programmes are essentially based on considerations of the time factor. Each of these approaches will have an importance to any study which is determined by the problem under investigation, but a combination will be required for many purposes. The establishment of a network of stations to be sampled during a monitoring programme will best be founded on the results of preliminary survey work and the choice of parameters to be monitored should be based on a knowledge of chemical behaviour, perhaps deduced from specific investigations.

In theory, monitoring requirements are most efficiently satisfied by using automated (continuous or time-discrete) sampling and, if possible, analytical and recording units. Similarly, surveying requirements are often greatly improved by, and may sometimes necessitate, the use of continuously recording devices which provide an immediate analytical output of the variables under investigation. Such techniques can greatly alleviate the cost in time and resources required to obtain chemical data of an extent and quality compatible with the requirements of many investigational programmes. However, although considerable advances in the development and operation of such techniques are being made, estuarine chemical investigations must continue for some time to rely on techniques involving discrete sampling at individual sampling points using limited transport facilites.

Whatever the approach or combination of approaches adopted, there are three essential practical decisions to be made, with full regard for the interdependent implications of each choice:

1. the sampling and analytical methodologies to be employed;
2. the number and positioning of sampling points;
3. the frequency of observations.

Other sections of this book are concerned with the various possible sampling and analytical techniques and their performance and limitations when applied to estuarine problems. Here, it must be emphasized that whereas optimum accuracy is an essential requirement, attempts to obtain maximum analytical precision, which are often accompanied by dramatically increased demands on instrumental costs and manpower utilization, are not always compatible with the turbulent nature of the estuarine mixing process and its imposition of transitory conditions at any single sampling point. For example, Mangelsdorf (1967) has pointed out that although techniques are available to determine salinity to a very high degree of precision, this precision is often superfluous to the requirements of estuarine studies and may introduce much useless noise and random fluctuations. Applications of salinity measurements in estuaries mainly demand convenience of measurement, for what normally is required is not ultimate precision, but more measurements in more places at more different times. Highly precise observations are justifiable only if commensurate precision is obtained for all other determinations and in operational procedures such as position fixing and timing; they must also be necessary for the purposes of the investigation.

The determination of any chemical constituent at a single sampling point within an estuary at a single time of sampling provides only a very limited amount of information, whatever analytical precision is obtained. There is no indication of how representative that determination is of average conditions or of the degree of fluctuation at that point, or in the immediate vicinity, nor can one deduce confidently the source, immediate history or future behaviour of that constituent with respect to transport and participation in *in situ* processes. Meaningful field investigation of any estuarine component or process must generally be based on a multiple sampling programme extended in both space and time.

The ideal strategy for investigations of the distributions of estuarine chemical components and their variability over any chosen time period is by repetitive synoptic sampling of the entire estuary, or a smaller region therein, utilizing a grid of sampling points, in one, two or three dimensions as appropriate. The distance between sampling points and the time intervals between samplings must be commensurate with the required resolutions in space and time, respectively. However, repetitive synoptic coverage of large estuarine areas will usually make inordinate demands on resources, especially in its requirement for multiple sampling facilities. Therefore, practical investigations will often necessarily entail successive rather than synoptic sampling. A practical approach to obtaining synoptic data using a single sampling vessel, at least for segments of smaller estuaries, has been demonstrated by Morris (1978). Sampling is carried out during rapid consecutive axial traverses yielding data in dimensions of time and space from which interpolations of synoptic states can be derived.

The number of sampling points and the frequency and number of observations control the rate at which analytical determinations and/or pre-storage treatments need to be carried out in order to keep pace with the flow of samples. Hence, a practical programme may be restricted either by the demands on resources necessary for producing sufficient analytical outputs of adequate precision or by the practicalities of maintaining the required sampling programme. Due consideration of this point is essential early in the planning. As for salinity determinations, utilization of more convenient methods of producing analytical output may be called for, even at the expense of potential precision. This should not, of course, be at the expense of the level of precision necessarily required for the purposes of the investigation. Careful consideration should be applied to the choice of sampling and analytical methodology with full recognition of the problem in hand.

Estuarine variability and the location and frequency of sampling

Consideration of the time factor is of paramount importance in practical studies of estuaries. The major factors which impose temporal variability in the composition of water pertaining at a fixed geographical position within an estuary are summarized in Table 1.3. The extent to which each facet of this variability is considered as an integral time-variant phenomenon, the effects of which must necessarily be characterized, is a function of the degree of time averaging appropriate to any individual investigation, and the time scale under consideration. For example, transient erratic fluctuations which are a function solely of the turbulent nature of the estuarine mixing processes are rarely of importance, except in studies of the mixing processes themselves, and effectively constitute unwanted noise about the required signal.

Tidally induced variability at individual sites within an estuary will be intrinsically important, for example, in studying the distribution of materials in the vicinity of an outfall or subsidiary fresh-water input, in water quality investigations at a mariculture establishment and in hydraulic studies of the transport of particulate materials. Full characterization will entail an investigation of the combined effects of tidal periodicity, variable input and metereorological conditions, which demands grid sampling for intervals covering one or more tidal cycles repeated at a number of appropriate times.

Whatever the frequency of variation of the phenomenon under investigation, the time interval between samplings must be carefully selected. Ideally, the demands on resources will be minimized by using the longest possible interval between samplings commensurate with adequate characterization of the phenomenon. To achieve this, the effort required for the statistical evaluation of the data must also be taken into account. Gunnerson (1966) has used power spectrum analysis of time-series data to illustrate this point. Using monthly data of dissolved oxygen in the tidal region of the Potomac River, Washington D.C., he demonstrated that a two-hour sampling interval was sufficient to resolve the frequencies of all the dominant factors that affected oxygen levels. In contrast, a more than sixfold increase in computational effort applied to data from a 0.2 hour sampling interval resulted in less resolution because of the increased amount of data processing required. He warns that 'it is even easier to perform useless statistics than to collect unnecessary samples'.

The design of rational sampling programmes which satisfy explicit investigational aims for rivers and effluents has been discussed thoroughly by Montgomery & Hart (1974). Their difficulties in deriving a 'blueprint' to cover all possible cases are even more accentuated for the estuarine

Table 1.3. *Factors which impose temporal variability on the composition of water at a fixed geographical position in an estuary*

Form of variability	Frequency	Process
Cyclic fluctuations about average conditions		
1. Small scale random fluctuations about mean level or trend	< Seconds to minutes	Turbulent eddy structure of water in mixing regime
2. Variability around mean level or trend	Minutes to hours	Eddying; incompletely mixed inputs; temporary isolation of water, e.g. in bays or over mud flats
3. Regular interruptions to mean level or trend	Often tidal	Intermittent discharge
4. Regularly cyclic	Usually $12\frac{1}{2}$ hours, with spring/neap variations in amplitude	Tidal advection
5. Regularly cyclic	Annual	Biological and/or climatic cycles
Intermittent fluctuations		
1. Irregular interruptions to mean level or trend	—	Irregular discharge
2. Intermittent significant change in water characteristics	Often annual, i.e. more probable at certain times of year	Climatic effects, e.g. exceptionally high or low fresh water run-off; storm surges; biological instability (plankton blooms)
3. Permanent discontinuity in water characteristics	—	Change in exploitation, e.g. new discharge. Natural phenomenon, e.g. morphological adjustment to estuarine bed form, rechannelling
Trend		
1. Persistent year to year trend	—	Change in exploitation, e.g. continuous increase or decrease in discharge. Natural estuarine evolution, e.g. continuing siltation

environment. Nevertheless, many of the general points raised are applicable to estuarine investigations provided the complications introduced by the oscillatory tidal motion and any vertical and lateral inhomogeneity are taken into account. Montgomery & Hart (1974) stress the different sampling strategy which is appropriate to the determination of summary values and frequency distributions of water quality indices in comparison with that required for the determination of actual values or sequences of values. Summary measures of water quality, for example, averages or frequency distributions of concentration require a sampling programme for any location which is unequivocally defined by the desired precision and confidence level of the result. Simple statistical evaluations of sampling frequency based on the assumption that the data are independent, randomly sampled and normally distributed have been used with success. The suitability of the application of such a technique to the estuarine environment remains to be tested although the monitoring of an effluent discharged directly into an estuary is potentially an appropriate application. However, considerable replication of sampling may be required for the detection of small differences between water samples when the inherent temporal and spatial variability of natural estuarine properties are considered, especially if those properties are a function of biological populations. For example, Carpenter *et al.* (1974) calculated that 88 replicate samples per station are required to detect a ± 5 per cent difference in chlorophyll *a*, reducing to 22 for a ± 10 per cent change and 6 for a ± 20 per cent change, in waters of eastern Long Island Sound.

Sampling at a geographically located fixed position over a time period is justifiable and efficient only when the variability of conditions at that fixed position are under investigation. For more general purposes, a reference frame based on the mixing pattern, usually quantified by salinity determinations, is appropriate. This applies particularly to dissolved constituents, whose correlation with salinity is unaffected by turbulent-induced fluctuations. In this case, considerations of time must be directed towards elucidating the extent of fluctuations brought about by the functioning of all processes occurring within the estuary over the time period of the investigation, and also towards the practical implications of sampling at different times in the tidal cycle, if synoptic sampling is not used.

Although it has been emphasized previously that the dynamic nature of estuaries introduces limitations in the extent to which steady-state descriptions of estuaries can adequately represent the true situation, the output of investigations are most often represented in this manner. Graphical representations of the geographical distribution of a chemical component within an estuarine system or the correlation between two chemical components, measured throughout the estuary, are typical, and often-utilized outputs (see Chapter 7 for examples). It is implicit, however, that these are synoptic or quasi-synoptic, which they must be if they are to be interpretable. To this end, the sampling grid on which they are based must be completed in as short a time as possible, for they necessarily assume a steady-state condition (apart from tidal oscillations) over the sampling period. The extent of the sampling period represents the major limiting factor in any attempt to approach synoptic coverage without unlimited facilities. Certainly, any quasi-synoptic representation should be based only on results collected over a time interval appreciably shorter than the turnover time for fresh water within the system and the time span within which any other potential perturbatory factors may be effective, unless the extent of their interference is fully characterized.

The essential requirement for the sampling to be confined within as short a period as possible will almost certainly entail sampling at individual stations at different times relative to the tidal period. For purposes requiring outputs determined relative to geographical coordinates, quasi-synoptic representations can be obtained by applying corrections to the actual sampling positions which allow for the tidal advection taking place between the actual time of sampling at each station and any arbitrary chosen fixed time (as explained in Chapter 7). For interpretative purposes, this fixed time is most often chosen appropriate to a specified tidal condition, for example, high or low water. This always entails a degree of misrepresentation because tidal wave progression within estuaries means that a synoptic representation of such tidal conditions is to a certain extent unreal. Similarly, when the investigational period extends in practice over more than one tidal cycle, changes in tidal excursion imply that even when applying advective corrections appropriate to the closest occurrence of the chosen time in the tidal cycle, some misrepresentation is unavoidable. Furthermore, without introducing complicated correction programmes, which are not easily substantiated, the influence of diffusive transport and wind stress and their fluctuations are not accounted for. The extent of discrepancies arising from these factors and their importance depend on each individual circumstance.

During the operational planning of an investigation, the practical implications of tidal movements must always be borne in mind. The actual sequence in which a grid of stations is sampled will be dependent on a number of factors including navigational considerations. However, if the inter-station distance is of the order of, or less than, the tidal excursion at that vicinity, care must be taken to ensure that sampling at adjacent stations is not carried out in a way such that one effectively samples the same parcel of water. For direct *in situ* studies of processes occurring over short time intervals and within restricted volumes of water, the complications introduced by tidal advection may be removed by following a 'labelled' parcel of water as in studies of chemical diffusion processes using dye patches (Talbot & Talbot, 1974).

Whatever the overall time scale of observations and the frequency of sampling at any position, it must be remembered that sampling at discrete times yields results which are characteristic only of those times. Restrictions on the use of operational craft in poor weather conditions may mean, for example, that perturbations to the system imposed by high wind stress are never investigated. It will often be advantageous to adjust the frequency of observations in accordance with the rate of change of the property under investigation. For example, a study of the ecological implications of the presence of the primary nutrient elements is likely to benefit from a greater sampling frequency during the period of enhanced phytoplankton production.

The choice of sampling positions is determined by the scale of the investigation and the gradients and fluctuations of properties of interest within the boundaries of the investigation. The positioning of individual sampling points and their spacing must be such that one can credibly interpolate between them. Any sharp fluctuations in properties of a scale, axial, transverse or depthwise, shorter than the distance between sampling points will remain undetected or poorly characterized. As noted earlier, characterization of the general estuarine condition or of processes occurring therein are most amenable to study using the salinity distribution as the primary reference frame rather than fixed geographical coordinates. Measurements at 1 ‰ salinity intervals, one-dimensionally and preferably at mid-channel along the estuarine length, is an appropriate guideline for general surveys, but extension into two or three dimensions and more closely spaced observations may be required for the study of *in situ* chemical processes or where inhomogeneity is accentuated around outfalls. It is advisable to extend the range into the fresh water. For economic reasons, sampling from the banksides, piers and bridges may appear favourable. However, extreme care must be taken, particularly for bank-side samples, to ensure that they are not subject to localized small-scale influences if they are to be utilized as indicators of general estuarine conditions. Also, restricted availability of, or access to, such sites will not generally allow a completely adequate spacing of sampling sites to be selected.

It is clear from earlier considerations that there will be a finite limit to the number of samples which can be collected, processed, analysed and

Table 1.4. *General guidelines to the operation of practical estuarine chemical investigations*

Type of investigation	Timing of investigation	Positioning and spacing of sampling points	Frequency
Water column			
1. 'Fire-brigade', e.g. accidental toxic discharge	Incident related	Determined by scale of incident	As necessary throughout period of occurrence of abnormal conditions and effects
2. Survey – whole estuary	Non-exceptional conditions with respect to tides, river flow, weather, season, etc. Synoptic or pseudosynoptic sampling	At geographical sites fixed according to the salinity gradient, e.g. at 1 or 2‰ salinity intervals through entire range, including fresh water sampling. Closer spacing at selected sites if required, e.g. around major inputs. One-, two- or three-dimensional sampling according to estuarine type and investigational aims	Repeated as appropriate to cover normal range of environmental conditions. Opportunistic sampling of abnormal conditions
3. Survey – regional, e.g. around point discharges	Non-exceptional conditions with respect to tides, river flow, weather, season, etc. Input must be representative, i.e. is the discharge regular or intermittent? Is it of constant composition? Synoptic sampling	Siting dependent on volume and rate of input of discharge. Increased spacing with distance from discharge. Sampling throughout depth range even in vertically homogeneous estuaries; one or two horizontal dimensions as appropriate to size of estuary	Hourly, as minimum, or more frequently over a tidal cycle. Repeated as appropriate to cover normal range of environmental conditions and variability in discharge. Opportunistic sampling of abnormal conditions
4. Survey – fixed site, e.g. mariculture establishment	Non-exceptional conditions with respect to tides, river flow, weather, season, etc.	Determined by scale of investigational area	Hourly, as minimum, or more frequently over a tidal cycle. Repeated as appropriate to cover normal range of environmental conditions. Opportunistic sampling of abnormal conditions

5. Monitoring	Defined by monitoring requirements. Regular and repetitive	Defined by monitoring requirements. May be based on preliminary survey results	Variable from continuous recording to discrete sampling with long interval. Frequency determined by precision of output required relative to environmental variability. Compositing of samples may be appropriate
6. Long-term studies, e.g. ecological investigations, model construction and testing, etc.	Regular and repetitive over period of study. Additional investigations for extremes of environmental conditions and regions of special interest. Synoptic or pseudosynoptic sampling	Either geographically or salinity related for entire estuary or region as appropriate; closer spacing at sites of special interest, e.g. major inputs. One-, two- or three-dimensional sampling according to estuarine type and purpose of study	Usually weekly to monthly; greater frequency at specified times, e.g. during spring bloom
7. Specific studies, e.g. characterization of interactive behaviour	Defined by aims of investigation	Determined by aims of investigation	Determined by aims of investigation
Sediments			
1. Survey – whole estuary or regional	Defined if related to proposed schemes	Regular grid throughout estuary or region, finer grid at sites of special interest or accentuated variability. Grid size reduced until required definition is obtained. Core sampling for historical record	Single survey
2. Monitoring	Defined by monitoring requirements. Regular and repetitive	Defined by monitoring requirements. May be based on preliminary survey results	Defined by monitoring requirements

reported on within the time schedule imposed by the requirements of the investigation; the actual rate determining factor being variable according to individual circumstances. Yet this limit must not reduce either temporal or spatial resolution to such an extent that the aims of the investigation are unattainable, i.e. there will be an essential minimum of resources necessarily required to carry out the task in hand. Thus the optimum procedure may well entail an irregularity in the spacing of sampling points such that closer spacing is utilized in vicinities of major importance with respect to the distributions and processes under consideration, with wider spacing elsewhere.

Sampling vehicles

The discussion has proceeded so far with little regard for the logistic problems associated with attempts to sample the estuarine environment, although they will often exert a considerable control over the investigational procedure adopted. Basically, a number of chosen sampling stations spread over an area or along a length of the estuary must be sampled within a restricted time interval. To achieve this, sampling vehicles must be considered according to the following criteria:

1. accessibility of chosen sampling positions and sufficient manoeuvrability under the tidal and meteorological conditions to be encountered;
2. sufficient speed to conform to the required time schedule;
3. adequate space and payload for sample equipment and its operation, personnel and collected samples;
4. availability of ancillary facilities such as electrical supply and position fixing devices.

The importance ascribed to each of these properties will be a function of each investigational problem, but it will often occur that none of the craft potentially suitable for estuarine work, conventional surface vessels, hydrofoils, hovercraft and helicopters, will be completely satisfactory on all counts. Utilization of more than one type of vehicle or the simultaneous use of multiple sampling vehicles may well necessarily be required for some estuarine research programmes (see Chapter 2 for details).

General guidelines for field studies

The foregoing discussions indicate in numerous ways the interdependent nature of the many considerations and decisions pertinent to the planning and execution of estuarine field studies. The general approaches which may be adopted for any particular purposes are outlined in Table 1.4.

This table must be considered only as a summary guideline, for the output requirements of any practical investigations are likely to be obtainable by a variety of tactical approaches. However, each approach will entail a specific demand on resources determined by the combination of operational procedures adopted and the limitations and constraints involved in their application within the estuarine environment. It is the principal task of the investigator to deploy the available resources with optimum efficiency.

References

Abdullah, M. I. & Royle, L. G. (1972). Heavy metal content of some rivers and lakes in Wales. *Nature, London*, **238**, 329–30.

Abdullah, M. I. & Royle, L. G. (1973). Chemical evidence for the dispersal of River Mersey run-off in Liverpool Bay. *Estuarine and Coastal Marine Science*, **1**, 401–9.

Allen, G. P., Salomon, J. C., Bassoullet, P., Du Penhoat, Y. & De Grandpré, C. (1980). Effects of tides on mixing and suspended sediment transport in macrotidal estuaries. *Sedimentary Geology*, **26**, 69–90.

Andelman, J. B. (1973). Incidence, variability and controlling factors for trace elements in natural fresh waters. In *Trace Metals and Metal–Organic Interactions in Natural Waters*, ed. P. C. Singer, pp. 57–88. Michigan: Ann Arbor Science.

Armstrong, F. A. J. & Butler, E. I. (1959). Chemical changes in sea water off Plymouth during 1957. *Journal of the Marine Biological Association of the United Kingdom*, **38**, 41–5.

Arons, A. B. & Stommel, H. (1951). A mixing length theory of tidal flushing. *Transactions. American Geophysical Union*, **32**, 419–21.

Aston, S. R. (1981). Estuarine chemistry. In *Chemical Oceanography*, ed. J. P. Riley & R. Chester, 2nd edn, vol. 7, pp. 361–440. London: Academic Press.

Bale, A. J. & Morris, A. W. (1981). Laboratory simulation of chemical processes induced by estuarine mixing: the behaviour of iron and phosphate in estuaries. *Estuarine, Coastal and Shelf Science*, **13**, 1–10.

Batley, G. E. & Gardner, D. (1978). A study of copper, lead and cadmium speciation in some estuarine and coastal marine waters. *Estuarine and Coastal Marine Science*, **7**, 59–70.

Bayne, D. R. & Lawrence, J. M. (1972). Separating constituents of phytoplankton populations by continuous particle electrophoresis. *Limnology and Oceanography*, **17**, 481–90.

Beck, K. C., Reuter, J. H. & Perdue, E. M. (1974). Organic and inorganic geochemistry of some coastal plain rivers of the south-eastern United States. *Geochimica et Cosmochimica Acta*, **38**, 341–64.

Bewers, J. M. & Yates, P. A. (1981). Behaviour of trace metals during estuarine mixing. In *River Inputs to Ocean Systems*, ed. J.-M. Martin, J. D. Burton & D. Eisma, pp. 103–5. Paris: UNEP/UNESCO.

Bowden, K. F. (1967). Circulation and diffusion. In *Estuaries*, ed. G. H. Lauff, pp. 15–36. Washington, D.C.: American Association for the Advancement of Science.

Bowden, K. F. (1980). Physical factors: salinity, temperature, circulation, and mixing processes. In *Chemistry and Biogeochemistry of Estuaries*, ed. E. Olausson & I. Cato, pp. 37–70. Chichester: J. Wiley & Sons.

Boyle, E., Collier, R., Dengler, A. T. Edmond, J. M., Ng, A. C. & Stallard, R. F. (1974). On the chemical mass-balance in estuaries. *Geochimica et Cosmochimica Acta*, **38**, 1719–28.

Brock, T. D. (1978). The poisoned control in biogeochemical investigations. In *Environmental Biogeochemistry and Geomicrobiology*, vol. 3, ed. W. E. Krumbein, pp. 717–25. Ann Arbor: Ann Arbor Science.

Bryan, G. W. & Hummerstone, L. G. (1973). Brown seaweed as an indicator of heavy metals in estuaries in south-west England. *Journal of the Marine Biological Association of the United Kingdom*, **53**, 705–20.

Buller, A. T. & McManus, J. (1979). Sediment sampling and analysis. In *Estuarine Hydrography and Sedimentation*, ed. K. R. Dyer, pp. 87–130 Cambridge: Cambridge University Press.

Carpenter, E. J., Anderson, S. J. & Peck, B. B. (1974). Copepod ana chlorophyll *a* concentrations in receiving waters of a nuclear power station and problems associated with their measurement. *Estuarine and Coastal Marine Science*, **2**, 83–8.

Chester, R. & Stoner, J. H. (1975). Trace elements in sediments from the lower Severn Estuary and Bristol Channel. *Marine Pollution Bulletin*, **6**, 92–6.

Conomos, T. J. & Gross, M. G. (1972). River–ocean suspended particulate matter relations in summer. In *The Columbia River Estuary and Adjacent Ocean Waters*, ed. A. T. Pruter & D. L. Alverson, pp. 176–202. Seattle: University of Washington Press.

D'Anglejan, B. F. & Smith, E. C. (1973). Distribution, transport and composition of suspended matter in the St Lawrence estuary. *Canadian Journal of Earth Science*, **10**, 1380–96.

Dillon, P. J. & Kirchner, W. B. (1975). The effects of geology and land use on the export of phosphorus from watersheds. *Water Research*, **9**, 135–48.

Duinker, J. C., Hillebrand, M. T. J., Nolting, R. F., Wellershaus, S. & Kingo Jacobsen, N. (1980). The River Varde Å: processes affecting the behaviour of metals and organochlorines during estuarine mixing. *Netherlands Journal of Sea Research*, **14**, 237–67.

Duinker, J. C. & Nolting, R. F. (1978). Mixing, removal and mobilization of trace metals in the Rhine Estuary. *Netherlands Journal of Sea Research*, **12**, 205–23.

Duxbury, A. C. (1972). Variability of salinity and nutrients off the Columbia River mouth. In *Columbia River Estuary and Adjacent Ocean Waters*, ed. A. T. Pruter & D. L. Alverson, pp. 135–50. Seattle: University of Washington Press.

Dyer, K. R. (1973). *Estuaries: A Physical Introduction*. London: John Wiley & Sons.

Dyer, K. R. & Taylor, P. A. (1973). A simple, segmented prism model of tidal mixing in well-mixed estuaries. *Estuarine and Coastal Marine Science*, **1**, 411–8.

Dyer, K. R. (1981). The measurement of fluxes and flushing times in estuaries. In *River Inputs to Ocean Systems*, ed. J.-M. Martin, J. D. Burton & D. Eisma, pp. 67–76. Paris: UNEP/UNESCO.

Edwards, A. M. C. & Liss, P. S. (1973). Evidence for buffering of dissolved silicon in fresh waters. *Nature, London*, **243**, 341–2.

Edzwald, J. K., Upchurch, J. B. & O'Melia, C. R. (1974). Coagulation in estuaries. *Environmental Science and Technology*, **8**, 58–63.

Eisma, D., Kalf, J. & Veenhuis, M. (1980). The formation of small particles and aggregates in the Rhine Estuary. *Netherlands Journal of Sea Research*, **14**, 172–91.

Elderfield, H. & Hepworth, A. (1975). Diagenesis, metals and pollution in estuaries. *Marine Pollution Bulletin*, **6**, 85–7.

Ewins, P. A. & Spencer, C. P. (1967). The annual cycle of nutrients in the Menai Straits. *Journal of the Marine Biological Association of the United Kingdom*, **47**, 533–42.

Fairbridge, R. W. (1980). The estuary: its definition and geodynamic cycle. In *Chemistry and Biogeochemistry of Estuaries*, ed. E. Olausson & I. Cato, pp. 1–35. Chichester: J. Wiley & Sons.

Festa, J. F. & Hansen, D. V. (1976). A two-dimensional numerical model of estuarine circulation: the effects of altering depth and river discharge. *Estuarine and Coastal Marine Science*, **4**, 309–23.

Festa, J. F. & Hansen, D. V. (1978). Turbidity maxima in partially mixed estuaries: a two-dimensional numerical model. *Estuarine and Coastal Marine Science*, **7**, 347–59.

Florence, T. M. & Batley, G. E. (1976). Determination of the chemical forms of trace metals in natural waters. *Talanta*, **23**, 179–86.

Forster, W. O. (1972). Radioactive and stable nuclides in the Columbia River and adjacent north-east Pacific Ocean. In *The Columbia River Estuary and Adjacent Ocean Waters*, ed. A. T. Pruter & D. L. Alverson, pp. 663–700. Seattle: University of Washington Press.

Foster, P. & Morris, A. W. (1974). Seasonal distribution of ultraviolet absorption in the surface waters of Liverpool Bay. *Estuarine and Coastal Marine Science*, **2**, 283–90.

Gallene, B. (1974). Study of fine material in suspension in the estuary of the Loire and its dynamic grading. *Estuarine and Coastal Marine Science*, **2**, 261–72.

Gameson, A. L. H. (ed.) (1973). *Mathematical and Hydraulic Modelling of Estuarine Pollution*. London: Her Majesty's Stationery Office.

Garvine, R. W. & Monk, J. D. (1974). Frontal structure of a river plume. *Journal of Geophysical Research*, **79**, 2251–9.

Gibbs, R. J. (1970). Mechanisms controlling world water chemistry. *Science, New York*, **170**, 1088–90.

Gibbs, R. J. (1973). Mechanisms of trace metal transport in rivers. *Science, New York*, **180**, 71–3.

Gobeil, G., Sundby, B. & Silverberg, N. (1981). Factors influencing particulate matter geochemistry in the St Lawrence Estuary turbidity maximum. *Marine Chemistry*, **10**, 123–40.

Gunnerson, C. G. (1966). Optimizing sampling intervals in tidal estuaries. *Proceedings of the American Society of Civil Engineers*, **92**, SA2, 103–25.

Hahn, H. H. & Stumm, W. (1970). The role of coagulation in natural waters. *American Journal of Science*, **268**, 354–68.

Hair, M. E. & Bassett, C. R. (1973). Dissolved and particulate humic acids in an east coast estuary. *Estuarine and Coastal Marine Science*, **1**, 107–11.

Hansen, D. V. & Rattray, M. Jr (1966). New dimensions in estuary classification. *Limnology and Oceanography*, **11**, 319–26.

Hess, K. W. (1976). A three-dimensional numerical model of the estuary circulation and salinity in Narragansett Bay. *Estuarine and Coastal Marine Science*, **4**, 325–38.

Hunter, K. A. (1980). Microelectrophoretic properties of natural surface-active organic matter in coastal seawater. *Limnology and Oceanography*, **25**, 807–22.

Hunter, K. A. & Liss, P. S. (1979). The surface charge of suspended particles in estuarine and coastal waters. *Nature*, **282**, 823–5.

Hunter, K. A. & Liss, P. S. (1982). Organic matter and the surface charge of suspended particles in estuarine water. *Limnology and Oceanography*, **27**, 322–34.

Ippen, A. T. (ed.) (1966). *Estuary and Coastline Hydrodynamics*. New York: McGraw-Hill.

Jefferies, D. F., Preston, A. & Steele, A. K. (1973). Distribution of caesium-137 in British coastal waters. *Marine Pollution Bulletin*, **4**, 118–22.

Johnson, R. G. (1974). Particulate matter at the sediment–water interface in coastal environments. *Journal of Marine Research*, **32**, 313–30.

Kennedy, V. C. (1971). Silica variation in stream water with time and discharge. In *Nonequilibrium Systems in Natural Water Chemistry*, ed. R. F. Gould, pp. 94–130. Washington, D.C.: American Chemical Society.

Kent, R. E. (1960). Diffusion in a sectionally homogenous estuary. *Proceedings of the American Society of Civil Engineers*, **86**, SA2, 15–47.

Ketchum, B. H. (1951). The exchange of fresh and salt waters in tidal estuaries. *Journal of Marine Research*, **10**, 18–38.

Ketchum, B. H. (1952). Distribution of coliform bacteria and other pollutants in tidal estuaries. *Sewage and Industrial Waters*, **27**, 1288–96.

Kharkhar, D. P., Turekian, K. K. & Bertine, K. K. (1968). Stream supply of dissolved silver, molybdenum, antimony, selenium, chromium, cobalt, rubidium and caesium to the oceans. *Geochimica et Cosmochimica Acta*, **32**, 295–8.

Kirchner, W. B. (1975). An examination of the relationship between drainage basin morphology and the export of phosphorus. *Limnology and Oceanography*, **20**, 267–70.

Knauer, G. A. & Martin, J. H. (1973). Seasonal variations of cadmium, copper, manganese, lead and zinc in water and phytoplankton in Monterey Bay, California. *Limnology and Oceanography*, **18**, 597–604.

Kranck, K. (1973). Flocculation of suspended sediment in the sea. *Nature, London*, **246**, 348–50.

Krone, R. B. (1978). Aggregation of suspended particles in estuaries. In *Estuarine Transport Processes*, ed. B. J. Kjerfve, pp. 177–90. Columbia: University of South Carolina Press.

Lewis, R. E. & Stephenson, R. R. (1975). Planning the pollution budget of an estuary. In *Pollution Criteria for Estuaries*, ed. P. R. Helliwell & J. Bossanyi, pp. 13.1–13.7. London: Pentech Press.

Li, Y-H. & Chan, L-H. (1979). Desorption of Ba and ^{226}Ra from river-borne sediments in the Hudson Estuary. *Earth and Planetary Science Letters*, **43**, 343–50.

Lion, L. W., Altman, R. S. & Leckie, J. O. (1982). Trace-metal adsorption characteristics of estuarine particulate matter: evaluation of contributions of Fe/Mn oxide and organic surface coatings. *Environmental Science and Technology*, **16**, 660–6.

Liss, P. S. (1969). Reactive silicate concentrations observed in the Irish Sea. *Journal of the Marine Biological Association of the United Kingdom*, **49**, 577–88.

Liss, P. S. (1976). Conservative and non-conservative behaviour of dissolved constituents during estuarine mixing. In *Estuarine Chemistry*, ed. J. D. Burton & P. S. Liss, pp. 93–130. London: Academic Press.

Liss, P. S. & Pointon, M. J. (1973). Removal of dissolved boron and silicon during estuarine mixing of sea and river waters. *Geochimica et Cosmochimica Acta*, **37**, 1493–8.

Liss, P. S. & Spencer, C. P. (1970). Abiological processes in the removal of silicate from sea water. *Geochimica et Cosmochimica Acta*, **34**, 1073–88.

Loder, T. C. & Reichard, R. P. (1981). The dynamics of conservative mixing in estuaries. *Estuaries*, **4**, 64–9.

Loring, D. H., Rantala, R. T. T., Morris, A. W., Bale, A. J. & Howland, R. J. M. (1983). The chemical composition of suspended particles in an estuarine turbidity maximum zone. *Canadian Journal of Fisheries and Aquatic Science*, **40** (Supplement 1), 201–6.

Luoma, S. N. & Bryan, G. W. (1981). Statistical assessment of the form of trace metals in oxidized estuarine sediments employing chemical extractants. *Science of the Total Environment*, **17**, 165–96.

Mackay, D. W. & Gilligan, J. (1972). The relative importance of freshwater input, temperature and tidal range in determining levels of dissolved oxygen in a polluted estuary. *Water Research*, **6**, 183–90.

Mantoura, R. F. C. & Morris, A. W. (1983). Measurement of chemical distributions and processes. In *Practical Procedures for Estuarine Studies*, ed. A. W. Morris, pp. 55–100. Swindon: Natural Environment Research Council.

Mantoura, R. F. C., Dickson, A. & Riley, J. P. (1978). The complexation of metals with humic materials in natural waters. *Estuarine and Coastal Marine Science*, **6**, 387–408.

Mangelsdorf, P. C. (1967). Salinity measurements in estuaries. In *Estuaries*, ed. G. H. Lauff, pp. 71–9. Washington, D.C.: American Association for the Advancement of Science.

Martin, J.-M., Thomas, A. J. & Van Grieken, R. (1978). Trace element composition of Zaire suspended sediments. *Netherlands Journal of Sea Research*, **12**, 414–20.

McCave, I. N. (1979). Suspended sediment. In *Estuarine Hydrography and Sedimentation*, ed. K. R. Dyer, pp. 131–85. Cambridge: Cambridge University Press.

Meade, R. H. (1972). Transport and deposition of sediment in estuaries. In *Environmental Framework of Coastal Plain Estuaries*, ed. B. W. Nelson, pp. 91–117. Memoir 133. Boulder: The Geological Society of America.

Meyers, P. A. & Quinn, J. G. (1974). Organic matter on clay minerals and marine sediments – effects on adsorption of dissolved copper, phosphate and lipids from saline solutions. *Chemical Geology*, **13**, 63–8.

Montgomery, H. A. C. & Hart, I. C. (1974). The design of sampling programmes for river and effluents. *Water Pollution Control*, **73**, 77–98.

Mook, W. G. & Koene, B. K. S. (1975). Chemistry of dissolved inorganic carbon in estuarine and coastal brackish waters. *Estuarine and Coastal Marine Science*, **3**, 325–36.

Morris, A. W. (1974). Seasonal variation of dissolved metals in inshore waters of the Menai Straits. *Marine Pollution Bulletin*, **5**, 54–9.

Morris, A. W. (1978). Chemical processes in estuaries: the importance of pH and its variability. In *Environmental Biogeochemistry and Geomicrobiology*, vol. 1, ed. W. E. Krumbein, pp. 179–87. Ann Arbor: Ann Arbor Science.

Morris, A. W. & Bale, A. J. (1975). The accumulation of cadmium, copper, manganese and zinc by *Fucus vesiculosus* in the Bristol Channel. *Estuarine and Coastal Marine Science*, **3**, 153–64.

Morris, A. W., Bale, A. J. & Howland, R. J. M. (1981). Nutrient distributions in an estuary: evidence of chemical precipitation of dissolved silicate and phosphate. *Estuarine, Coastal and Shelf Science*, **12**, 205–16.

Morris, A. W., Bale, A. J. & Howland, R. J. M. (1982*a*). Chemical variability in the Tamar Estuary, South-west England. *Estuarine, Coastal and Shelf Science*, **14**, 649–61.

Morris, A. W., Bale, A. J. & Howland, R. J. M. (1982*b*). The dynamics of estuarine manganese cycling. *Estuarine, Coastal and Shelf Science*, **14**, 175–92.

Morris, A. W., Loring, D. H., Bale, A. J., Howland, R. J. M. & Woodward, E. M. S. (1982*c*). Particle dynamics, particulate carbon and the oxygen minimum in an estuary. *Oceanologica Acta*, **5**, 349–53.

Morris, A. W., Mantoura, R. F. C., Bale, A. J. & Howland, R. J. M. (1978). Very low salinity regions of estuaries: important sites for chemical and biological reactions. *Nature, London*, **274**, 678–80.

Mouvet, C. & Bourg, A. C. M. (1983). Speciation (including adsorbed species) of copper, lead, nickel and zinc in the Meuse River. *Water Research*, **17**, 641–9.

Myers, V. B., Iverson, R. L. & Harriss, R. C. (1975). The effect of salinity and dissolved organic matter on surface charge characteristics of some euryhaline phytoplankton. *Journal of Experimental Marine Biology and Ecology*, **17**, 59–68.

Neal, V. T. (1972). Physical aspects of the Columbia River and its estuary. In *The Columbia River Estuary and Adjacent Ocean Waters*, eds A. T. Pruter & D. L. Alverson, pp. 19–40. Seattle: University of Washington Press.

Neihof, R. A. & Loeb, G. A. (1972). The surface charge of particulate matter in sea water. *Limnology and Oceanography*, **17**, 7–16.

Neihof, R. & Loeb, G. (1974). Dissolved organic matter in sea water and the electrical charge of immersed surfaces. *Journal of Marine Research*, **32**, 5–12.

Neu, H. J. A. (1982*a*). Man-made storage of water resources – a liability to the ocean environment? Part I. *Marine Pollution Bulletin*, **13**, 7–12.

Neu, H. J. A. (1982*b*). Man-made storage of water resources – a liability to the ocean environment? Part II. *Marine Pollution Bulletin*, **13**, 44–7.

Nihoul, J. C. J. (ed.) (1975). *Modelling of Marine Systems*. Elsevier Oceanography Series, No. 19. Amsterdam: Elsevier.

Officer, C. B. (1979). Discussion of the behaviour of non-conservative constituents in estuaries. *Estuarine and Coastal Marine Science*, **9**, 91–4.

Officer, C. B. (1980). Discussion of the turbidity maximum in partially mixed estuaries. *Estuarine and Coastal Marine Science*, **10**, 239–46.

O'Kane, J. P. (1980). *Estuarine Water-Quality Management*. Boston: Pitman.

Ongley, E. D., Bynoe, M. C. & Percival, J. B. (1982). Physical and geochemical characteristics of suspended solids, Wilton Creek, Ontario. *Hydrobiologica*, **91**, 41–57.

Park, P. K., Osterberg, C. L. & Forster, W. O. (1972). Chemical budget of the Columbia River. In *The Columbia River Estuary and Adjacent Ocean Waters*, ed. A. T. Pruter & D. L. Alverson, pp. 123–34. Seattle: University of Washington Press.

Pearson, C. R. & Pearson, J. R. A. (1965). A simple method for predicting the dispersion of effluents in estuaries. In *New Chemical Engineering Problems in the Utilization of Water*. American Institution of Chemical Engineers and Institution of Chemical Engineers Symposium No. 9, pp. 50–6. London: Institution of Chemical Engineers.

Perkins, E. J. (1974). *The Biology of Estuaries and Coastal Waters*. London: Academic Press.

Peterson, D. H., Conomos, T. J., Broenkow, W. W. & Scrivani, E. P. (1975). Processes controlling the dissolved silica in San Francisco Bay. In *Estuarine Research*. Vol. 1. *Chemistry, Biology and the Estuarine System*, ed. L. E. Cronin, pp. 153–87. New York: Academic Press.

Pierce, J. M. & Siegel, F. R. (1979). Particulate materials suspended in estuarine and oceanic waters. *Scanning Electron Microscopy*, **1**, 555–62.

Pomeroy, L. R., Smith, E. E. & Grant, G. M. (1965). The exchange of phosphate between estuarine water and sediments. *Limnology and Oceanography*, **8**, 50–5.

Postma, H. (1967). Sediment transport and sedimentation in the estuarine environment. In *Estuaries*, ed. G. H. Lauff, pp. 159–79. Washington: American Association for the Advancement of Science.

Preston, M. R. & Riley, J. P. (1982). The interactions of humic compounds with electrolytes and three clay minerals under simulated estuarine conditions. *Estuarine, Coastal and Shelf Science*, **14**, 567–76.

Pritchard, D. W. (1955). Estuarine circulation patterns. *Proceedings of the American Society of Civil Engineers*, **81**, 1–11.

Pritchard, D. W. (1967). Observations of circulation in coastal plain estuaries. In *Estuaries*, ed. G. H. Lauff, pp. 37–44. Washington, D.C.: American Association for the Advancement of Science.

Pritchard, D. W. (1969). Dispersion and flushing of pollutants in estuaries. *Proceedings of the American Society of Civil Engineers*, **95**, 115–24.

Rattray, M. Jr & Officer, C. B. (1979). Distribution of a non-conservative constituent in an estuary with application to the numerical simulation of dissolved silica in the San Francisco Bay. *Estuarine and Coastal Marine Science*, **8**, 489–94.

Rattray, M. Jr & Officer, C. B. (1981). Discussion of trace metals in the waters of a partially-mixed estuary. *Estuarine, Coastal and Shelf Science*, **12**, 251–66.

Rattray, M. Jr & Uncles, R. J. (1983). On the predictability of the ^{137}Cs distribution in the Severn Estuary. *Estuarine, Coastal and Shelf Science*, **16**, 475–87.

Rowe, G. T., Clifford, C. H., Smith, K. L. & Hamilton, P. L. (1975). Benthic nutrient regeneration and its coupling to primary productivity in coastal waters. *Nature, London*, **255**, 215–17.

Sayles, F. L. & Mangelsdorf, P. C. Jr (1979). Cation-exchange characteristics of Amazon River suspended sediment and its reaction with sea water. *Geochimica et Cosmochimica Acta*, **43**, 767–79.

Schubel, J. R. (1968). Turbidity maximum of the northern Chesapeake Bay. *Science, New York*, **61**, 1013–15.

Schubel, J. R., Wilson, R. E. & Okubo, A. (1978). Vertical transport of suspended particles in upper Chesapeake Bay. In *Estuarine Transport Processes*, ed. B. J. Kjerfve, pp. 161–75. Columbia: University of South Carolina Press.

Sholkovitz, E. R. (1976). Flocculation of dissolved organic and inorganic matter during the mixing of river water and sea water. *Geochimica et Cosmochimica Acta*, **40**, 831–45.

Sholkovitz, E. R. (1979). Chemical and physical processes controlling the chemical composition of suspended material in the River Tay Estuary. *Estuarine and Coastal Marine Science*, **8**, 523–45.

Sholkovitz, E. R. & Price, N. B. (1980). The major-element chemistry of

suspended matter in the Amazon Estuary. *Geochimica et Cosmochimica Acta*, **44**, 163–71.
Sigleo, A. C. & Helz, G. R. (1981). Composition of estuarine colloidal material: major and trace elements. *Geochimica et Cosmochimica Acta*, **45**, 2501–9.
Smith, R. (1980). Buoyancy effects upon longitudinal dispersion in wide well-mixed estuaries. *Philosophical Transactions of the Royal Society* A, **296**, 217–24.
Stommel, H. (1953). Computation of pollution in a vertically mixed estuary. *Sewage and Industrial Wastes*, **25**, 1065–71.
Talbot, J. W. & Talbot, G. A. (1974). Diffusion in shallow seas and in English coastal and estuarine waters. *Rapports et Procès-Verbaux des Réunions. Conseil Permanent International pour l'Exploration de la Mer*, **167**, 93–110.
Turekian, K. K., Harriss, R. C. & Johnson, D. G. (1967). The variation of Si, Cl, Na, Ca, Sr, Ba, Co and Ag in the Neuse River, North Carolina. *Limnology and Oceanography*, **12**, 702–6.
Ulanowicz, R. E. & Flemer, D. A. (1978). A synoptic view of a coastal plain estuary. In *Hydrodynamics of Estuaries and Fjords*, ed. J. C. J. Nihoul, pp. 1–26. Elsevier Oceanography Series, No. 23. Amsterdam: Elsevier.
Vuceta, J. & Morgan, J. J. (1978). Chemical modelling of trace metals in fresh waters: role of complexation and adsorption. *Environmental Science and Technology*, **12**, 1302–9.
Whitfield, P. H. & Schreier, H. (1981). Hysteresis in relationships between discharge and water chemistry in the Fraser River Basin, British Columbia. *Limnology and Oceanography*, **26**, 1179–82.
Westall, J. (1980). Chemical equilibrium including adsorption on charged surfaces. In *Particles in Water. Characterization, Fate, Effects and Removal*, ed. M. C. Kavanaugh & J. O. Leckie, pp. 33–44. Washington, D.C. : American Chemical Society.
Wilson, T. R. S. (1974). Caesium-137 as a water movement tracer in the St George's Channel. *Nature, London*, **248**, 125–7.

General reading

A selection of physical, chemical and biological texts covering theoretical and experimental observations pertinent to the study of estuarine chemical systems.

Burton, J. D. & Liss, P. S. (eds) (1976). *Estuarine Chemistry*. London: Academic Press.
Church, T. M. (ed.) (1975). *Marine Chemistry in the Coastal Environment*. ACS Symposium Series, No. 18. Washington, D.C.: American Chemical Society.
Cronin, L. E. (ed.) (1975). *Estuarine Research*, vol. 1, *Chemistry, Biology and the Estuarine System*. New York: Academic Press.

Cronin, L. E. (ed.) (1975). *Estuarine Research*, vol. 2, *Geology and Engineering*. New York: Academic Press.

Dyer, K. R. (1973). *Estuaries: A Physical Introduction.* London: John Wiley & Sons.

Dyer, K. R. (ed.) (1979). *Estuarine Hydrography and Sedimentation.* Cambridge: Cambridge University Press.

Faust, S. D. & Hunter, J. V. (eds) (1971). *Organic Compounds in Aquatic Environments.* New York: Marcel Dekker.

Gameson, A. L. H. (ed.) (1973). *Mathematical and Hydraulic Modelling of Estuarine Pollution.* London: Her Majesty's Stationery Office.

Hamilton, P. & Macdonald, K. B. (eds) (1980). *Estuarine and Wetland Processes with Emphasis on Modelling.* Marine Science Series, No. 11. New York: Plenum Press.

Jenne, E. A. (ed.) (1979). *Chemical Modeling in Aqueous Systems: Speciation, Sorption, Solubility and Kinetics.* ACS Symposium Series, No. 93. Washington, D.C.: American Chemical Society.

Lauff, G. H. (ed.) (1967). *Estuaries.* Washington, D.C.: American Association for the Advancement of Science.

Martin, J.-M., Burton, J. D. & Eisma, D. (eds) (1981). *River Inputs to Ocean Systems.* Paris: UNEP/UNESCO.

Morris, A. W. (ed.) (1983). *Practical Procedures for Estuarine Studies.* Swindon: Natural Environment Research Council.

Nelson, B. W. (ed.) (1973). *Environmental Framework of Coastal Plain Estuaries.* Memoir 133. Boulder: The Geological Society of America.

Nihoul, J. C. J. (ed.) (1975). *Modelling of Marine Systems.* Elsevier Oceanography Series, No. 19. Amsterdam: Elsevier.

Nihoul, J. C. J. (ed.) (1978). *Hydrodynamics of Estuaries and Fjords.* Elsevier Oceanography Series, No. 23. Amsterdam: Elsevier.

Officer, C. B. (1976). *Physical Oceanography of Estuaries (and Associated Coastal Waters).* New York: Wiley Interscience.

Olausson, E. & Cato, I. (eds) (1980). *Chemistry and Biogeochemistry of Estuaries.* Chichester: J. Wiley & Sons.

Perkins, E. J. (1974). *The Biology of Estuaries and Coastal Waters.* London: Academic Press.

Pruter, A. T. & Alverson, D. L. (eds) (1972). *The Columbia River Estuary and Adjacent Ocean Waters.* Seattle: University of Washington Press.

Rheinheimer, G. (ed.) (1977). *Microbial Ecology of a Brackish Water Environment.* Ecological Studies, vol. 25. New York: Springer-Verlag.

Stevenson, L. H. & Colwell, R. R.. (eds) (1973). *Estuarine Microbial Ecology.* Belle W. Baruch Library in Marine Science, No. 1. Columbia: University of South Carolina Press.

Stumm, W. & Morgan, J. J. (1981). *Aquatic Chemistry: An Introduction Emphasizing Chemical Equilibria in Natural Waters*, 2nd edn. New York: Wiley Interscience.

UNESCO (1978). *Biogeochemistry of Estuarine Sediments.* Paris: UNESCO.

2

Operations in the field

T. M. LEATHERLAND

Field work is the essential link between theoretical plans and laboratory studies. There is no way by which even scrupulously careful or precise analysis in a shore laboratory can restore a sample which has deteriorated since its collection, been contaminated during the collection procedure, or simply become lost or muddled with another because of inadequate labelling. It is also much harder to derive statistical confidence limits for errors which may arise in the field than for those arising during standard chemical analyses. Every effort must therefore be made to eliminate errors arising out of field work. Successful field work is promoted by forethought, organization and perhaps most important of all, practice at the techniques employed.

Practice is also a significant factor in increasing the safety of sampling operations. It is important that strict safety regulations are adhered to at all times during field-work operations. British estuaries and coastal waters can be inhospitable and dangerous, particularly during the winter months, and even during the early summer when sea-water temperatures are still low, especially on the east coast. Every group carrying out operations on or by the sea should have its own set of safety regulations, which must be known and understood by all concerned.

The availability of life-jackets which can be worn uninflated, but which inflate either automatically when immersed in water, or when a trigger is pulled to release compressed CO_2 from a small cylinder, now makes the essential compulsory wearing of life-jackets at all times in small boats much less restrictive and more comfortable.

Time can also have an important part to play; the costs of field surveys have probably increased at least as fast as other items in the scientific budget in recent years, and it is therefore necessary to be time-efficient while sampling. The aim of completing certain surveys during the course of one particular phase of a tide, tidal cycle or during daylight hours may also make time a significant factor. The speed at which field work can be undertaken is also increased by efficient organization and practice.

Field transport

Transport between sampling sites may be by land, water or air. Travelling by road in conjunction with shoreline sampling is not really to be recommended if it can be avoided. From the shoreline it is usually difficult or impossible to take a sample representative of the main water body, the water there will tend to have a different salinity and higher turbidity than that further offshore. However, some piers or jetties may be suitable for taking surface or near-surface samples, although even these rarely extend far enough to provide the means for taking good depth profiles. Shoreline sampling can also be more dangerous than working from a boat. Boatwork will never be carried out with less than two people, but with a single person operating from the shore, there is a possibility of the lack of an alarm signal should he fall from a remote jetty or stumble into deep water from a rocky shore. Seaweed can often make obtaining a good sample from a rocky shore difficult, and sedimentary beaches will usually have the greater turbidity at the shoreline.

The use of helicopters for sampling is a comparatively recent innovation. The helicopter's speed enables almost synoptic sampling to be carried out over a relatively large area, and they should be able to travel at the speed of tidal progression in almost any estuary. They also provide rapid transport of samples back to a laboratory. Helicopter sampling has been used successfully not only in America, but also in the Bristol Channel and Mersey Estuary, in Great Britain; and Port Phillip Bay, in Australia (Anon., 1973). Particularly for occasional large surveys, the costs of such sampling, using a hired machine, may not be too unfavourable. However, the downdraught from a helicopter as it hovers is very considerable, and will dramatically disturb the surface water layers. This may not be too significant in a turbulent, well-mixed area, but in stratified waters it could have an undesirable effect on any near-surface samples taken or on results obtained from lowered probes.

Small hovercraft have been used for taking chemical samples. Their unique amphibious abilities may give them a permanent niche for surveys in particular areas where access is very difficult because of extensive soft mudbanks with only narrow or shallow meandering water channels. However, they tend to have only very restrictive space and payload carrying capacity, and they are very noisy and more expensive to operate and maintain than conventional boats of similar capacity.

Most routine surveys will probably continue to be made by boat, and it is about boat work that this section is mainly concerned. Survey boats may perhaps be most conveniently considered in three different size categories. At the smallest end, there are those which can be manhandled and launched almost anywhere. Secondly, there are those in the approximate length range 4–6 m which can be conveniently transported on a trailer and launched from suitable slipways, or from a firm sloping beach. Thirdly, there are the larger vessels, generally over 6 m in length which will usually spend most of their time in the water.

In the smallest size category, inflatable dinghies will probably generally be found to offer the greatest number of advantages. They tend to be more stable and to have a greater weight-carrying capacity than conventional boats of a similar size, although space tends to be somewhat restricted. In this smallest size category, the inflatable boats also tend to be more seaworthy than glass-fibre hulls. The largest size of engine which can be reasonably conveniently manhandled and carried on and off a small boat will have a power of around 25 hp, but this is sufficient to propel a reasonably loaded 3.5 m long inflatable at about 20 knots. This sort of speed should be adequate for many chemical sampling operations. In most instances it will enable sampling along the length of an estuary to be carried out more rapidly than sampling from the shore with driving between the fixed stations.

If facilities are available to allow the boat to be carried on a trailer, with the engine semi-permanently in place, and to be launched from a slipway or firm sloping beach, then the choice of viable boat types is widened considerably. The upper end of this size category will probably be limited by the combined size and weight of the boat, engine and trailer. If a Land-Rover type vehicle is to do the towing, then the maximum practical boat length will be about 6 m. The carrying capacity, sea-worthiness and versatility of boats increases very rapidly with increasing size. Towards the upper end of this size group, it will even be possible to fix some sort of derrick arrangement and manual winch for sampling from depth, if required. Glass-fibre boats with trihedral hulls (relatively flat-bottomed) and inflatables with a rigid V-shaped hull are both available and have particular advantages in this size range. Both can be operated at quite high speeds in fairly choppy water, and provide a stable working platform when at rest. With a sufficiently powerful engine, the semi-rigid inflatable may have the better ultimate performance and sea-worthiness, and a flooding V-hull will reduce windage when not under power, but the rigid trihedral hull types will provide more useful working space, and their solid construction makes the provision of a shelter for electronic instruments and crew possible, as well as providing more hard points for the lashing and stowage of gear. The ride of flatter-bottomed boats is hard at speed in anything but smooth conditions, but modified trihedral hulls are now available which are claimed to overcome this slight disadvantage.

An engine of 50 hp should propel a 5.5 m long hull with an empty weight of 500 kg at 20 knots, even when the boat is loaded with 2–3 scientists and equipment.

Boats above 6 m in length clearly fall into a different category. Besides requiring a greater depth of water for safe navigation, they are getting a little large to be powered by an outboard motor and tend to be too heavy to be easily and routinely towed around on roads. The necessary capital outlay and maintenance costs of larger boats also increase very dramatically with increasing size and at some point it is likely that a choice between owning and hiring will have to be made. Beyond a certain size, the speed of larger survey vessels is also likely to be less than that of smaller craft, as the hull form tends towards that of a displacement hull. The power requirement and fuel consumption of propelling a 15 m hull at 20 knots are likely to be prohibitive.

Much could be written about the qualities obviously desirable in larger boats, such as seaworthiness, area of working space, clean areas, quietness and freedom from vibration. However, the availability of these features generally increases with hull size, which is likely to be cost-limited. All vessels over 6 m length will usually have some sort of cabin accommodation and power supply, which will enable much more ambitious wet chemical manipulations to be carried out than is possible in smaller open boats. A survey vessel with some cabin accommodation also facilitates longer surveys. Fixed point 12.5 h tidal cycle surveys can be carried out using small open boats, but in this case they tend to become less pleasant unless arrangements can be made for the relief or changing of the operators at some stage during the work.

If widespread areas have to be covered, the possibility of using a relatively small (road-towable) boat operating from a small motorized (dormobile-type) laboratory should also be considered.

Depending upon the frequency of boat usage, hiring may give access to a larger size of boat than owning for any given budget, but owning can provide very considerable advantages unless suitable boats are readily available locally for hire. Ownership enables the permanent fitting of facilities which are unlikely to be found on a hired vessel, permits much more control over cleanliness (which is of particular importance in trace analysis) and also ensures availability.

Position fixing

The method chosen for fixing any sampling position will depend upon the scale of the survey to be undertaken, the mode of transport used, the equipment available and the presence or absence of natural or artificial navigational features. The form of the survey is also important, a vessel may be stopping to take discrete samples from one or a series of fixed positions, or sampling continuously while navigating by either a series of running fixes or positions surveyed at regular intervals.

For some estuarine surveys, the salinity of the samples taken will be at least as significant as their geographical position, and the sampling interval may then be dictated by the readout from a portable salinometer. Nevertheless, it is still desirable to fix the isohalines derived into a geographical framework so that the significance of any intermediate inputs can be assessed. The time at which every sample is taken should also be noted so that corrections for tidal movement can be made if necessary. (See Chapter 7).

Whatever the circumstances, the station or sampling positions should, as far as strategic limitations permit, be easy to locate or fix within the required limits of error so as to minimize delay to the sampling programme. Some of the methods available are considered below under two subheadings, for those using either optical or electronic techniques.

More detailed discussion of position-fixing techniques can be found in the EBSA handbook on estuarine hydrography and sedimentation (Dyer, 1979).

Optical

In calm, sheltered water very precise position fixing of even the smallest open boat can be achieved using two sextant angles between accurately fixed or surveyed landward features. These may, if necessary, be quite easily set up for one specific survey close to some feature of interest such as an outfall. Specially made station pointers are virtually essential to translate sextant angles to a position on a chart; the alternative geometric construction needed to find the position of the observer is both complicated and slow.

Sextant fixing is more useful when not using preselected station positions; the series of angle readings obtained can then be used to calculate positions after the field work is completed. Even using station pointers it is a slow procedure to plot out a series of sextant fixes in order to 'home in' on a predetermined point. Ideally, for sextant work, two observers and two sextants are required, so that simultaneous readings can be taken. However, with clear sighting objects, a single practised observer can fix the two required angles within a period of about 15 seconds, even from a gently moving platform. This degree of accuracy should be sufficient for chemical sampling work.

The use of cross bearings, especially in conjunction with natural transects, although not as accurate, is much more convenient than sextant work. They are much more useful for 'homing in' on predetermined stations. Cross bearings are more easily made with a hand-held liquid-filled prismatic compass than with a boat's steering compass, even if the latter is fitted with sight vanes or an azimuth mirror. All bearings obtained must be corrected for magnetic variation (i.e. subtract $\sim 8°$ in the British Isles) and any known deviation before being transferred to a chart. In small boats, compass readings should be taken from as far away as possible from potentially magnetic objects such as engines, to minimize the introduction of unknown deviations. Sighting objects should be chosen so that the angle between bearings is somewhere around 90°. The use of very acute or obtuse angles results in a lower precision. Predetermined station positions fixed by cross bearings are best approached by sailing along one of the bearing lines until intersecting with another.

Where they are available, the finding of station positions is facilitated by the use of transect lines and one bearing, or, better still, two transect lines. The latter make the most convenient station positions, particularly for surveys which are going to be regularly repeated. Such positions can be rapidly found by sailing along a transect line until either the required bearing to another object is attained, or until the course intersects with a second transect line. It is easier to maintain a course along a transect line than along a bearing to a single object, particularly where there are substantial tidal currents. Experienced coxswains are usually able to accurately reproduce station positions using transects formed by a surprisingly wide variety of features, provided that there is adequate visibility.

If continuous sampling is undertaken and no electronic navigational aids are available, running fixes will be needed, requiring some knowledge of the boat's speed. The construction of such fixes is described in many readily available elementary texts such as that of Lund (1966).

Many estuaries are regularly navigated, and navigational aids such as buoys, perches or lights, which are widely distributed, can be very useful for position fixing. Buoys are inevitably subject to some movement around their moorings, but many channel marking buoys are very tightly moored and are subject to only minimal movement, especially around high water. However, their position should be checked before they are regularly used; they may also be replaced in a somewhat different position after maintenance. If samples are taken from depth, stations should not be made too close to any buoy because of the danger of fouling their mooring lines with the sampling wire, especially if the boat is liable to drift.

In more confined waters, shoreline or landward features such as posts, walls and bridges may also be very convenient for station fixing. The precise positions of suitable features can generally be determined from Admiralty Charts or Ordnance Survey maps of a suitable scale. It is particularly easy to locate sampling points fixed by such features even in poor weather. No special equipment is required, the precision attainable will probably be adequate for most applications, and their use can be recommended in many circumstances.

All optical position-fixing techniques demand reasonable visibility, but in a small boat without electronic navigation equipment, surveys should not anyway be undertaken in conditions of bad visibility.

Electronic

Electronic navigational aids are becoming increasingly important and they may now be fitted on quite small survey vessels, provided that they have an adequate power supply. At night they are virtually essential even if a good system of illuminated navigation buoys is available, and in fog they become absolutely essential. However, the continuance of survey work in very poor visibility in coastal waters which are often confined, shallow and hazardous is generally not worth the risk inherent in sailing in such conditions.

Radio beacons which are marked on navigational charts are not really useful for fixing sampling positions. The Decca Navigator system is far more versatile and precise and is now very widely used around the whole of the British Isles, and elsewhere, by large commercial ships, fishermen and scientific investigators. A series of Decca chains cover the whole of the British Isles coastal waters.

The dial units of a Decca receiver, which are essentially phasemeters, may be read to one or two hundredths of a lane, and lanes are generally 0.5–1.0 km wide, so a relative precision of ± 10 or 20 m can theoretically be obtained. However, absolute precision is not as good as this. There are various sources of fixed and variable error, such as the distortion of radio waves by hills or other massive objects, some of which can be allowed for, which reduce the normal operating accuracy to more like ± 50–250 m. Close inshore, this accuracy is not as good as can be obtained by optical fixes. In some estuaries, and especially in areas bounded by large hills or mountains, of which there are many on the west coast of Scotland, the resulting distortion of the radio signals may effectively render the system completely useless. However, where the system is applicable, it provides a convenient, rapid and reasonably precise means of position fixing for either discrete or continuous water sampling.

Radar can also be a useful aid to position fixing, perhaps particularly in confined waters where the Decca System may be less reliable. Movable range-finding rings, with a digital readout, are desirable for precise ranging, but may not be available on all makes of radar equipment. Once the range calibration of a radar set has been checked (e.g. from a position accurately surveyed under calm conditions) two radar ranges to prominent features, or one range and a bearing may be used to fix a position. However, a third range or bearing is always desirable as an additional check.

Further offshore, large survey vessels can now accurately fix their position by satellite navigation.

Water depth can often provide useful confirmation of a boat's position. Small echo-sounders of adequate accuracy are now available which may be fitted to almost any boat. These instruments are particularly useful in instances where the sampling stations are at naturally deep points – which in turn are often of hydrographic or chemical interest. It is in any event good practice to record the prevailing water depth at sampling stations which are regularly visited.

Recently, various other position-fixing systems have been devised, some of which are extremely sophisticated. Some of these have been developed for the off-shore oil industry, and many provide a degree of accuracy far better than is required for chemical sampling surveys. However, if they are already available for other purposes, then they may be usefully employed by chemists. Underwater acoustic installations are perhaps a good example of these systems. Acoustic transponders are moored to the seabed and are interrogated by a shipboard underwater transducer which receives from the transponders a reply to its coded signals and hence, via a receiver unit, drives a position display.

Once a sampling position has been arrived at, by whatever means, it has to be maintained until sampling has been completed. This may not always be easy, but whether or not it is desirable to anchor on station will depend upon several variables such as water depth, current speed, wind speed and the sampling time involved. If a sampling hydroline is to be lowered and the prevailing current is strong, then it may be necessary to allow the survey boat to drift for a short time in order to keep the hydroline vertical so that sampling depths can be accurately measured. On some surveys, particularly those on which samples are being taken for trace metal or specific organic analysis, it is desirable to avoid potential contaminants which may emanate from the sampling vessel. Since it is impossible to sample both upwind and upstream of the boat if it is drifting sideways in the wind, the boat should be headed into the wind/current and samples taken from as far as possible away from the boat's side.

Water sampling

Particularly in estuaries, it is now possible to obtain a considerable amount of data about a water body without actually taking any water from the water mass. The development of electronic instruments has made it possible to determine at least pH, salinity, temperature, oxygen and turbidity with an accuracy and precision which is adequate for many purposes, by simply lowering suitable probes into the water. These instruments often do not require mains electricity and can therefore be used from even the smallest boats. An example of the survey work which can be carried out using only submersible probes is described by Morris *et al.* (1982).

Although much current research work is directed towards the development of other remote sensing heads in the form of ion-selective electrodes, (see Chapter 6), these devices are not yet all fully suitable for widespread use in the field. The polluted water present in some estuaries presents particular problems for some electrodes, and all electronic instruments may be subject to some shortcomings (Pennak, 1973). The physical taking of water samples from a water mass is therefore still an essential prerequisite for many chemical determinations. Every effort must be made to try to ensure that the water sample taken is representative of the water mass sampled.

The range of sampling devices available may be conveniently classified according to the section of the water column they are designed to sample:

1. surface microlayer;
2. subsurface;
3. mid-depth;
4. near-bottom.

The surface microlayer

The importance of the chemistry of the surface microlayer is being increasingly recognized in connection with air–sea interaction studies. However, the sampling of this layer, particularly under field conditions which may be anything but flat calm, presents enormous difficulties. No entirely satisfactory sampling device has yet been marketed, most microlayer samplers are still one-off devices constructed in laboratory workshops. Samplers working on four widely used principles have been compared in laboratory tests by Hatcher & Parker (1974), and yet another method, using glass-fibre cloth, has been employed by Rittall (1974). All current techniques for sampling the surface microlayer probably also recover a larger proportion of more normal subsurface water.

The chemistry and sampling of the sea surface microlayer has been extensively reviewed by Liss (1975). Its effects are most likely to be significant in estuaries or close to coasts where there are large inter-tidal areas or other sources of organic material. Some recent work suggests that it is even more significant as a concentrator of hydrophobic organohalogens (Platford *et al.*, 1982) than for trace metals (Lion & Leckie, 1982).

Subsurface sampling

Subsurface sampling is perhaps the most widely used technique in estuaries; it may be particularly expedient where it has been established that there is no vertical stratification. Into this sampling category must come the ubiquitous, carefully cleaned, bucket. However, the use of a bucket for sampling cannot always be recommended because of the inclusion of an unknown part of the surface microlayer. This may contain a sufficiently elevated concentration of trace metals, or hydrophobic organic compounds to influence the overall concentration found in the bulk sample. The concentrations of other components, such as major ions and plant nutrients are less likely to be influenced by the surface microlayer, and bucket sampling is apparently suitable for these parameters.

Other methods for sampling near-surface waters overlap with those used for mid-depth sampling, and include the displacement water sampler, pumping systems and sampling bottles triggered when only just below the surface. Displacement samplers, (see Fig. 2.1), which have been standard equipment in fresh-water work for many years (Rainwater & Thatcher, 1960) are now available in an all plastic-coated configuration. They are both cheaper and more practical than their older copper/brass equivalents. They are also preferable to pumped water systems for taking samples which are to be analysed for dissolved gases.

Polypropylene bodied displacement samplers are now commercially available in a range of sizes from about 1.4 l (suitable for filling a 250–500 ml bottle) to 7 l. For specific applications, other sizes can be fairly readily constructed.

Pumped water systems may be used for either discrete or continuous water sampling. In the continuous sampling mode, the depth of immersion of the intake will be restricted by the speed of the boat, but if the boat is stationary the sampling tube may be lowered to a considerable depth. In this way, large volume discrete samples, of up to at least 1000 l, can be readily brought to the surface within a relatively few minutes by a pump of only a few hundred watts power. On a smaller scale, pumps of only a few watts, perhaps operated from a 12 V car battery or the lighting circuit of an outboard motor, are perfectly adequate for bringing in samples of up to a few litres from just below the water surface. Either standard or Terylene-reinforced PVC tubing is suitable for most applications, but possibly not for taking samples to be analysed for some organic compounds. All tubing should be thoroughly flushed through with the sample water before an aliquot is taken for analysis. The volume of water required to achieve this increases as the sampling depth and length of tubing used is increased.

Despite these various mechanical aids, the most satisfactory method of subsurface sampling, particularly for trace component analysis, is still the simple (manual) submersion of the sample bottle underwater. The top can then be opened, and the bottle allowed to fill before re-capping. This simple method has been extensively used as it should eliminate any possible

Fig. 2.1. Displacement sampler: *a*. Being used to take a subsurface sample. *b*. Ready to take a mid-depth sample.

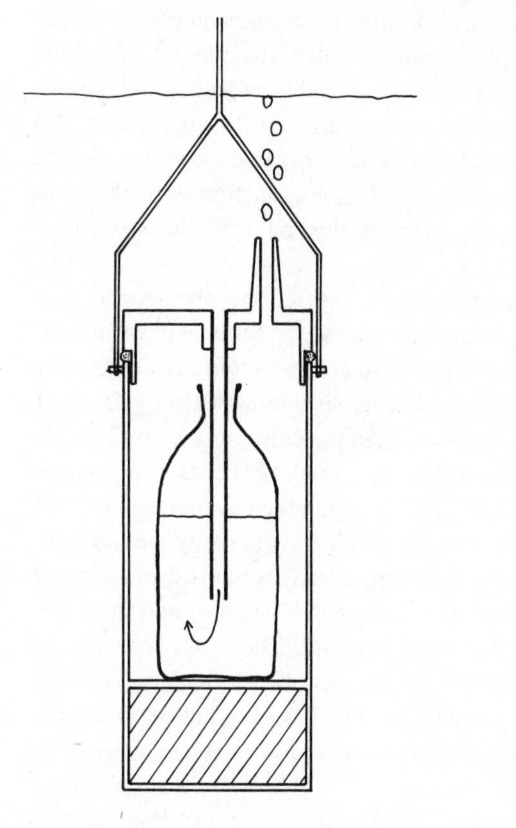

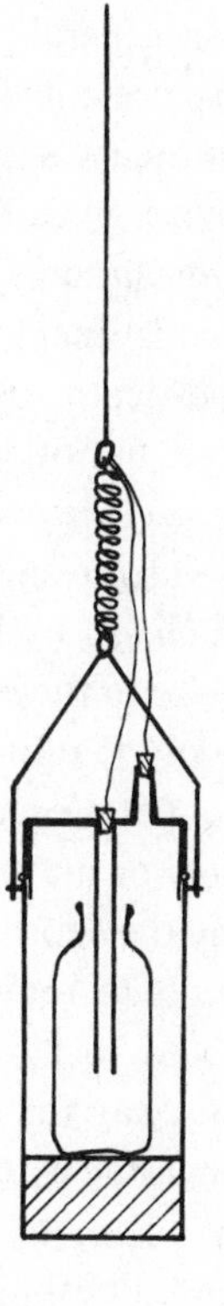

contamination. It also prevents possible adsorptive losses on to the internal surfaces of a sampling device. It can easily be done over the side of a small boat moving slowly into the wind/stream under its own power, or from a tender tied alongside a larger vessel. If the water is rough, or if no small boat is available to bring the water surface into arm's reach, then the sample bottle can be tied into a weighted framework (which need not be rigid, and should be quite readily constructed). This can then be lowered to just below the water surface, preferably from close to the forward end of the survey vessel, again as it steams slowly into the wind. When this sampling procedure is adopted, it may be necessary to use a sufficiently large sampling bottle to allow subsamples to be taken subsequently for salinity or other determinations which may be required on the same sample.

Mid-depth sampling

A very wide range of mid-depth sampling devices is available. This, combined with the wide range of analyses which may be carried out on estuarine samples means that it is not possible to recommend one particular sampler or even type of sampler which is well suited for all mid-depth sampling. In general, the choice is wider for samples to be analysed for constituents present at high concentrations and much more limited for trace constituents such as many metals and organic compounds. The general advice given here should be used in conjunction with the more specific information given in the chapters dealing with the analysis of particular constituents.

Most mid-depth discrete water samplers are of the open-ended free-flushing type, but other designs remain suitable for some applications. In particular, displacement water samplers are generally more easily handled and operated from small open boats. For use as mid-depth samplers, both orifices are closed by bungs which are attached either to a second line to the surface, or to the main lowering line by non-elastic cords which bypass a spring link in that line (Fig. 2.1). The bungs can then be removed and the sample taken at a predetermined depth which is most easily measured by regular knots or markings on the lowering line. This method of sampling is limited by the depth at which the sampler may be crushed and the bungs made difficult to remove by the water pressure. They can, however, be routinely used to a working depth of at least 12 m which is probably adequate anyway for small boat surveys. The lowering line should be of relatively inelastic nylon or polyester material, and of sufficient diameter to be easily handled.

One type of Niskin sampler operates by expanding and thus filling a polyethylene (or similar) bag when lowered to the required depth and triggered by a messenger. This has the advantage that there is no possibility of contaminating the walls of the sampling bag with material adsorbed from the surface or from intermediate water layers. Principally because the sample bags can be sterilized they have been most widely used in bacteriological studies, and they have been relatively little used for chemical investigations. An alternative approach to sampling shallow waters with a device incorporating a compressed air-operated valve to open and close a sampling/storage bottle at the required depth has been recently described by Fletcher & Polson (1982).

Various designs of open-ended water bottle, which are free-flushing as they are lowered through the water column, and closed by tightly fitting end caps when triggered by a messenger have been marketed. The types used include those based on designs by Munro–Ekman, Nansen, Van Dorn and Fjarlie. The last two of these are now most widely used. Munro–Ekman and Nansen bottles may not only close when triggered, but also detach from the hydrographic wire at their top end, fall through 180° and thus invert reversing thermometers clamped to them. Standard sized Fjarlie bottles (NIO bottles, made and marketed by the Institute of Oceanographic Sciences; formerly the National Institute of Oceanography, of the United Kingdom), and some but not all makes of Van Dorn bottle, have reversing frames, which hold and invert reversing thermometers when the end caps are triggered shut. NIO bottles and most Van Dorn bottles are of almost entirely plastic construction and are available in a range of sizes, 1.3–7.5 l for NIO bottles and typically 2–30 l for the Van Dorn types. Larger water sampling bottles of the same general type, but with a capacity of 200 l or more have been constructed for radiochemical work (Gerard & Ewing, 1961). Bottles with a capacity of 7 l or more tend to become cumbersome and may need more than one person to bring them in-board off a hydrographic line. Some of the commercially available mid-depth samplers, along with addresses of the manufacturers, are listed in Batley & Gardner (1977). Sampling devices have also been compared by Bewers & Windom (1982).

Most of the bottles marketed are designed so that when they are triggered shut by a messenger dropped down the hydrographic line they release a further messenger to trigger the next bottle on the line. In this way several samples can be taken from different depths on just one cast. Even with bottles not built with a second messenger release mechanism, a reliable alternative can usually be improvised via a cord to the operating arm.

Particularly in estuaries, stations may be close together, requiring a high turn-round speed for the bottles. Small NIO and Van Dorn bottles are, with practice, convenient to handle, empty and reset for another cast but, especially when used many times a day, some of the plastic operating parts are prone to failure. A reasonable stock of spare parts is therefore required. Although free-flushing, sampling bottles gradually acquire a dirty, greasy film, especially when used in polluted waters. This contamination arises principally from passage through the surface film. Regular cleaning of the internal surfaces with a suitable detergent or inert solvent is therefore desirable. Although available standard equipment is suitable for many applications, for some trace analyses much more stringent precautions are necessary during sampling operations to avoid contamination or loss. The problem is perhaps particularly severe with hydrophobic trace organics and organohalogens – relatively few analyses of these compounds in water samples from mid-depths have even been attempted.

For the more important trace metals, most of the sampling problems were recognized and overcome during the 1970s (e.g. Bender & Gagner, 1976). A considerable amount of consistent trace metal data has since been produced by different sets of workers (see, e.g. Bruland *et al.*, 1978), even using different types of sampling apparatus (e.g. Magnusson & Rasmussen, 1982). In relatively shallow estuarine areas, we have the advantage that the use of a totally synthetic hydroline (e.g. Kevlar) is a more realistic proposition than at sea, where theoretically slightly less satisfactory PVC coated wire and clean stainless steel hydrolines have also been used. All metal parts on the water samplers should be eliminated or plastic coated, and ordinary rubber replaced by silicone rubber for seals or to spring shut the end caps of the fairly widely used Niskin-type Van Dorn samplers (Bewers & Yeats, 1978). This group also use large messengers made of solid PVC. Besides the modified Niskin samplers, General Oceanics 'Go-Flo' type samplers and Hydro-Bios TPN samplers have been quite widely employed (Danielsson, 1980; Duinker & Nolting, 1982).

Especially in relatively shallow estuarine areas, pumped water sampling systems are now being more widely employed for collecting mid-depth samples. Both peristaltic, and centrifugal pumps with impellers unlikely to introduce contamination, are becoming more widely available. Such systems have much to recommend them, they can be very convenient to use on some types of survey, and may also be less expensive to buy than a set of discrete water samplers. A fast pumping rate is desirable to minimize the time required to flush the sampling tube.

Near-bottom samples

The taking of near-bottom water samples requires yet another sampling method. The water bottles just described require to be clamped to a hydrographic line above a weight used to keep that wire taut. Any approach of the water bottle closer than ~1 m from the bottom is liable to result in the weight touching and stirring up the bottom sediment, and thus spoiling the water sample. For deep-water work, an acoustic pinger at the end of the hydroline is often used, in conjunction with an echo-sounder, to judge the approach of the hydroline to the seabed. Bottom water samplers are designed to be lowered and left resting on the sediment surface until all disturbed sediment has resettled. The samples are then taken. Although commercial samplers of this type may not be available, successful designs have been published (Smith, 1971; Joyce, 1973).

Although sediments are outside the main scope of this book, reference should be made to the monograph by Holme & McIntyre (1984) which describes many of the grabbing devices commonly used to obtain sediment samples. Coring devices are also useful for taking samples for more specialized chemical work. In particular, the Craib corer (Craib, 1965) is useful for obtaining short, undisturbed cores, usually complete with overlying water which can be sucked off.

Sampling depth

On small boats, probes and sampling devices will normally be lowered manually using a weighted calibrated line to which any electrical wires attached to the probes will be taped. We have adopted this practice even when the cables supplied with probes are supposed to be load bearing.

On boats with a fitted winch, sampling depths are most quickly and conveniently measured by running the hydroline out over a metre block; attempts to calibrate a hydrographic wire directly are not usually very successful; markings tend to either wear off or jam pulley systems. Suitable measuring blocks have a precisely known circumference and are fitted with dials to display the length in metres of line paid out.

When a hydrographic winch is in use, an echo-sounder is also desirable to determine the depth of water available, so that the bottom can be avoided. Quite apart from stirring up sediment, if a sampling bottle hits the bottom, its firing mechanism may be upset so that it is not closed when struck by a messenger.

Determining accurately the depth at which samples are actually taken may not be easy in an estuary if there are strong tidal currents. If, under these conditions, the sampling boat is anchored, or maintains way to remain stationary with respect to fixed reference points, then the sampling wire will stream out at an angle. This angle can be reduced only by letting the sampling vessel drift, or by using an increasingly massive streamlined weight or depressor on the end of the hydroline. The error introduced in this way is small for moderate wire angles (3.5 per cent for a mean wire angle of 15° from the vertical), but increases more rapidly at greater wire angles. Oceanographic techniques for recording sampling depths, such as the use of protected and unprotected reversing thermometers and barographic recordings usually do not have an accuracy much better then ± 5 m. Although this is good in the context of oceanic depths, it is probably not precise enough to be useful in even the deepest estuaries, so recording the length of hydrographic line paid out remains the most widely used and expedient technique of determining sampling depths.

Storage bottles

Once a water sample has been taken, it usually has to be transferred to one or more bottles prior to pre-treatment, storage or analysis. The problems inherent in storing samples prior to analysis, which are discussed more fully in Chapters 3, 4 and 5, usually result in the need for each sample to be split into several fractions which will be treated differently prior to analysis. This proliferation of the number of bottles required to be on hand at each sampling station, although it leads to better conditions for each fraction involved and is often very convenient when back in the laboratory, can become inconvenient in the field, especially if several depths are sampled at one station. It has been found to be good practice to have bottles of markedly different physical characteristics for each fraction, all clearly labelled with the sample number. When several samples are taken on one station, wooden or plastic bottle racks or crates are very useful to keep all bottles in order. Particularly in rougher weather, loose bottles fall over, roll around and get lost or dirty. For fractions stored in glass bottles the risk of breakage is ever present. Felt-tipped pens have revolutionized bottle numbering, but they will not write on wet surfaces, so prenumbering should be carried out as far as possible.

Plastic, especially polyethylene or polypropylene bottles are generally preferred for shipboard use, because they are less liable to break. However, for chemical reasons, they are not always suitable for all types of samples. Plastic bottles are particularly advantageous when deep-freezing is used as a means of storage. Bottles whose tops have integral inner sealing flanges are preferable, those with separate washers of card, cork or other material should be avoided. It is also worth noting that very good quality machining or moulding is required to produce a fully watertight polypropylene bottle top, and consequently some makes of the harder polypropylene bottles are not as watertight as their polyethylene equivalents.

Samples for most trace metal analysis are best transferred to acid-cleaned, low density polythene bottles and acidified to ~pH 2 for storage (Bewers *et al.*, 1981). Water samples may be pressure filtered straight from the sampling devices into storage bottles using in-line filters (e.g. Bewers & Yeats, 1978), but this may not be practicable in more turbid estuarine waters. Samples for trace metal analysis must not be acidified prior to filtration.

Glass bottles remain essential for some samples, and are generally used for the storage of samples for mercury analysis. A totally impermeable container is also required for water samples which are to be returned to a laboratory for precise salinity determination. Polyethylene does not meet this requirement. Glass bottles with a good rubber seal incorporated into the top are therefore generally used for these samples. Most glass is suitable, but new soda glass may leach some ionizable material into sea water, and if such bottles are to be used, they should be 'aged' by soaking in sea water for several weeks before being brought into routine use. Samples taken for precise salinity determination are generally considered to be stable for long periods without any special preservative measures, but they are best kept in the dark to prevent algal growth, and protected from large temperature fluctuations.

If dissolved oxygen is to be determined by the Winkler technique, a glass bottle will also be required for this fraction. Suitable bottles have a tapered neck and stopper which can be inserted to exclude all air bubbles. Samples for oxygen determination will be the first to be run off from each sampling bottle (see page 97 for details).

Table 2.1. *Possible fractions into which a water sample may be split, and treatment required in the field*

Parameter	Field filtration?	Recommended container	Field treatment
Salinity	No	Glass	—
Dissolved oxygen	No	Glass	Manganous sulphate plus alkaline iodide fixation
BOD	No	Glass	Keep in dark
Major nutrients	Yes	Polyethylene	Freezing (and/or bacteriocide)
Suspended solids	Preferable	Polyethylene	Bacteriocide if not filtered
Most trace metals	Preferable	Low density polyethylene	Acidification
Mercury	Preferable	Glass	Acidified oxidant
Organics	No	Glass	Cool and dark,
Chlorophyll	Yes, and add $MgCO_3$*	Dessicate and cool filter and keep in dark	

* See Chapter 5.

Glass containers are also required for the transport and storage of many samples to be analysed for specific organic components. Plastics and the associated 'fillers' used in their manufacture can interfere with some determination methods.

Glass bottles have also often been recommended for the storage of samples to be analysed for phosphate, but this may reflect a problem of sterility with plastic bottles in routine use, rather than chemical absorption. Even in an unfrozen state, a filtered and sterilized water sample in a sterile polythene bottle has been known to retain its phosphate concentration unchanged for six months (Leatherland, unpublished).

A summary of possible fractions required from each water sample taken is given in Table 2.1. Preservative measures employed during storage are subject to considerable controversy and are discussed more fully in Chapters 3, 4 and 5.

Sample pre-treatment

If determinations are not to be made in the field, it may be essential to carry out work on some samples as soon as possible after they are taken, to prevent changes in their chemical composition brought about by microbiological activity or physico–chemical processes. Commonly used techniques include cooling, fitration, freezing or the addition of a biocide to minimize microbiological activity, and acidification to prevent metal precipitation or adsorption reactions. Filtration, or centrifugation, is also necessary where it is desired to separately determine the 'dissolved' and 'particulate' fractions of any determinand.

The cooling of samples, usually to approximately 4 °C, at the same time keeping them in the dark, is fairly general practice and may be adequate to minimize undesirable changes in the sample if the time between collection and analysis, or other further treatment, can be kept short. If this is not possible, then more drastic measures are required.

Filtration

Filtration not only removes phytoplankton and most bacteria from samples, thus minimizing biological activity, but also removes turbidity which can interfere in commonly used colorimetric determinations. For many applications vacuum filtration is sufficiently fast. It can be undertaken in most field conditions using a Buchner flask and a suitable filter funnel assembly. Vacuum may be supplied either by a hand pump, or, more conveniently and rapidly, by an electric pump. Various suitable designs of filter funnel, based on the Steffli–Cole principle, are available. The filter support may be either sintered glass or stainless steel mesh. There is a smaller pressure differential across the mesh supports, which therefore give faster filtration, but they may not be suitable for all applications.

Cellulose acetate membrane filters with a pore size of 0.45 μm and Whatman GF/C glass-fibre filters have both been widely used in estuarine work. The glass-fibre filters probably have a larger effective mean pore size, and they are much less rapidly clogged by the high loads of suspended material found in many estuaries. A more retentive, but slower glass-fibre paper, the GF/F has not become as widely used in estuarine studies. See Chapter 5 for further information about filter types and performance.

Whichever system and filters are employed, it is necessary to wash through both the apparatus and the filter with an aliquot of the sample, prior to filtration of the main sample. Washing of the filter is required to leach out potentially interfering or contaminating ions, principally phosphate from membrane filters and silicate from GF/Cs.

When many relatively small samples are being handled for nutrient analysis, disassembling the Buchner apparatus twice for each sample is inconvenient. A neat way of overcoming this difficulty has been described by Folkard (1968), using a separating funnel in place of the usual Buchner flask (Fig. 2.2).

Fig. 2.2. Filtration assembly which avoids the necessity of disassembling the Buchner apparatus twice for each sample (after Folkard, 1968). Funnel assembly expanded.

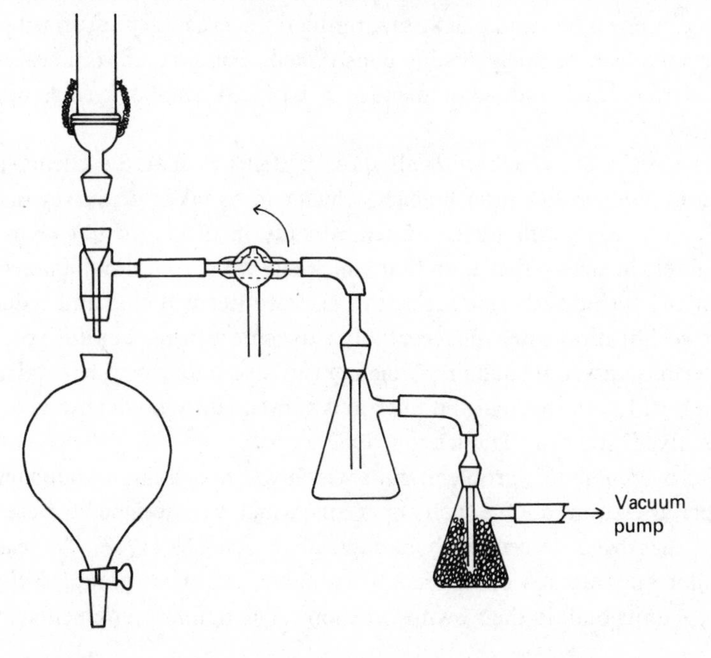

If large or particularly turbid samples are to be filtered, then pressure filtration, which can be made very much more rapid than vacuum filtration, may be necessary. Most commercially available pressure filtration units employ stainless steel in their construction. This is perfectly satisfactory for some organic, major element and nutrient analyses, but is not acceptable for samples to be analysed for trace metals. If pressure filtration is employed, a clean filtered air supply is also needed; this is normally supplied either from a hand or foot pump, or, more conveniently, from cylinders of compressed air or nitrogen.

The filtration of samples to be analysed for many trace components demands particular care to exclude contaminants which may be found in the atmosphere on or around a boat. For some applications, filtration may not be necessary at all; it should certainly not be employed unless it is essential. The filtration of water samples to be analysed for trace metals demands extreme care and attention to fine details to avoid both contamination and adsorptive losses of metals. Pre-cleaned filters will usually be used and some form of totally enclosed (pressure) filtration system is essential. Adequate checks should be made to ensure that the supplies of compressed air or gas do not contain detectable amounts of the constituents to be determined. On larger boats, apparatus may be installed in portable clean air cabinets. The use of in-line filters attached directly to a pressurized sampler has already been referred to, but this will only be practicable for samples of low suspended solid content. Other neat systems using in-line filters and plastic separating funnels as the pressurized sample reservoirs can be fairly readily constructed. For smaller volumes, Burrell (1974) has used and recommended a basically similar syringe-operated system.

The main drawback with all these systems is that the cheap, plastic demountable in-line filter holders, which can be taken on survey sealed in polythene bags with washed filters already in place, do not seem to be available in sizes larger than that which will take a 47 mm diameter filter. In many turbid estuarine areas, this size of filter will clog and reduce the rate of filtration to an unacceptably slow rate before a useful volume of water has passed through it. Clogging can also influence the metal species which will pass through the filter and subsequently be determined in the 'dissolved' fraction (Danielsson, 1982).

To overcome this problem, units which will take a larger diameter filter paper are required. One such apparatus, which was designed for use in the field, has been described by Skougstad & Scarbro (1968). At least one similar apparatus is commercially available, but most groups still prefer to use units built in their own workshops. The main requirements are that they should be free from any potential contamination, easy to keep clean and reasonably simple to use.

Larger diameter filters, unfortunately, also represent a potentially greater source of contamination. This is worse for cellulose acetate membrane filters than for Nucleopore filters, but the larger sizes of both types will probably require to be acid-washed and thoroughly rinsed with ultrapure water prior to use. After roasting, glass-fibre filters are suitable for the filtration of samples to be analysed for mercury or organics, but are not suitable for samples on which most other metals will be determined.

After filtration, samples for trace metal determination are usually further stabilized, against sorption or precipitation processes, by the addition of acid.

Continuous centrifugation has been used in oceanographic studies (e.g. Chester & Stoner, 1975), but offers less control over the particle size fraction removed, particularly in more polluted areas where organic colloids and aggregates may be abundant (see also page 133).

Chemical stabilization

The chemical stabilization of samples may be carried out as an additional stage following filtration, where filtration is not required or, although infrequently as it is not usually good practice, prior to filtration. Its aim is to prevent or minimize any microbiologically or biologically induced changes in the sample and/or to prevent the adsorption or precipitation of chemical species of interest.

Reagents can be quite easily and rapidly added to samples in the field under most conditions. Calibrated dropping pipettes are adequate for many applications, but there are now available reasonably robust semi-automatic dispensers which are more rapid, more precise and unlikely to introduce any contamination. Such dispensers can be strapped into place and are easy to use even in rough weather conditions. They are also particularly useful for large reagent additions.

One of the more common applications of chemical stabilization is the acidification, after filtration, of samples for subsequent trace metal analysis. This lowering of pH, usually to approximately pH 2 has been shown to effectively stabilize the sample against biological or chemical changes even during quite prolonged storage.

Samples for the subsequent extraction and determination of hydrocarbons can be stabilized against biological changes by the addition of isopropanol, which can remain in the sample as it will be part of the mixed solvent used to extract the hydrocarbons from the water. This example illustrates the general point that the stabilizing reagent added should be compatible with the subsequent analysis, in order to minimize sample manipulation and hence also the risk of contamination.

If a biocide is to be added prior to eventual filtration, then a suitable chemical should be selected which is unlikely to lead to the rupture of living cells in the sample. The release of such cellular material may lead to changes in the dissolved components of the sample, as well as reducing the mass of material eventually retained by the filter. Mercuric chloride has been used prior to nutrient analysis, but the use of such an environmentally hazardous material can hardly be recommended.

Some other examples of chemical stabilization involve more dramatic alteration of the sample, and will be considered later in connection with simple field chemistry.

Freezing

Deep-freezing of water samples, usually to −20 °C, has also become a fairly widely used technique for the preservation of samples during storage. In the field, rapid freezing may conveniently be achieved using an alcohol or glycol bath cooled by solid carbon dioxide. If polyethylene bottles are used, acetone should not be used as the heat transfer medium, as this solvent does slowly diffuse through polyethylene. Care must always be taken of deep-frozen samples as even polyethylene becomes quite brittle at very low temperatures and may shatter if dropped or crushed.

Freezing should be carried out as rapidly as possible after filtration or sampling. It effectively halts biological activity, but does give rise to some chemical changes, which are not necessarily reversible, especially if the freezing is carried out prior to filtration (Burton *et al.*, 1970).

As with any other storage procedure adopted, its effectiveness and possible influence on the chemical species of interest should be rigorously investigated prior to routine use.

Field measurements

It is generally desirable to complete as many determinations in the field as can be managed without impeding or interrupting the sampling programme. Measurements made in the field reduce the number of samples which have to be transported back to the laboratory and also eliminate the need for storage, thus avoiding that particular problem. There may also be an associated saving of analytical time because, besides saving the time spent treating samples prior to storage, there is in some cases work to be done on a sample after storage to bring it back to its original condition before it can be analysed. Some parameters, such as temperature, can only be satisfactorily measured in the field.

Temperature

Numerous instruments are available for measuring temperature. In oceanographic work, reversing thermometers have been very widely used. These instruments also give very good results in estuarine waters, but they are relatively delicate and can only be used with discrete water sampling bottles with suitable reversing frames. In deeper coastal areas they may be almost essential, but unless accuracy to the second decimal place is required, they cannot be recommended for routine shallow estuarine applications. Here, the generally less accurate and less precise alternatives, which may be cheaper and more robust, will probably be adequate.

Electrical means of measuring temperature, which generally employ some form of thermistor, the resistance of which changes with temperature, can be made at least as precise as the best expanding fluid devices. They are often combined with other sensing probes, especially those for salinity or dissolved oxygen, which require temperature compensation. Some oceanographic devices such as the bathythermograph are not intended for use in shallower coastal waters, but other towed oceanographic instruments, which will also measure salinity and depth, may be useful.

An example of an instrument designed for estuarine use is the portable temperature and salinity measuring bridge built to an Institute of Oceanographic Sciences (IOS) specification. This combines a thermistor and carbon conductivity electrodes in a moulded epoxy-resin measuring probe This probe is connected to its control unit by a cable up to 100 m long. It can be lowered through the water column and temperature (and salinity) read off after equilibration at the required depths, with a typical precision for temperature of ± 0.1 °C. Similar thermistors are incorporated in the probes of many dissolved oxygen meters, and they can usually be read with a similar precision. However, in any such instrument, the accuracy will probably not be as good as the precision, and this should therefore occasionally be checked in the laboratory against a reference thermometer.

If discrete water samples are taken from depth, and reversing thermometers are not available, temperature measurements sufficiently accurate for some purposes can still be obtained by immersing a mercury-in-glass thermometer into the water sample as soon as possible after collection. Unless the discrete sample taken is very large, probes incorporating thermistors should not be used in this way, as their greater mass and thus thermal capacity is more likely to perturb the temperature of the sample. Most modern plastic water sampling bottles are relatively good thermal insulators, and using field instruments it has consistently been shown (Leatherland, unpublished) that provided the sampling bottles are well-equilibrated with the water temperature at their sampling depth, are brought rapidly to the surface, and temperature is measured as soon as possible after return to the surface, then results within ±0.25 °C of those recorded by reversing thermometer can be achieved even when samples are taken from below a marked thermocline. Under normal conditions it is sufficient to allow two minutes for polypropylene bottles to attain thermal equilibrium at their sampling depth, i.e. about the same length of time as required by reversing thermometers. Some older makes of water sampling bottles have a concentric tube construction to retain an insulating layer of sample water around the main sample, and were therefore particularly suited to this retrospective method of temperature measurements.

One disadvantage of pumped water sampling systems is that, unless the rate of flow is high and the residence time of water in the pump and pipe is small, a substantial temperature change may take place before it can be measured at the outlet. Temperature rises of 0.5 °C have been regularly found in one such system. A thermistor fixed in the water stream close to the inlet point should overcome this difficulty.

Salinity

The determination of salinity is central to virtually any estuarine survey, but the accuracy and precision required of the determination is likely to depend upon the particular type of survey being carried out (see Chapter 1).

The present definition of salinity and its history, are given in Chapter 7. It will be seen from this definition that salinity is not readily determined directly and most methods of salinity determination rely on the constant relative composition of major ions in sea water. This constancy of composition gradually breaks down towards the head of estuaries, where dilution with river water of variable composition is very great, but it can be assumed for most practical purposes down to a salinity of about 1–2‰. At lower concentrations of sea water, measurements of chlorinity or conductivity (corrected for temperature) may be more useful. Assuming constancy of composition, allows salinity to be determined from the temperature compensated conductivity. Most laboratory salinometers now utilize the principle of inductive coupling between two coils to determine the conductivity of sea water. This overcomes the main problems associated with the fouling of electrodes used in instruments which measure conductivity directly. However, because of problems of direct calibration, the inductively coupled laboratory instruments do not measure absolute conductivity or salinity, but are designed to give a conductivity ratio relative to a standard sea water. They are regularly calibrated against either IAPSO (International Association of Physical Sciences of the Oceans) standard sea water, or bulk secondary standard which has been calibrated against the IAPSO standard. IAPSO standard sea water, which is calibrated against a potassium chloride conductivity standard, is supplied worldwide by the IOS, Brook Road, Wormley, Godalming, Surrey, GU8 5UB, England, to ensure that all oceanographers work to a common salinity standard.

Many field salinometers, with probes which can be lowered into the sea, work on the same inductive coupling principle, but are calibrated to give a reading directly in salinity. However, in order to ensure accuracy equal to that specified by the manufacturers, frequent calibration of these instruments against either an IAPSO standard or secondary standard is necessary. Bulk secondary standards, stored in impermeable containers and protected from evaporation or dilution are stable for many weeks. They may be stored in glass carboys, with a take-off tube near the base, under a layer of liquid paraffin. When used in estuaries, the linearity of field instruments should also be checked, preferably against a good laboratory instrument.

Other field salinometers, especially less expensive ones, such as those built to an original IOS specification, do work on the direct conductivity principle, and have carbon electrodes embedded in a moulded epoxy-resin probe which also houses a thermistor for temperature compensation and measurement. Even when the electrodes are regularly cleaned, such instruments are more prone to fluctuations in calibration and it is good practice to regularly take a few samples of water, the salinity of which has been determined in the field, for subsequent accurate analysis in the laboratory. If the best results are required from a field instrument, this should be done on every survey.

Field salinometers may be used either by lowering the probe, usually attached to a weighted line, down through the water column from a stationary vessel, or by submerging it is a stirred sample of water brought aboard a survey vessel by a discrete sampler or pumped sampling system. In any application, time must be allowed for equilibration of the temperature compensating thermistor. Some conductivity, temperature and depth (CTD) probes, conceived primarily for oceanographic use, are designed to be towed from a moving vessel. However, their pressure sensors, for determining depth, may not be sensitive enough for use in shallow water.

Despite the availability of a variety of conductivity instruments, the determination of salinity via chloride titration remains useful in some situations. Suitable methods for general estuarine applications are fully described in Strickland & Parsons (1972). These methods may be particularly useful if only a few salinity measurements are to be made, if only a very small volume of sample is available (e.g. of sediment pore water), or for calibration purposes in the absence of a laboratory salinometer.

Dissolved oxygen

The classical Winkler titrimetric technique for the determination of dissolved oxygen (see Chapter 3 for details) does not lend itself to use in the field although the field fixation of samples for later determination in the laboratory is an integral part of its application to environmental samples. However, with the development of transportable microprocessor controlled titration systems (Williams & Jenkinson, 1982) the possibility of using the titrimetic techniques in the field, for example in estuarine productivity studies, now exists.

The majority of field oxygen determinations which are often extremely useful in pollution studies will continue to be made by means of oxygen sensing electrodes which are generally of two basic types: galvanic, e.g. the Mackereth electrode or polarographic, e.g. the Clark electrode. A good description of these and their use is given in HMSO (1980). In attempting to measure oxygen concentrations in estuaries it is important to remember that the electrodes respond to the *in situ* partial pressure of the gas and that this is influenced by temperature, pressure and salinity. Thus meters calibrated in per cent oxygen saturation will give equally accurate values in fresh and saline waters whereas those with a concentration readout but without an in-built facility for salinity correction will only be accurate in fresh waters. Such meters can be used in saline waters but the ambient salinity must be measured to enable the values to be corrected. In general, by careful attention to calibration procedures and determinations of the necessary corrections for temperature, pressure and salinity it is possible to obtain data which are of similar precision to all but the most careful titrimetric determinations.

Before use in the field, the performance of an oxygen electrode under controlled conditions in the laboratory should be determined by measuring the oxygen concentration of a set of solutions of known oxygen content with the electrode and by the Winkler method. These results may be used to prepare a calibration graph for the particular electrode. When using the instrument in the field, its performance should be checked by a single point measurement of air-saturated water or if it has been shown to be appropriate for the particular electrode of water-saturated air. Full details of calibration procedures are given in HMSO (1980). During the course of a set of field measurements, individual water samples should be collected and fixed as described in Chapter 3. The oxygen concentration of these samples as determined by the Winkler technique is used as a further check of the electrode performance and thus the validity of the instrumental data.

If it is necessary to replace the electrode membrane, it will be necessary to re-calibrate it. When not in use it is generally recommended that the electrodes should be stored in moist conditions but the individual manufacturer's recommendations should be followed. The response of oxygen sensors may be displayed in concentration units or per cent saturation. For most estuarine applications it is useful to know both these quantities and it will therefore be necessary to have simultaneous information of the salinity and temperature of the water. A measurement of the atmospheric pressure is also necessary for the most accurate work although the effect of atmospheric pressure is much smaller than those of temperature and salinity. Where correction to allow for atmospheric pressure is required it should be carried out as set out in HMSO (1980). Corrections for temperature and salinity effects are described in Chapter 3.

Transparency

Water transparency, or light absorption is also best determined in the field. Light absorption is affected both by dissolved and particulate material. The relative importance of each component will vary between different situations, but in most places the loading of particulate material is likely to be the dominant factor.

An estimate of the transparency of the surface layer can be made using a Secchi disk. This is a matt white disk usually 30 cm in diameter which is weighted and fixed in the horizontal plane. It is lowered on a calibrated cord until it is no longer visible. The 'Secchi depth' recorded is the mean of the depths at which it disappears when lowered and reappears when raised. For a somewhat subjective method, quite surprisingly consistent results can be obtained even by different workers; unsuitable weather conditions put the biggest constraint upon the accuracy attainable. Although 'Secchi depth' is not only dependent on the particulate loading of a water mass, reasonable correlation between the two may well be obtained in localized situations. Along the length of the Firth of Clyde and Clyde Estuary an equation of the form:

$$Z_{SD} = k/(SS)^n$$

where Z_{SD} is the Secchi depth in metres and (SS) the concentration of suspended solids in mg l^{-1} has been found to fit experimental data from over 100 samples, taken throughout the year, with a correlation coefficient of 0.93.

Some indication of light penetration can also be obtained by measuring incident radiation at the sea surface with a photocell, and then lowering it through the water column. The results obtained by this method may not be very meaningful because of the differential absorption of light of different wavelengths.

This problem of changing absorption at different wavelengths, which may be particularly variable in the estuarine environment, may also affect the results obtained from true light transmission measurement instruments. These employ a suitable shielded light source and photocell mounted some distance apart so that light attenuation over the known path length between the source and detector can be measured. They measure the combined effects of dissolved and particulate material. Problems may arise with such instruments, especially in polluted estuarine waters, because of the soiling of optical windows.

If it is desired to get information about concentrations of suspended solids, then instruments working on a nephelometric (light scattering) basis may be more suitable if they are sufficiently sensitive. Such instruments can be quite readily designed to automatically compensate for any soiling of optical surfaces. Even with such instruments, some calibration against solids' concentrations determined gravimetrically is necessary for each separate situation because of the natural variation in the composition of the solids, which may give rise to different degrees of light scattering from any given concentration of material.

Neither transmission nor nephelometric instruments have yet come into widespread use in estuaries, but designs are continually being improved and they should have an important part to play in coastal monitoring and survey work.

The principles and applications of other, potentiometric field measurements such as those of pH, Eh and ions determined via ion-selective electrodes are discussed in Chapter 6.

Simple field chemistry

In a few instances, the concept of chemically stabilizing samples in the field can be quite readily extended to field chemistry. The dividing line may not be clear.

The determination of oxygen by the Winkler method (see Chapter 3) is undoubtedly the best known example. Some Winkler determinations of oxygen are carried out on many surveys, even when dissolved oxygen meters are used for the majority of measurements. As explained in Chapter 3 it is still good practice to carry out such checks of the calibration of oxygen meters under field conditions.

Unless taken by a displacement sampler, the sample for dissolved oxygen determination will be the first to be run off from the water sampling bottle, via a tube leading to the bottom of the bottle which is being filled. The dissolved oxygen is then chemically fixed by the sequential addition of aliquots of manganous sulphate and alkaline iodide solutions, with thorough mixing to disperse the precipitate. After fixation in this way, the samples should be stable in subdued light for at least a few days provided they are not subjected to temperature fluctuations likely to cause the ingress of small air bubbles. The fixed samples can be stored under water to prevent any such ingress. After return to a laboratory, they can be acidified and titrated in the usual way.

One important nutrient which is amenable to field fixation, and which is also difficult to preserve satisfactorily, is ammonia. The method of Solórzano (1969) is readily employed under field conditions, and has the additional advantage that where turbidity is relatively low, accurate results can even be obtained from unfiltered samples. The time taken for complete colour development is quite long, but the indophenol blue colour produced is then stable for many hours. The samples are therefore usually ready for colour measurement when, or soon after, they have been returned to a laboratory. Suitable standards and blanks must be prepared beforehand, and treated in the same way as the samples. Proposed modifications to this method (Liddicoat *et al.*, 1975; HMSO, 1982) may improve the blank, but make the colour development phase UV light sensitive, and may not be as practicable for use under field conditions. However, the recent modification described by Mantoura & Woodward (1983) specifically for use in estuaries should be suitable for use in the field as it does not require exposure to a UV light source.

The manual silicate method described by Strickland & Parsons (1972) may also be applied in a similar way, although in this case samples must be filtered prior to chemical treatment. Both the ammonia and silicate reactions can be conveniently carried out in 50 or 100 ml polyethylene bottles.

For reagent additions in the field, piston-operated automatic pipettes with disposable tips are very convenient, especially on boats where suitable spring-clip rack space can be provided. Since the piston does not come into contract with the liquid being dispensed, they are particularly useful for alkaline reagents.

If some sort of sheltered temporary base (or better) is available, the next extension of chemistry into the field might be the use of a spectrophotometer. If mains power is not available, some spectrophotometers can be modified to run off a 12 V DC source. Other smaller instruments are also available which operate off their own internal batteries, but many of these offer a lower sensitivity and precision, which will not be acceptable in many instances.

On larger vessels, with laboratory space and mains power, normal laboratory equipment and instruments such as autoanalysers may be advantageously employed.

Chemistry and other measurements carried out in the field remove the difficulties and uncertainty of sample storage, and any methodological adaptations which can be made to facilitate their use should be considered, provided that there is no significant loss of precision.

Survey organization

Most of the material included in this final section will probably seem very obvious, but much of it has been verified by personal experience, and it remains worth emphasizing. Even a minor accident or mishap may result in the loss of vital sample, or spoil the results of an otherwise well-planned survey. Attention to apparently minor details is particularly important in small boats where space is strictly limited.

However simple or sophisticated the work carried out in the field is to be, forethought and the provision of special racks, boxes or clamps for virtually everything to be used, can make surveys go more smoothly and prevent damage to equipment or the loss of valuable samples. Particularly in small dinghies, the maintenance of ropes and cables in an orderly and knot-free state is essential and can be facilitated by the use of suitable spools or wooden formers onto which they can be wound every time they are brought in-board. Figure-of-eight winding is particularly useful because it minimizes cable twisting. Between stations, shorter lengths of cable may be conveniently dumped, only loosely coiled, in an open (laundry-type) basket.

The availability of some sort of workshop facilities to make up special items, such as winding formers, is very useful. Small items, such as polyethylene pipettes can be made from tubing and calibrated in any laboratory.

Verbal communications over engine noise and vibration can become difficult, especially if weather conditions are unpleasant. Therefore, before survey work is started, all members of a team should be very clear of their exact task, how they will do it, and where the required equipment is stowed. Checklists should also be prepared and gone through to ensure that no items are overlooked. Soft pencils are useful for writing on damp data sheets when the weather is very wet. The making of notes in the field is often useful, they may later prove valuable when interpreting results or to enable improvements to be made for a subsequent survey. Survey books made. from rag paper are sufficiently water-resistant for most purposes, but plastic 'paper' is now available for use in particularly extreme conditions. Soft pencils are also suitable for writing on this material.

As far as is practicable, laboratory-style cleanliness and contamination avoidance procedures must be maintained in the field. This is often difficult, but it is useless to have a scrupulously clean, contamination-free laboratory procedure if samples are spoilt before reaching the laboratory. A lot of sampling gear is designed to minimize cross-contamination, but it can still become dirty, both when left unused for a period of time, or when used in polluted waters. Equipment used in polluted areas gradually acquires a thin coating of dirty, mainly organic material; the acquisition of such a coating can also upset the accuracy of conductivity or potentiometric measuring probes. It can usually be removed with either a solvent or detergent wash. Equipment to be used for taking samples for trace metal analysis should be given an acid wash and thorough rinsing prior to use, and will normally be stored and taken out on surveys sealed in clean polyethylene bags.

Finally, the weather, an ever-present factor, but one which must not be ignored, particularly when working from a small boat. Although good weather should not be regarded as prerequisite, it is certainly very helpful, and conducive to successful survey work. Cold hands and a continual downpour may eventually dampen enthusiasm, and it is under such conditions of stress, or when afflicted by seasickness that muddles or mistakes are most likely to occur. Particularly at such times, familiarity with the techniques being used, and convenient placing of all equipment will help minimize errors. However, a line must be drawn somewhere, the temptation to undertake or continue a survey under dangerously adverse conditions must be resisted, even if it does result in a gap in a set of data.

A very useful catalogue, which gives the addresses of, and describes equipment produced by many American manufacturers of oceanographic equipment is the 'Ocean Master' catalogue. This is available from: Charles Kerr Enterprises Inc., 129 North Main Street, New Hope, PA 18938, U.S.A.

References

Anon. (1973). *Environmental Study of Port Phillip Bay, Report on Phase One, 1968–73.* Melbourne: Melbourne and Metropolitan Board of Works/Fisheries and Wildlife Department of Victoria.

Batley, G. E. & Gardner, D. (1977). Sampling and storage of natural waters for trace metal analysis. *Water Research*, **11**, 745–56.

Bender, M. L. & Gagner, C. (1976). Dissolved copper, nickel and cadmium in the Sargasso Sea. *Journal of Marine Research*, **34**, 327–39.

Bewers, J. M., Dalziel, J., Yeats, P. A. & Barron, J. L. (1981). An

intercalibration for trace metals in seawater. *Marine Chemistry*, **10**, 173–93.
Bewers, J. M. & Windom, H. L. (1982). Comparison of sampling devices for trace metal determinations in sea water. *Marine Chemistry*, **11**, 71–86.
Bewers, J. M. & Yeats, P. A. (1978). Trace metals in the waters of a partially mixed estuary. *Estuarine and Coastal Marine Science*, **7**, 147–62.
Bruland, K. W., Knauer, G. A. & Martin, J. H. (1978). Cadmium in northeast Pacific waters. *Limnology and Oceanography*, **23**, 618–25.
Burrell, D. C. (1974). *Atomic Spectrometric Analysis of Heavy Metal Pollutants in Water*. Ann Arbor: Ann Arbor Science.
Burton, J. D., Leatherland, T. M. & Liss, P. S. (1970). The reactivity of dissolved silicon in some natural waters. *Limnology and Oceanography*, **15**, 473–6.
Chester, R. & Stoner, J. H. (1975). Trace elements in total particulate material from surface seawater. *Nature, London*, **225**, 50–1.
Craib, J. S. (1965). A sampler for taking short undisturbed cores. *Journal du Conseil Permanent International pour l'Exploration de la Mer*, **30**, 34–9.
Danielsson, L. G. (1980). Cadmium, cobalt, copper, iron, lead, nickel and zinc in Indian Ocean Water. *Marine Chemistry*, **8**, 199–215.
Danielsson, L. G. (1982). On the use of filters for distinguishing between dissolved and particulate fractions in natural waters. *Water Research*, **16**, 179–82.
Duinker, J. C. & Nolting, R. F. (1982). Dissolved copper, zinc and cadmium in the Southern Bight of the North Sea. *Marine Pollution Bulletin*, **13**, 93–6.
Dyer, K. R. (ed.) (1979). *Estuarine Hydrography and Sedimentation.* Cambridge: Cambridge University Press.
Fletcher, W. K. & Polson, D. (1982). A simple device for sampling shallow waters. *Limnology and Oceanography*, **27**, 188–9.
Folkard, A. R. (1968). *An apparatus for the filtration of sea water.* International Council for the Exploration of the Sea Interlaboratory Report. Information on Techniques and Methods for Sea Water Analysis No. 2.
Gerard, R. & Ewing, M. (1961). A large-volume water sampler. *Deep-Sea Research*, **8**, 298–301.
Hatcher, R. F. & Parker, B. C. (1974). Laboratory comparisons of four surface microlayer samplers. *Limnology and Oceanography*, **19**, 162–5.
HMSO (1980). *Dissolved Oxygen in Natural and Waste Waters* 1979. Methods for the examination of waters and associated materials. London: Her Majesty's Stationery Office.
HMSO (1982). *Ammonia in Waters* 1981. Methods for the examination of waters and associated materials. London: Her Majesty's Stationery Office.
Holme, N. A. & McIntyre, A. D. (eds) (1984). *Methods for the Study of Marine Benthos*. International Biological Programme Handbook No. 16, 2nd edn. Oxford: Blackwell Scientific Publications.

Joyce, J. R. (1973). An improved bottom-water sampler. *Journal of the Marine Biological Association of the United Kingdom*, **53**, 741–4.

Liddicoat, M. I., Tibbitts, S. & Butler, E. I. (1975). The determination of ammonia in sea water. *Limnology and Oceanography*, **20**, 131–2.

Lion, L. W. & Leckie, J. O. (1982). Accumulation and transport of Cd, Cu and Pb in an estuarine salt marsh surface microlayer. *Limnology and Oceanography*, **27**, 111–25.

Liss, P. S. (1975). Chemistry of the sea surface microlayer. In *Chemical Oceanography*, ed. J. P. Riley & G. Skirrow, 2nd edn, vol. 2, pp. 193–244. London: Academic Press.

Lund, C. A. (1966). *Coastal and Deep Sea Navigation for Yachtsmen.* 6th edn. Glasgow: Brown, Son & Ferguson.

Magnusson, B. & Rasmussen, L. (1982). Trace metal levels in coastal sea water. *Marine Pollution Bulletin*, **13**, 81–4.

Mantoura, R. F. C. & Woodward, E. M. S. (1983). Optimization of the indophenol blue method for the automated determination of ammonia in estuarine waters. *Estuarine, Coastal and Shelf Science*, **17**, 219–24.

Morris, A. W., Bale, A. J. & Howland, R. J. M. (1982). Chemical variability in the Tamar Estuary, south-west England. *Estuarine, Coastal and Shelf Science*, **14**, 649–61.

Pennak, R. W. (1973). Water chemistry and the realiability of electronic field instruments. *Limnology and Oceanography*, **18**, 811–12.

Platford, R. F., Carey, J. H. & Hale, E. J. (1982). The environmental significance of surface films: Part I, octanol–water partition coefficients for DDT and hexachlorobenzene. *Environmental Pollution, Series B*, **3**, 125–8.

Rainwater, F. H. & Thatcher, L. L. (1960). *Methods for Collection and Analysis of Water Samples.* United States Geological Survey Water Supply Paper No. 1454. Washington, D.C.: United States Government Printing Office.

Rittall, W. F. (1974). Surface slicks and films – a need for control. In *Proceedings of a Seminar on Methodology for Monitoring the Marine Environment*, Seattle, October 1973, pp. 55–71. Washington, D.C.: United States Environmental Protection Agency.

Skougstad, M. W. & Scarbro, G. F. (1968). Water sample filtration unit. *Environmental Science and Technology*, **2**, 298–301.

Smith, K. L. (1971). A device for sampling immediately above the sediment–water interface. *Limnology and Oceanography*, **16**, 675–7.

Solórzano, L. (1969). Determination of ammonia in natural waters by the phenolhypochlorite method. *Limnology and Oceanography*, **14**, 799–801.

Strickland, J. D. H. & Parsons, T. R. (1972). *A practical handbook of seawater analysis. Bulletin of the Fisheries Research Board of Canada*, No. 167, 2nd edn.

Williams, P. J. leB. & Jenkinson, N. W. (1982). A transportable micro-processor-controlled precise Winkler titration suitable for field station and shipboard use. *Limnology and Oceanography*, **27**, 576–84.

3

Salinity, dissolved oxygen and nutrients

P. C. HEAD

The measurement of these parameters forms the classical repertoire of the marine chemist. Originally studied for their biological and hydrographic significance, in the estuarine context, they were frequently employed as indicators of pollution and of the natural processes restoring polluted waters.

The term 'nutrients' usually refers to the dissolved inorganic forms of nitrogen, phosphorus and silicon utilized by photosynthetic organisms in the formation of organic matter. Since the process of photosynthesis releases dissolved oxygen, and nutrients are regenerated during the oxidation of organic matter, chiefly through bacterial respiration, nutrient concentrations and the concentration of dissolved oxygen often show opposing trends. In the complex estuarine environment, however, such simple relationships are easily masked. Salinity, on the other hand, may be said to define the very extent of estuarine waters.

Salinity

The determination and definition of salinity, which approximates very closely to the total dissolved solids content of a water sample, is something which has, over the years, occupied a considerable amount of time of marine chemists and physicists. The salinity of marine waters is a fundamental property which can be used to determine much about the previous mixing and chemical history of the waters (see Chapter 7 for the definition of salinity). In the open ocean and even in coastal waters, variations in salinity are small and thus it is necessary to use very precise methods to determine the extent of real differences. In estuaries, differences in salinity are typically much greater and, as explained in Chapter 1, the use of high precision methods is often counter-productive in that they can introduce much unwanted background noise which may obscure the information sought. Until 1982 salinity determinations were related to a primary international standard by means of their chloride content. With the adoption in 1982 of the Practical Salinity Scale 1978 this relationship has been superseded by one involving the relative conductivities of standard sea waters and a standard solution of potassium chloride. (See Chapter 7 for details; and Parsons (1982), Sharp & Culberson (1982) and Gieskes (1982) for a discussion of the implications to biologists and chemists of the adoption of the Practical Salinity Scale 1978.)

Sampling and storage

Where discrete samples are required for salinity determinations they may be obtained by a wide variety of methods as the probability of sample contamination is low. Most of the sampling devices described in Chapter 2 may be used to collect salinity samples. Where non-flow-through devices are being used, the only precaution to take is to ensure that the sampler is well-rinsed with the sample so that traces of previous samples or fresh water are removed.

Once the sample has been collected, the type of storage container required is in some ways dependent on the subsequent precision of the salinity determination, although it is always good practice to try to minimize both systematic and random changes to the sample prior to analysis. For very precise work it is usual to store salinity samples in 200–250 ml glass bottles with impermeable stoppers. Such bottles when sealed with wax-impregnated corks should maintain the salinity of samples for many years, as should beer-bottle types with a flip-lock fitted with a silicone rubber seal. For most estuarine work, bottles sealed with ordinary rubber bungs should be adequate as samples are unlikely to be stored for extended periods before analysis. Where the relatively low precision methods are employed, any glass bottle with a stopper providing a virtually airtight seal may be employed. In all cases the storage bottles should be well-rinsed with the sample to ensure that all traces of previous samples are removed.

Salinity determination

Where precise salinities are required they are best determined by laboratory-based inductively coupled salinometers. Such salinometers will provide salinity data precise enough for the good quality density information required in water movement studies and may be used to calibrate batches of sub-standard sea water used for checking less precise field instruments. Where access to such instruments is not available, precise determinations of chlorinity are possible by titrimetric methods but as is explained in Chapter 7 chlorinity is no longer related directly to Practical Salinity and they should not now be used to calculate salinity.

The principles and use of laboratory-based conductimetric salinometers are fully described by Wilson (1975) and Grasshoff *et al.* (1983). These instruments must always be standardized against International Association of Physical Sciences of the Oceans (IAPSO) standard sea water. This standard sea water, which is now calibrated against a potassium chloride conductivity standard is supplied by the Institute of Oceanographic, Sciences (IOS), Brook Road, Wormley, Godalming, Surrey GU8 5UB, England.

Details of the high precision titrimetric methods for the determination of chlorinity are given by Strickland & Parsons (1972); Wilson (1975) and Grasshoff *et al.* (1983). Before attempting to use such methods it is important that investigators are clear in their own minds about the present and past relationships between chlorinity, chlorosity and salinity (see Chapter 7).

For most estuarine investigations, relatively low precision salinity determinations are usually adequate. These may be obtained from field salinometers or laboratory titrations to determine chlorinity as described by Strickland & Parsons (1972). For these latter titrations, the subtleties of the chlorosity, chlorinity, salinity relationship may be ignored, although there is a strong case for not attempting to calculate salinities but to report the data in terms of chlorinity or chlorosity. Thus where information is being sought on the relationship between a constituent and an index of fresh-water content, adequate data should be obtainable by the careful use of field salinometers or low precision chloride titrations. Where reactions in the very low salinity regions of estuaries are being investigated, chloride data, either obtained with a chloride electrode, or by titration, are to be preferred (Gieskes, 1982).

With a laboratory-based inductive salinometer, a standard deviation of about ± 0.01 should be routinely achievable at salinities near to that of the standard sea water (Practical Salinity 35.00). For samples with a much greater fresh-water component, relatively less precise figures will be obtained. Where the relatively low precision titration method is used, chlorinity or salinity figures correct to between 0.05 and 0.1‰ should be achievable for salinities of between about 4 and 40‰ (chlorinities of about 2–22‰).

Dissolved oxygen

Sampling and storage

Samples for oxygen determination are easily contaminated by atmospheric oxygen if suitable care is not taken in transferring them from sampling devices to storage bottles. The use of displacement samplers as described in Chapter 2 provides a way round these problems as the sample is taken in the bottle in which it is subsequently stored. Where displacement samplers are not employed, the transfer of a portion of a sample for oxygen determination should be carried out as soon as possible after the sampler has been recovered. Grasshoff *et al.* (1983) recommend the use of a special type of glass bottle with a sample volume of about 50 ml. This bottle has a relatively wide, flat-bottomed glass stopper which can be easily inserted without introducing any atmospheric oxygen. However, the volume of these bottles is too small for precise titration without specialized apparatus and almost any small glass bottle with a volume of up to 250 ml, a tapered neck and glass or high density plastic stopper may, with care, be satisfactorily used to store samples. The transfer of the sample from the sampler to the storage bottle is accomplished by means of a tube, preferably transparent, which is inserted to the bottom of the bottle and slowly withdrawn after flushing the bottle with about twice its volume of sample.

If the dissolved oxygen concentration is not to be determined immediately, either titrimetrically or instrumentally, the sample may be fixed for subsequent titrimetric determination by the addition of the manganous salt and alkaline iodide reagents. These should be introduced below the surface of the sample by means of narrow-tipped pipettes or a purpose-built reagent dispenser. Following the addition of the fixing reagents the stopper is inserted carefully, to ensure that no air is entrapped, and the contents mixed thoroughly by repeated vigorous inversion. Once fixed, the sample should be stable for up to about 24 h providing it is stored in the dark and not subject to large temperature changes.

Titrimetric method

The Winkler method (Winkler, 1888) has stood the test of time for almost a century as the standard analytical procedure for the determination of oxygen in aqueous solutions. It is based on a series of redox reactions. Manganous hydroxide is precipitated in the sample by the addition of two solutions, one containing a manganous salt and the other an alkali metal hydroxide. The oxygen dissolved in the sample oxidizes an equivalent amount of the precipitated Mn(II) to higher oxidation states (+3 and +4) which, on subsequent acidification in the presence of iodide, liberates an equivalent amount of iodine. This is titrated against standard thiosulphate. The basic technique has been well described by Carpenter (1965), Riley (1975), HMSO (1980) and Grasshoff *et al.* (1983).

The method is simple, capable of a high degree of precision and gives accurate results in clean estuarine waters. The large number of modifications which have been suggested to overcome interferences from a variety of natural substances and pollutants have been fully discussed by Phillips (1973). In many cases only simple precautions are necessary, but tests for interferences should be carried out whenever unfamiliar samples are to be analysed.

The main drawback to the Winkler method is that it involves handling corrosive reagents, which can be particularly hazardous on a sampling vessel. Syringe pipettes are to be preferred for reagent transfer and Öström (1973) has described the use of sealed ampoules which are crushed after addition to the sample. The titration can be automated to reduce fatigue and enhance sensitivity and routine precision (Bryan *et al.*, 1976; Williams & Jenkinson, 1982), but such techniques have not improved on the throughput of at least twenty samples per hour attainable by an experienced analyst.

If dissolved oxygen concentrations are required in terms of per cent saturation, allowance must be made for the salinity and *in situ* temperature of the sample. Allowances for changes in oxygen solubility at different atmospheric pressures are generally small enough to be neglected for most investigations and, for most estuaries, sampling does not go down deep enough for the effects of hydrostatic pressure to become significant. Tables for determining the solubility of oxygen in waters of varying salinity have been published by UNESCO (UNESCO, 1973) and convenient summaries are given by Riley & Skirrow (1975), Grasshoff *et al.* (1983) and HMSO (1980). The solubility of oxygen in water within the range of temperatures and chlorosities usually encountered in estuaries is illustrated in Fig. 7.1 (page 286).

The accuracy, precision and limit of detection obtainable with this method are very much dependent on the exact methodology and composition of the samples. Using the method suggested by Grasshoff *et al.* (1983) a standard deviation of about 0.001 mmol l^{-1} (0.03 mg l^{-1}) should be achievable, but for routine investigations, standard deviations nearer to 0.002 mmol l^{-1} (0.06 mg l^{-1}) are probably more realistic (Strickland & Parsons, 1972). The routine standard deviation of the automated methods is about an order of magnitude smaller than that obtainable manually.

Instrumental methods

Because of the high probability that the dissolved oxygen concentrations of unfixed samples will significantly alter during transport from the sampling location to a shore-based laboratory, instrumental methods of oxygen determination are confined to field use and are discussed in Chapter 2.

Nutrients

Half a century ago the sensitivity and specificity of their nutrient analyses placed marine chemists at the forefront of analytical chemistry but inter-laboratory comparison exercises have repeatedly shown that even long-established procedures must never be taken for granted. The price of consistent results is continual vigilance. Most chemical methods used in estuaries are adapted from methods developed for sea water, and are therefore more than adequate in terms of sensitivity. However, there are several characteristics of estuarine waters that can upset the typically spectrophotometric methods used for the determination of nutrients in saline water. These include wide ranges of concentration, varying salinity, turbidity and colouration. Most of these problems can be overcome by suitable blank and calibration procedures.

Spectrophotometric nutrient analyses are readily adapted to discrete or continuous autoanalytical systems. The initial development of continuous autoanalytical systems involved air-segmented methodology but recently methods employing flow-injection analysis (Růžička & Hansen, 1978, 1981) have been developed. A good introduction to the principles and problems of continuous automatic analysis is given by Petts (1980) and its use for determining nutrient concentrations in saline waters by Folkard (1978) and Grasshoff *et al.* (1983). As discussed in Chapter 1 the data obtained from such systems are extremely useful in investigating short-term or spatially isolated phenomena in estuaries and the use of a continuous system to determine nutrients on board a small sampling vessel has been described by Morris *et al.* (1981).

Although there is a certain amount of agreement amongst marine and estuarine chemists on the general methods for nutrient analyses in saline water, it is not yet possible to recommend methods for all the various nutrient forms which are suitable for even the majority of samples. In the following sections an attempt has been made to introduce the reader to the methods available and to indicate those which are generally applicable. However, when choosing any method whether previously used widely or on a more limited scale, it is the responsibility of the analyst to evaluate it for the particular situation and ensure that sensible results are produced.

Sampling, filtration and storage

The sampling of estuarine waters for nutrient analysis presents no particular problems if normal standards of cleanliness are maintained. However, compliance with this apparently simple injunction requires additional effort for any field work, and especially on a moving platform such as a boat.

All sampling equipment (buckets, water bottles, tubing, etc.) should be of plastic and reserved from new for this purpose only. Before first use their internal surfaces should be thoroughly leached with distilled water and left open to drain and dry out naturally in a dust-free atmosphere. On returning from a survey, the exterior of water bottles should be hosed off with copious tap water to remove salt which might otherwise corrode metal fittings. Their interiors and other sampling equipment should be rinsed with distilled water and left to drain as before. In the case of grease or other resistant contamination, soaking in a solution of phosphate-free detergent is permissible, but mechanical abrasion should be kept to a minimum.

In the field, samplers should be well flushed with water similar to that to be sampled before each sample is taken. This occurs automatically in the case of flow-through water bottles and pump sampling systems. Water bottles should not be allowed to linger at the water surface where surface films may adhere to them. Indeed, it may be advisable to use the bottles in the same relative positions on the wire to reduce the risk of cross-contamination. These precautions are particularly important in estuaries, where major changes in composition can occur over relatively short distances and depth ranges.

As a result of the shallow depths involved in most estuarine work, samples usually spend a relatively short time in water bottles and so contamination does not normally arise from this source. However it can easily pass unnoticed. If this problem is suspected, the bottle should be flushed with water typical of the area, closed, and samples drawn off for analysis at intervals. In this way absorption of phosphate and reduction of nitrate was confirmed in a nickel-plated Nansen bottle.

All vessels with which the sample will come into contact prior to analysis (beakers, filtration apparatus, sample bottles, etc.) should be cleaned in advance by rinsing with a little concentrated sulphuric acid followed by several washes with distilled water, and dried by shaking. Open vessels, such as beakers, should be covered with aluminium foil until required.

For most estuarine investigations, relatively high concentrations of particulate matter will be encountered and it will be necessary to filter the sample before analysis. If analysis is postponed, filtration should be carried out as soon as possible after collection, rather than just prior to analysis, to minimize changes during storage. Samples should be filtered using an apparatus of the type described in Chapter 2. Various investigations have been made into the effects of filtration on nutrient concentrations (see Riley, 1975; Eaton & Grant, 1979; Grasshoff *et al.*, 1983; for details). From these it is apparent that no one type of filter is ideal for all nutrient determinations but glass-fibre filters probably represent the best compromise for estuarine work. Glass-fibre filters with a nominal pore size of 1.2 μm (e.g. Whatman GF/C) have been widely used although more retentive ones are available. For estuarine waters a pore size of 1.2 μm probably represents the best compromise between flow rate and retentivity. Further information about the structure and performance of the various types of filter available is given in Chapter 5. The filters should be rinsed *in situ* with distilled water and the first portion of each sample filtered should be discarded. The filtrate will be essentially free of living material and optically clear from the point of view of subsequent spectrophotometry.

Ideally, analysis should be carried out immediately after filtration, and this can be achieved on large vessels with laboratory facilities. However, it is in the nature of most estuarine studies that they are undertaken too close to a shore laboratory for such facilities to be justified, and so samples must be stored at least until the return to land. A number of approaches have been made to the problem of preserving the samples' composition unchanged during this period. Filtration itself is an important method of temporary preservation since, by removing most living material and suspended particles, it arrests the exchange of nutrients by absorption or metabolism. The most generally effective method of preservation is rapid freezing by placing the samples in an insulated box containing solid carbon dioxide. Obviously plastic bottles are more suitable for this treatment, but glass medical flats can be used if they are filled no more than two-thirds full and allowed to freeze at an angle of about 45°. Polypropylene bottles of about 300 ml capacity are adequate for most purposes but glass bottles are to be preferred for phosphate samples unless an iodization procedure is employed to prevent the absorption of phosphorus (see HMSO, 1981*a* for details). Samples intended for silicate analysis should not be stored in glass bottles. The only drawbacks to freezing as a method of preservation occur in low-salinity samples. Hard waters may form a precipitate which removes phosphate from solution and a fraction of silicate is rendered unavailable for a short period after thawing. In any case, samples should be completely thawed before removing aliquots for analysis because dissolved components are fractionated between the solid and liquid phases during freezing.

The chief alternative to freezing is the use of chemical preservatives, such as chloroform and mercuric chloride. Here the risk is that of interference with analytical procedures. For example, mercurous chloride may be precipitated under the reducing conditions of the most commonly used phosphate method and mercury also interferes in ammonia determinations, perhaps through the formation of complexes. Nevertheless, chemical preservatives can be useful if freezing facilities are unavailable or as an additional safeguard. Their possible effects on analytical results should always be checked first, however. The importance of filtration prior to storage can now be appreciated: the death of plankton cells by rupture during freezing or poisoning by chemical preservatives would release their contents irrecoverably into solution.

Phosphorus

Essentially all the phosphorus present in the aquatic environment is in the +5 oxidation state as various forms of phosphate. There have been many complex schemes put forward to classify the fractions which can be distinguished by a combination of filtration and analytical procedures (see Burton, 1973; HMSO, 1981*a*; Strickland & Parsons, 1972 for details). The significance of many of these fractions is unclear and for most investigations only one or two of the four main fractions need be considered. These four fractions are distinguished operationally on the basis of filtration and oxidation. Filtration distinguishes between the dissolved and particulate fractions and oxidation between the inorganic (largely orthophosphate) and organic forms. For all the fractions, the final analytical stage involves the measurement of the concentration of orthophosphate ions by the formation of a reduced phosphomolybdenum blue complex in an acidic solution containing molybdic acid, ascorbic acid and trivalent antimony. The most popular of the methods relying on this reaction, which was developed by Murphy & Riley (1962), is that given by Strickland & Parsons (1972). As pointed out by Burton (1973), the reaction conditions, particularly in terms of acidity, for this variation are significantly different from those of the original method. In these methods the reagents are mixed into a single solution of limited stability for addition to the sample and the optical density determined at 885 nm after ten minutes. Similar reaction conditions are used in a variation of the method described by Grasshoff *et al.*, (1983) in which two more stable reagent solutions are used and the optical density measured within 30 minutes to reduce any possible interference from arsenate. There is no salt error with these methods which are capable, with care, of detecting as little as 0.01 μmol l^{-1} if 100 mm cuvettes are used in the spectrophotometer. Such a detection limit is unlikely to be required for routine investigations of most estuaries and therefore more convenient shorter cuvettes may be employed.

Various descriptions of automated phosphate procedures based on that of Murphy & Riley (1962) have been described (see for example Chan & Riley 1966; Folkard, 1978; Grasshoff *et al.*, 1983). To ensure adequate colour development, fairly long mixing coils or a heating step were usually employed. Most of these methods have not been systematically compared with the manual method as described by Strickland & Parsons (1972), which is generally accepted as the most suitable for saline samples. However, the automated techniques described by Atlas *et al.* (1971) and Tréguer & Le Corre (1975) have been shown to give a good agreement with this manual method and their use is, therefore, to be recommended. This latter method has recently found favour with a number of British laboratories as it gives a very stable baseline and good sensitivity and reproducibility. As it is not easy to obtain a copy of the method, an English translation is included as an appendix to this chapter. When using any automated phosphate method, it is important to ensure that the reduced molybdenum blue complex is confined to glass tubing to minimize absorption on to the tubing walls, and that for saline samples due allowances are made for any possible errors associated with refractive index effects (Froelich & Pilson, 1978). For autoanalytical methods detection limits similar to those of the manual methods should be achievable. A flow injection method which is capable of determining phosphate at a rate in excess of ninety samples per hour has been developed by Johnson & Petty (1982).

The most probable source of interference in phosphate determinations is from arsenate which also forms a molybdenum blue complex under the reaction conditions. However, because the concentration of arsenate in estuarine waters is unlikely to normally exceed 0.01 μmol l^{-1} and its molybdenum blue complex absorption maximum occurs at a lower wavelength, interference is rarely likely to be significant.

Where it is necessary to measure phosphate concentrations of less than about 0.1 μmol l^{-1}, it is worth considering an extraction procedure to concentrate the molybdenum blue complex. This may be accomplished by the use of isobutyl acetate (Cescon & Scarazzato, 1973) or *n*-hexanol (HMSO, 1981*a*). Alternatively the yellow phosphomolybdic acid may be extracted into isobutyl acetate and then reduced to molybdenum blue with a mixed reagent containing ascorbic acid and stannous chloride (Pakalns & McAllister, 1972). By these methods it should be possible to determine as little as 0.015 μmol l^{-1} with 40 mm cuvettes (HMSO, 1981*a*) or 0.005 μmol l^{-1} with 100 mm cuvettes (Grasshoff *et al.*, 1983).

Reactive phosphate

The generally employed method for orthophosphate determination described above involves acidic conditions, principally to prevent interference from silicon, which could hydrolyse some polyphosphates and the more labile organophosphates which may be present in the sample. For this reason it is usual to refer to all the species responding to the molybdenum blue procedure as reactive phosphate and not simply as phosphate or orthophosphate.

Total phosphorus

In the course of its incorporation in biological tissue, much phosphorus is incorporated into poly- and organophosphates. To determine the total phosphorus concentration of a sample, an oxidation procedure is required to bring the poly- and organophosphates into a reactive state. The most convenient way of doing this is by heating with an oxidizing agent such as persulphate or by exposure to ultraviolet (UV) radiation from a mercury vapour lamp in the presence of hydrogen peroxide (see Chapter 5 for details).

With the UV irradiation precedure, relatively short exposure times result in the breakdown of most organophosphates but not of polyphosphates. Total phosphorus may be determined by prolonged irradiation or more conveniently by hydrolysis of the irradiated samples. Although this method requires the construction of a special irradiator and silica reaction vessels, its simplicity and elegance make it well worth considering if routine organic or total phosphorus determinations are to be made.

Polyphosphate

Large amounts of the inorganic phosphorus in algae occur as polyphosphates which will not react with the acid molybdate reagent (Spencer, 1975). The death and decay of algae is therefore a potential source of polyphosphate. Although not thought to be present in most unpolluted marine and estuarine waters in very large quantities under normal circumstances, Solórzano & Strickland (1968) have reported appreciable quantities of polyphosphate in waters of a red-tide bloom. Also the use of polyphosphates in detergents means that there is a potential source in the waste discharges made to many estuaries. Determination of polyphosphate may be made either by acid hydrolysis or UV irradiation procedures as described by Solórzano & Strickland (1968) and Grasshoff *et al.* (1983).

Particulate phosphorus

Particulate reactive inorganic phosphorus may be determined by difference from reactive phosphorus analysis of unfiltered and filtered portions of a sample. This fraction will include calcium and ferric phosphates and any inorganic phosphate associated with cellular material. Analysis of particulate material for reactive phosphate after oxidation, as discussed in Chapter 5, will give a measure of its total phosphorus content. It is probable that this will mainly be organic phosphorus but inorganic species unreactive to acid molybdate will also be included.

Nitrogen

In contrast to phosphorus the dissolved nitrogen present in the aquatic environment occurs in many of the nine oxidation states from -3 to $+5$. However only a few of the many possible nitrogen forms occur in appreciable amounts. Molecular nitrogen occurs in water at or near its saturation value and is thus the dominant form of the element with a concentration of about 600 μmol l^{-1} as against a maximum concentration of about 200 μmol l^{-1} for combined nitrogen forms in estuaries. However, it is thought to be virtually inert and, of the combined forms, only ammonia, nitrite, nitrate and organically bound nitrogen are usually involved to any great extent in biological processes and thus determined routinely. These species may be determined separately or in various combinations. In addition to the conversion of inorganic to organic nitrogen by phytoplankton and other microorganisms there exists a number of specialized groups of microorganisms which are capable of interconverting most of the usually found inorganic species. Thus bacteria of the genus *Nitrosomonas* are capable of oxidizing ammonia to nitrite and of the genus *Nitrobacter* nitrite to nitrate. Other bacteria are capable of reducing nitrate to molecular nitrogen.

The determination of nitrite has acquired a special significance, not because it is particularly abundant, but because other forms of nitrogen can often be readily converted to nitrite, which can be measured with great sensitivity.

Ammonia

The acid ionization of ammonium ions represented by the equation

$$NH_4^+ + H_2O \rightleftharpoons NH_3 + H_3O^+$$

is a reaction which is very dependent on temperature, salinity and pH. The reaction is very sensitive to the normal range of pH found in estuarine waters, and particularly to changes in the range pH 8.0–9.5 as is illustrated in Fig. 3.1. The proportion of ammonia present in the more toxic un-ionized form decreases with increasing salinity but increases with increasing temperature. Methods for the determination of ammonia determine both the ionized and un-ionized forms and thus the term ammonia is used here to include both free dissolved ammonia gas (NH_3) and ammonium ions (NH_4^+). Where necessary the proportions of the two species in any particular sample may be calculated from the total ammonia concentration using the data of Whitfield (1974) (see Hampson, 1977 for details of a computer program to do this).

The development of a suitable method for the determination of ammonia at the low concentrations, typically $< 3\ \mu mol\ l^{-1}$, at which it occurs in marine waters has proved difficult. Methods involving distillation or oxidation to nitrite, have proved impracticable for routine use and the classical Nessler method lacks sensitivity and is subject to interference from the alkali earth metals present in saline samples. A fluorometric method has been developed by Gardner (1978).

The presently favoured methods rely on modifications of the indophenol blue technique first proposed by Berthelot (1859). The initial application of this method to sea water by Solórzano (1969) involved the use of nitroprusside as a catalyst and a sodium hypochlorite solution as the chlorine source. Liddicoat *et al.* (1975) found that, when using this method, they were troubled by high and erratic blanks and variations in the intensity of the indophenol blue colour for standards. Low and reproducible blanks were obtainable by substituting potassium ferrocyanide as the catalyst and using UV light to stabilize colour development. As the original method was developed in Southern California, samples would have been naturally

Fig. 3.1. The effect of pH and salinity on the ionization of ammonia. Redrawn from Hampson (1977).

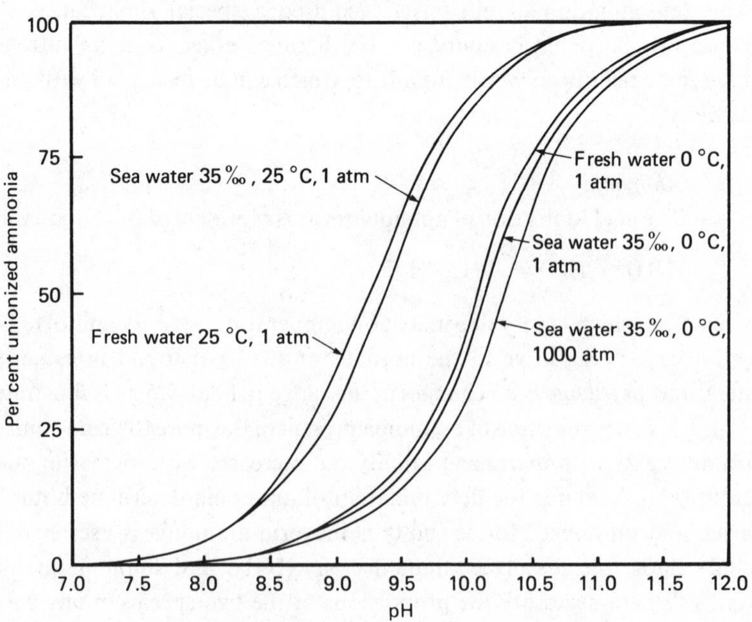

exposed to fairly high levels of UV light. Liddicoat *et al.* (1975) also favoured the use of sodium dichloroisocyanurate as the chlorine source. With minor modifications in the standardization procedure to take account of interference by nitrite and various primary and secondary amines as described by HMSO (1982*a*), this procedure is probably the best chemical method presently available for determining ammonia in estuarine and marine waters, although an adaptation of the automated method, developed specifically for estuaries by Mantoura & Woodward (1983), may prove to be more versatile as it requires no exposure to UV light and should suffer from no salt effects. If an automated method is required, that of Mantoura & Woodward (1983) seems to be the most appropriate as it is capable of giving consistent results over the range of salinities encountered in estuaries.

Although the use of ammonia sensing electrodes for the detection of the typically low levels of ammonia present in estuarine and sea water has not yet gained widespread acceptance, their use is possible and may offer definite advantages in that they do not suffer from the interference effects of the chemical methods (see Chapter 6).

Whatever the method used for determining ammonia in unpolluted estuarine waters, great care must be taken to ensure that samples, blanks and standards are not contaminated during the course of analysis. The inadvertent introduction of containers of concentrated ammonia solutions into, or near to, a laboratory where ammonia determinations are in progress can easily ruin a batch of expensively collected samples.

Nitrite

Nitrite is present in estuarine waters in generally low concentrations as an intermediate product of microbial reduction of nitrate or oxidation of ammonia and as an excretion product of plankton. In estuaries subject to large discharges of treated or untreated sewage, higher than usual concentrations are to be expected. In most waters, nitrite concentrations seldom account for more than 10 per cent of total oxidized nitrogen. All methods for the determination of nitrite in saline waters are based on the classical Griess–Ilosvay diazotization technique as adapted for sea water by Bendschneider & Robinson (1952). Full details are given by Strickland & Parsons (1972), HMSO (1982*b*), and Grasshoff *et al.* (1983) and for a flow injection method by Johnson & Petty (1983). Morris *et al.* (1981) report negligible refractive index effects when using an automated method involving this reaction in estuarine waters. The detection limit of the method is about 0.02 μmol l^{-1}.

Nitrate

In spite of its importance as the most abundant nitrogenous nutrient present in saline waters and a considerable amount of analytical development work there is as yet no satisfactory method for the direct determination of nitrate in waters of appreciable salinity. However, suitable methods for the quantitative reduction of nitrate to nitrite have been developed and thus the determination of total oxidized nitrogen is possible. The concentration of nitrate may be established by difference after a separate nitrite determination on the same sample, As, however, in most situations, nitrite concentrations are very low in relation to nitrate, unless knowledge of the nitrite concentration is particularly required, as for example when studying nitrification, it is often sufficient to determine total oxidized nitrogen alone. Such determinations are often loosely referred to as nitrate although in reporting them the distinction should be pointed out. Compared with the development of methods for the determination of phosphorus and silicon, it was not until comparatively recently that a satisfactory nitrate method using cadmium or copperized cadmium to achieve a reduction to nitrite in saline waters was developed (Grasshoff, 1964; Morris & Riley, 1963; Wood *et al.*, 1967). For manual methods this reduction is usually achieved by passing the sample through a column of copperized cadmium granules either by gravity (Strickland & Parsons, 1972) or by aspiration (Grasshoff *et al.*, 1983). However, Lambert & DuBois (1971) have described a method whereby the samples are shaken in the presence of copperized cadmium filings. With the reductor column methods considerable care is required to achieve consistent results as unquantifiable variations, particularly associated with the unpredictable loss of activity by the columns, are encountered. This source of variation is much more controllable in automated systems as all the samples and standards pass through the same reductor column and its performance is more easily monitored.

Early attempts to automate the cadmium reduction technique involved the passage of the sample through columns of similar character to those used in the manual methods (Strickland & Parsons, 1972; Grasshoff, 1976). Although satisfactory reduction could be achieved with such columns, problems were often encountered as a result of the mixing of samples within the column and uneven dissolution of the reductor material. These problems have been overcome by the adoption of a simpler type of reductor involving a single strand of copperized cadmium wire as suggested by Stainton (1974). Full details of the method are given by Folkard (1978) and HMSO (1982*b*). Errors associated with refractive index effects appear to be negligible (Morris *et al.*, 1981). A flow injection method for nitrate has been developed by Johnson & Petty (1983). Both manual and automated methods should be capable of detecting about 0.1 μmol l^{-1} of total oxidized nitrogen.

Total nitrogen

Nitrogen in living material is incorporated into a vast range of compounds ranging from simple amino acids containing a single nitrogen atom to complex proteins containing many thousands. On death or decay these compounds are released to the surrounding water and become part of the organically bound nitrogen fraction of natural waters. As with phosphorus, an oxidation step is required to render this fraction reactive and the choice again lies between wet oxidation and UV irradiation. The general principles of both of these methods are given in Chapter 5. Until relatively recently most total nitrogen determinations have relied on the well-established Kjeldahl treatment in which organic nitrogen is converted to ammonia by heating with sulphuric acid (see Strickland & Parsons, 1972). However, this method is not without problems when applied to saline samples and the alkaline digestion procedure developed by Koroleff (see Grasshoff *et al.*, 1983) seems to offer considerable advantages. In this method, samples are heated under pressure for about 30 minutes in the presence of NaOH and persulphate to convert organically bound nitrogen to nitrate which is then determined by the normal reduction method. That this method results in a single oxidation product, nitrate, gives it some advantages over UV irradiation procedures where a combination of nitrate, nitrite and ammonia can be produced.

Particulate nitrogen

Ammonia, nitrite and nitrate associated with particulate material may, if required, be determined by difference from analyses of filtered and unfiltered samples. The determination of the more important organically bound particulate organic nitrogen is discussed in Chapter 5.

Silicon

Unlike nitrogen, silicon does not normally undergo oxidation/reduction reactions in the aquatic environment and is found in natural waters in a stable +4 oxidation state. The predominant form in saline waters is probably undissociated silicic acid $Si(OH)_4$. Polymeric forms of silicic acid have not been found in significant amounts in saline waters although they may be present in fresh water. In the course of uptake by diatoms, radiolaria and some sponges, silicic acid is converted into silica (SiO_2) and incorporated into skeletal material. With the death of such organisms, particulate silica is released to the water to become part of a particulate silicon fraction also consisting of authigenic and weathered crustal alumino-silicate material. The particulate material may be subsequently dissolved or become incorporated into coastal or oceanic sediments. The relative importance of biological and abiological processes in controlling the silicon concentration of natural waters is still a matter of controversy although the short-term effects of biological activity are readily demonstrable.

The majority of methods presently used for determining silicon in saline waters rely on measuring the absorbance of the intensely coloured molybdenum blue complex resulting from the reduction of β-silicomolybdic acid formed by the addition of an acidic molybdate reagent. Interference from phosphorus and arsenic is eliminated by the addition of oxalic acid which decomposes any phospho- and arsenomolybdate complexes formed during the initial addition of the molybdate reagent. As only relatively short-chain polymers of silicic acid can also react with the molybdate reagent at an appreciable speed the quantity determined by the method is usually termed 'reactive silicate'. It is unlikely that this quantity differs significantly from total dissolved silicic acid for most saline samples. For the initial development of the method, stannous chloride was used for the reduction step but as reaction conditions have to be closely controlled to achieve consistent results, subsequent workers have preferred to use metol (*p*-methylamino phenol sulphate) or ascorbic acid. Present workers appear to favour the use of ascorbic acid as described by Grasshoff *et al.* (1983) and also HMSO (1981*b*). Full details of a method involving metol as the reductant are given by Strickland & Parsons (1972). Both variations of the method are subject to a discernible salt error and this must be taken into account for precise work in estuarine waters.

Silicate determination in saline waters has been successfully adapted to continuous autoanalytical systems using both ascorbic acid (Folkard, 1978), metol (Grasshoff *et al.*, 1983) and stannous choride (Strickland & Parsons, 1972) as the reductant. In the case of the latter method, the closer control of analytical conditions achievable with the automatic system allows better precision than can be achieved by manual analysis. Refractive index effects as described by Froelich & Pilson (1978) need to be taken into account. The detection limits of both manual and automated methods are about 0.2 μmol l^{-1}.

Total Silicon

Most of the silicon occurring in saline waters which is unavailable to the molybdate reagent will be in the form of particulate material associated with organisms or silicate minerals. As most silicate minerals are relatively stable materials, fairly extreme extraction methods involving fusion with sodium carbonate in platinum crucibles are necessary to bring the silicon into a reactive state. Details of a suitable method are given in Grasshoff *et al.* (1983). Grasshoff *et al.* (1983) also describe an alternative method involving alkaline persulphate oxidation which will recover organically associated silicon and much of that associated with inorganic suspended material. For this method polytetrafluoroethylene (PTFE) heating vessels are required.

References

Atlas, E. L., Hager, S. W., Gordon, L. I. & Park, P. K. (1971). *A Practical Manual for use of the Technicon AutoAnalyzer in Sea-water Nutrient Analyses*. Revised Technical Report, No. 215, Reference 71–22, Oregon State University.

Bendschneider, K. & Robinson, R. J. (1952). A new spectrophotometric method for the determination of nitrite in seawater. *Journal of Marine Research*, **11**, 87–96.

Berthelot, M. E. P. (1859). *Répertoire de Chimie Appliquée*, p. 284.

Bryan, J. R., Riley, J. P. & Williams, P. J. leB. (1976). A Winkler procedure for making precise measurements of oxygen concentration for productivity related studies. *Journal of Experimental Marine Biology and Ecology*, **21**, 191–7.

Burton, J. D. (1973). Problems in the analysis of phosphorus compounds. *Water Research*, **7**, 291–307.

Carpenter, J. H. (1965). The accuracy of the Winkler method for dissolved oxygen analysis. *Limnology and Oceanography*, **10**, 135–40.

Cescon, B. S. & Scarazzato, P. G. (1973). Determination of low phosphate concentrations in seawater by an isobutyl acetate extraction procedure. *Limnology and Oceanography*, **18**, 499–500.

Chan, K. M. & Riley, J. P. (1966). The automatic determination of phosphate in seawater. *Deep-Sea Research*, **13**, 467–71.

Eaton, A. D. & Grant, V. (1979). Sorption of ammonium by glass frits and filters: implications for analyses of brackish and fresh water. *Limnology and Oceanography*, **24**, 397–9.

Folkard, A. R. (1978). *Automatic Analysis of Sea-water Nutrients*. Fisheries Research Technical Report No. 46. Lowestoft: Ministry of Agriculture, Fisheries and Food.

Froelich, P. R. & Pilson, M. E. Q. (1978). Systematic absorbance errors with Technicon AutoAnalyzer II Colorimeters. *Water Research*, **12**, 599–603.

Gardner, W. S. (1978). Microfluorometric method to measure ammonium in natural waters. *Limnology and Oceanology*, **23**, 1069–72.

Gieskes, J. M. (1982). The practical salinity scale 1978: a reply to comments by T. R. Parsons. *Limnology and Oceanography*, **27**, 387–9.

Grasshoff, K. (1964). Zur Bestimmung von Nitrat in Meeres- und Trinkwasser. *Kieler Meeresforschungen*, **20**, 5–11.

Grasshoff, K. (1976). *Methods of Seawater Analysis*. Weinheim: Verlag Chemie.

Grasshoff, K., Erhardt, M. & Kremling, K. (eds) (1983). *Methods of Seawater Analysis*, 2nd edn. Weinheim: Verlag Chemie.

Hampson, B. L. (1977). Relationship between total ammonia and free ammonia in terrestrial and ocean waters. *Journal du Conseil Permanent International pour l'Exploration de la Mer*, **37**, 117–22.

HMSO (1980). *Dissolved Oxygen in Natural and Waste Waters* 1979. Methods for the Examination of Waters and Associated Materials. London: Her Majesty's Stationery Office.

HMSO (1981*a*). *Phosphorus in Waters, Effluents and Sewages* 1980. Methods for the Examination of Waters and Associated Materials. London: Her Majesty's Stationery Office.

HMSO (1981*b*). *Silicon in Waters and Effluents* 1980. Methods for the Examination of Waters and Associated Materials. London: Her Majesty's Stationery Office.

HMSO (1982*a*). *Ammonia in Waters* 1981. Methods for the Examination of Waters and Associated Materials. London: Her Majesty's Stationery Office.

HMSO (1982*b*). *Oxidised Nitrogen in Waters* 1981. Methods for the Examination of Waters and Associated Materials. London: Her Majesty's Stationery Office.

Johnson, K. S. & Petty, R. L. (1982). Determination of phosphate in sea water by flow-injection analysis with injection of reagent. *Analytical Chemistry*, **54**, 1185–7.

Johnson, K. S. & Petty, R. L. (1983). Determination of nitrate and nitrite in sea water by flow-injection analysis. *Limnology and Oceanography*, **28**, 1260–6.

Lambert, R. S. & DuBois, R. J. (1971). Spectrophotometric determination of nitrate in the presence of chloride. *Analytical Chemistry*, **43**, 955–7.

Liddicoat, M. J., Tibbitts, S. & Butler, E. I. (1975). The determination of ammonia in seawater. *Limnology and Oceanography*, **20**, 131–2.

Mantoura, R. F. C. & Woodward, E. M. S. (1983). Optimization of the indophenol blue method for the automated determination of ammonia in estuarine waters. *Estuarine, Coastal and Shelf Science*, **17**, 219–24.

Morris, A. W. & Riley, J. P. (1963). The determination of nitrate in seawater. *Analytical Chimica Acta*, **29**, 272–9.

Morris, A. W., Bale, A. J. & Howland, R. J. M. (1981). Nutrient distributions in an estuary: evidence of chemical precipitation of dissolved silicate and phosphate. *Estuarine, Coastal and Shelf Science*, **12**, 205–16.

Murphy, J. & Riley, J. P. (1962). A modified single solution method for the determination of phosphate in natural waters. *Analytica Chimica Acta*, **27**, 31–6.

Öström, B. (1973). Expendable ampoules for oxygen determination. *Marine Chemistry*, **1**, 323–7.

Pakalns, P. & McAllister, B. R. (1972). Determination of phosphate in sea water by an isobutyl-acetate extraction procedure. *Journal of Marine Research*, **30**, 305–11.

Parsons, T. R. (1982). The new physical definition of salinity: biologists beware. *Limnology and Oceanography*, **27**, 384–5.

Petts, K. W. (1980). *Air Segmented Continuous Flow Automatic Analysis in the Laboratory*, 1979. Methods for the Examination of Waters and Associated Materials. London: Her Majesty's Stationery Office.

Phillips, A. J. (1973). The Winkler method and primary production studies under special conditions. In *A Guide to the Measurement of Marine Primary Production under Some Special Conditions*, pp. 48–54. Monographs on Oceanographic Methodology, No. 3. Paris: UNESCO.

Riley, J. P. (1975). Analytical chemistry of seawater. In *Chemical Oceanography*, ed. J. P. Riley & G. Skirrow, 2nd edn., vol. 3, pp. 143–514. London: Academic Press.

Riley, J. P. & Skirrow, G. (eds) (1975). *Chemical Oceanography*, 2nd edn., vol. 1, pp. 561–2. London: Academic Press.

Růžička, J. & Hansen, E. H. (1978). Flow-injection analysis. *Analytica Chimica Acta*, **90**, 37–76.

Růžička, J. & Hansen, E. H. (1981). *Flow-Injection Analysis*. London: Wiley Interscience.

Sharp, J. H. & Culberson, C. H. (1982). The physical definition of salinity: a chemical evaluation. *Limnology and Oceanography*, **27**, 385–7.

Solórzano, L. (1969). Determination of ammonia in natural waters by the phenolhypochlorite method. *Limnology and Oceanography*, **14**, 799–801.

Solórzano, L. & Strickland, J. D. H. (1968). Polyphosphate in seawater. *Limnology and Oceanography*, **13**, 515–18.

Spencer, C. P. (1975). The micronutrient elements. In *Chemical Oceanography*, ed. J. P. Riley & G. Skirrow, 2nd end., vol. 2, pp. 245–300. London: Academic Press.

Stainton, M. P. (1974). Simple, efficient reduction column for use in the automated determination of nitrate in water. *Analytical Chemistry*, **46**, 1616.

Strickland, J. D. H. & Parsons, T. R. (1972). A practical handbook of seawater analysis. *Bulletin of the Fisheries Research Board of Canada*, No. 167, 2nd edn.

Tréguer, P. & Le Corre, P. (1975). Phosphates. In *Manuel d'Analyse des Sels Nutritif dans l'Eau de Mer* (*Utilisation de l'AutoAnalyzer II, Technicon*), 2nd edn., Laboratoire d'Oceanologie Chimique. Brest: Universite de Bretagne Occidentale.

UNESCO (1973). International Oceanographic Tables, vol. 2. Paris: National Institute of Oceanography/UNESCO.

Whitfield, M. (1974). The hydrolysis of ammonia ions in seawater – a theoretical study. *Journal of the Marine Biological Association of the United Kingdom*, **54**, 565–80.

Williams, P. J. leB. & Jenkinson, N. W. (1982). A transportable micro-processor-controlled precise Winkler titration suitable for field station and shipboard use. *Limnology and Oceanography*, **27**, 576–84.

Wilson, T. R. S. (1975). Salinity and the major elements of seawater. In *Chemical Oceanography*, ed. J. P. Riley & G. Skirrow, 2nd edn., vol. 1, pp. 365–413. London: Academic Press.

Winkler, L. W. (1888). Die Bestimmung des im Wasser gelösten Sauerstoffes. *Berichte der Deutschen Chemischen Gesellschaft*, **21**, 2843–54.

Wood, E. D., Armstrong, F. A. J. & Richards, F. A. (1967). Determination of nitrate in sea water by cadmium–copper reduction to nitrite. *Journal of the Marine Biological Association of the United Kingdom*, **47**, 23–31.

APPENDIX TO CHAPTER 3

An automated method for the determination of phosphate in sea water

Translated from 'Tréguer, P. & Le Corre, P. (1975). *Manual d'Analyse des sels Nutritif dans l'Eau de Mer* (*Utilisation de l'AutoAnalyzer II, Technicon*, 2nd edn., Brest: Université de Bretagne Occidentale'.

Principle of the method

Dissolved inorganic phosphorus in sea water is essentially present in the form of orthophosphate ions (principally HPO_4^{2-} and PO_4^{3-} (Kester & Pytkowicz, 1967)). In unpolluted ocean waters the concentrations of mineral pyrosphosphates are very small (in the order of 0.1 μmol l^{-1}). According to Solórzano & Strickland (1968) polyphosphate originating from marine phytoplankton is soon re-utilized and thus does not accumulate.

The reaction described by Deniges (1920) has been generally used for the determination of dissolved inorganic phosphate. The Mo^{VI} ion (in the form of ammonium molybdate) reacts under acid conditions ($pH < 1$) with orthophosphate to form a heteropolyanion with the probable formula $(NH_4)_3P(Mo_3O_{10})_4$.

This complex is subsequently reduced with stannous chloride under the conditions described by Atkins (1923), Harvey (1948), Robinson & Thompson (1948) or Wooster & Rakestraw (1951) and measured spectrophotometrically. The chemistry of the reaction is not well understood and the intensity of the colour produced is dependent on the experimental conditions. Modifications were suggested by Burton & Riley (1956) and Murphy & Riley (1958, 1962) to reduce variability associated with the speed of the reaction and salinity of the sample.

In the most commonly used oceanographic method described by Strickland & Parsons (1972) orthophosphate ions react with an acidic reagent (0.25 N H_2SO_4) containing Mo^{VI}, ascorbic acid (reductant) and Sb^{VI} ions (catalyst) to form a well-defined complex with P, Mo and Sb present in the ratio 1:12:1. The optical density is independent of salinity and the Beer–Lambert law is obeyed.

Reductions using stannous chloride and ascorbic acid give different results and the differences are dependent on the nature of the sea water analysed (Armstrong, 1965; Jones, 1966; Piton & Voituriez, 1967).

Of the possible interfering ions capable of forming complexes at the pH employed arsenate causes the greatest effect. However, the speed of formation of the arsenate complex is slow in sea water and the concentrations encountered are small (less than 0.1 μmol l^{-1} in open oceanic water) and thus the interference is usually negligible. High concentrations of nitrite can also cause interference as can concentrations of copper greater than 500 μg l^{-1} when stannous chloride is used.

According to Murphy & Riley (1962), Armstrong (1965) and Jones (1966) the ascorbic acid method is, in general, less sensitive to interferences.

Finally, reaction with polyphosphates is possible. Hydrolysis of polyphosphates is generally achieved by heating to 100 °C with 0.1 N HCl and a partial hydrolysis is possible by heating at the natural pH of sea water (Solórzano & Strickland, 1968). In a relatively acidic medium (0.2–0.5 N H_2SO_4) suitable for the determination of orthophosphate no hydrolysis of polyphosphate occurs, provided the temperature does not exceed 30 °C. However, the possibility of interference from metaphosphate cannot be excluded (Souchay, 1969).

The experimental conditions determine, to a large extent, the precision of orthophosphate measurements in sea water.

At a concentration of 0.3 μmol l^{-1} the precision given by Strickland & Parsons (1972) is $\pm$0.02 μmol l^{-1} for a single determination.

For concentrations less than 0.3 μmol l^{-1} Proctor & Hood (1954) and Stephens (1963) have used concentration procedures involving the extraction of the reduced phosphomolybdic complex into isobutanol, and Pakalns & McAllister (1972) used extraction with isobutyl acetate prior to reduction. At a concentration of 0.2 μmol l^{-1} the standard deviation for 12 samples was $\pm$0.005 μmol l^{-1}.

The determination of inorganic phosphate has been adapted to a continuous flow autoanalytical system (Technicon 1971, 1973). Hager *et al.* (1968) and Coote *et al.* (1970) have used stannous chloride reduction but most investigators have preferred to automate the method of Murphy & Riley (1962). Various procedures are described by Molof *et al.* (1965), Brewer *et al.* (1966), Grasshoff (1965), Bernhard & Macchi (1966), Chan & Riley (1966), Armstrong *et al.* (1967), Grasshoff (1969), Revel (1969), Coste (1971), Atlas *et al.* (1971), Mangelsdorf (1972), Friederich & Whitledge (1972) and Hager *et al.* (1973) employing a wide range of conditions as described in the various forms of the manual method. The acidity of the final solution varied from 0.25 to 0.8 N and the temperature from 35 to 80 °C. The authors suggest that the degree of interference was similar to the manual methods.

Hager *et al.* (1973) compared their automated method with the manual method described by Strickland & Parsons (1972) for 46 samples collected from one oceanographic cruise. The correlation between the two methods was good ($r = 0.998$) and the precisions were comparable (standard deviation $\pm 0.02\ \mu mol\ l^{-1}$ at a concentration of $1\ \mu mol\ l^{-1}$).

Grasshoff (1969) described a method for the 'Technicon CSM_6 Auto-Analyzer' which used very small sample volumes and was capable of determining 15 samples per hour. Technicon (1971) proposed an adaptation of this method for the 'AutoAnalyzer II' but, because of the instability of the reagent and problems in determining the base line, its use for oceanographic work is difficult.

The method which we have developed for the 'AutoAnalyzer II' takes account of these problems and allows the determination of 0 to $3\ \mu mol\ l^{-1}$ PO_4–P, without significant interferences, with a detection limit of $0.01\ \mu mol\ l^{-1}$ and a maximum error of $0.02\ \mu mol\ l^{-1}$.

Apparatus

The arrangement of pump tubes and mixing coils is shown in Fig. 3A.1. Plastic tubes and joints must not be used after the sample and reagent have been brought into contact. It is essential to reduce the number of joints to the minimum and to use glass-to-glass joints between individual components. A single plastic joint is employed at the entry to the flow-through cuvette to allow sufficient flexibility between the colorimeter and the analytical coils and tubing.

Reagents

The reagents should be prepared with analytical grade chemicals and distilled water.

Reagent 1

500 ml 4.9 N sulphuric acid (136 ml of concentrated H_2SO_4 made up to 1000 ml with distilled water).

150 ml of a solution of ammonium molybdate (40 g of $NH_4Mo_7O_{24}.H_2O$ made up to 1000 ml with distilled water).

50 ml of a solution of potassium antimonyl tartrate (3 g $K(SbO)C_4H_4O_6.0.5H_2O$ made up to 1000 ml with distilled water, stored in deep-freeze and thawed immediately before use).

Reagent 2
Solution of ascorbic acid (9 g of $C_8H_8O_6$, made up to 500 ml with distilled water. This solution is stable for several days if stored in a refrigerator).

Distilled water*
1 ml of Levor IV (Technicon) per litre of distilled water.

Standards

Working standards should be made up in sea water with a low phosphate content.

Stock solution

Dissolve 68 mg of KH_2PO_4 in distilled water and make up to 1000 ml with distilled water. Add 1 drop of chloroform and store in a

Fig. 3A.1. Flow diagram for the determination of phosphate phosphorus in sea water. Normal sampling rate 20 samples per hour. See text for description of distilled water.*

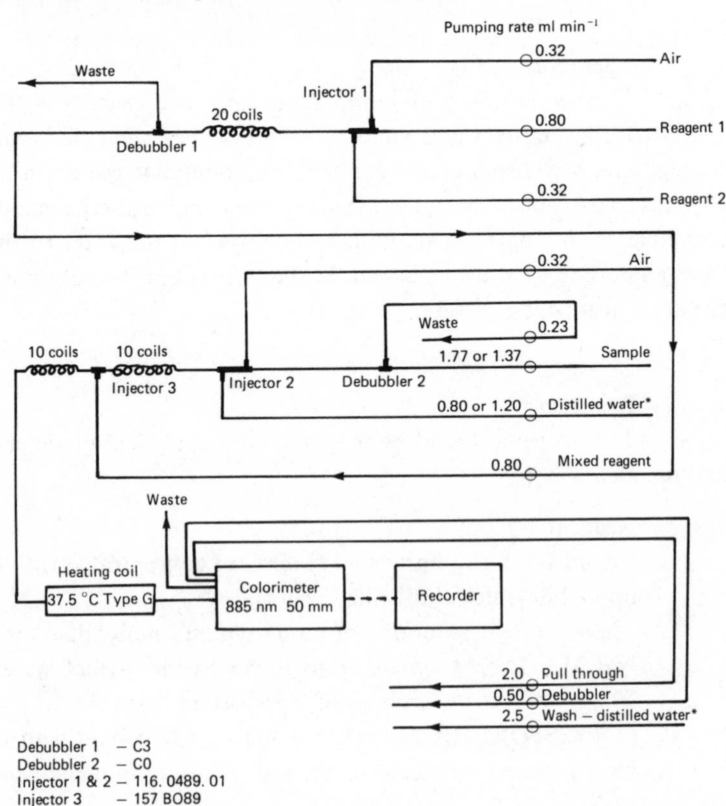

refrigerator in the absence of light. The stock solution is stable for several months.

$$1 \text{ ml} = 0.5\ \mu\text{mol PO}_4\text{—P}$$

Method

The colorimeter

The colorimeter should be switched on several hours before the samples are to be measured and should be adjusted to give a full-scale deflection for an optical density of 0.18.

The baseline

The baseline should be set using distilled water* rather than phosphate-poor water so as to provide a rigorously defined constant reference point. The response of distilled water without Levor IV is the same. During the pumping of distilled water* the digital display is set at 0.00.

Standardization of the digital display

Sea water of low phosphate concentration (SW_0) and a standard ($SW_0 + 1\ \mu mol\ PO_4—P\ l^{-1}$) are sampled successively from the sample cups and the concentration of phosphate in both SW_0 and $SW_0 + 1$ is determined. The corresponding values are entered on the digital display (calibration factor 300).

Sampling

The samples are pumped from 4.5 ml plastic cups previously washed with acid and distilled water. A standard ($SW_0 + 1$) is analysed at the beginning and after every ten samples.

Washing and sampling rate

The sampling rate should be 20 samples per hour with 1 minute sampling and 2 minutes of washing with distilled water*. This will allow the signal to return to the base line. It is possible to sample at 30 or 40 samples per hour without the signal returning to the base line.

Washing with distilled water between samples causes a disturbance to the signal when measuring low concentrations. The peak height should be measured once a plateau has been reached.

Dilution of the sea water

To prevent the formation of a precipitate which produces an irregular signal it is necessary to dilute the sea water (pumped at

1.37 ml min^{-1}) with distilled water (1.20 ml min^{-1}). This will cause a diminution in the optical density recorded. If the maximum sensitivity is required, it is possible to employ a smaller dilution (sea water 1.77 ml min^{-1} and distilled water 0.80 ml min^{-1}) (Fig. 3A.1).

Turbidity

It is necessary to allow for the optical density associated with any turbidity in the sample by comparing it with distilled water under acidic conditions. The turbidity is determined using the following reagents:

Reagent 1 is replaced by 500 ml of 4.9 N H_2SO_4 to which has been added 200 ml of distilled water.

Reagent 2 is replaced by distilled water.

A new base-line is obtained when the sampler is pumping the washing solution and the turbidity is the difference between this new base line and the signal obtained for the sample. The turbidity varies linearly with salinity and for sea water with a salinity of 35‰ filtered through a 0.45 μm

Fig. 3A.2. Response of phosphate-poor sea water (SW_0) spiked with a series of phosphate standards (0.25 to 3.0 μmol l^{-1}). Dilution 1.37:1.20, 20 samples per hour.

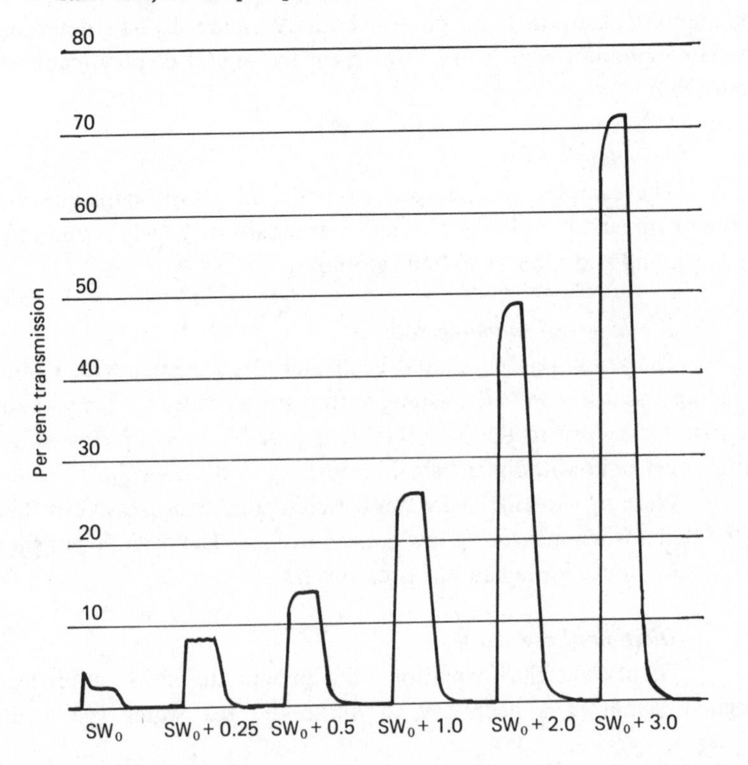

filter is equivalent to 0.12 μmol l^{-1} PO_4—P. A similar value is obtained for filtration through a 200 μm filter. Thus a sea-water sample free of phosphate will give a response equivalent to 0.12 μmol l^{-1} PO_4—P.

Standardization

The response of standards of 0.075 to 3.0 μmol l^{-1} PO_4—P is shown in Fig. 3A.2 and the corresponding calibration curve in Fig. 3A.3. Taking into account the dilution and temperature differences, the molar extinction coefficient obtained is similar to that for the manual method. The Beer–Lambert Law is obeyed for the concentration range usually found in sea water.

Precision

At a level of 0.4 μmol l^{-1}, the standard deviation for 8 samples of water filtered through a 200 μm filter is less than 0.01 μmol l^{-1}. This value is valid for phosphate concentrations of between 0 and 3 μmol l^{-1}. The

Fig. 3A.3. Calibration curve for a series of phosphate standards of between 0.1 and 3.0 μmol l^{-1}.

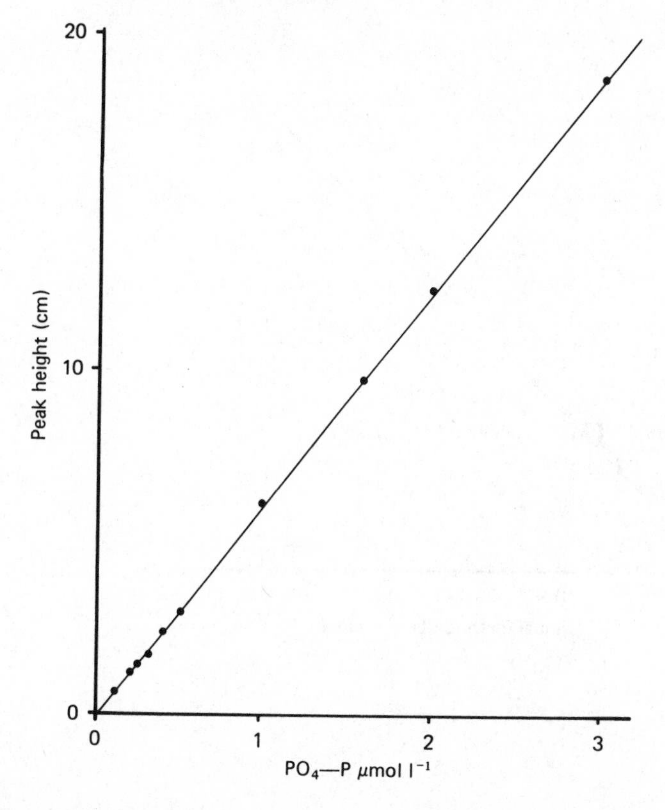

Fig. 3A.4. Reproducibility of eight 1 μmol l^{-1} phosphate standards. Dilution 1.77:0.80, 30 samples per hour.

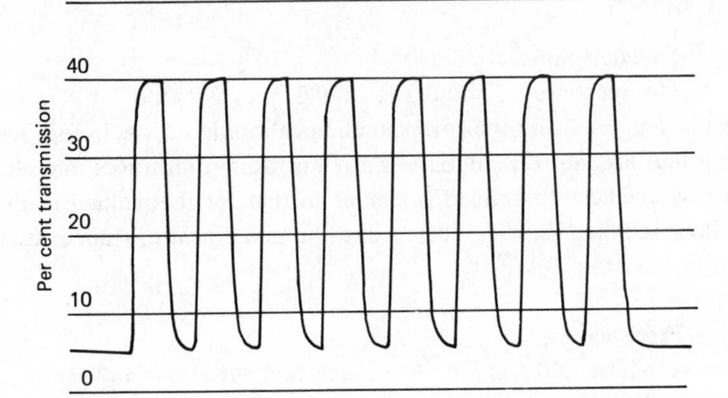

Fig. 3A.5. Comparison of the automated method with the manual method of Strickland & Parsons (1972). The ideal relationship is indicated.

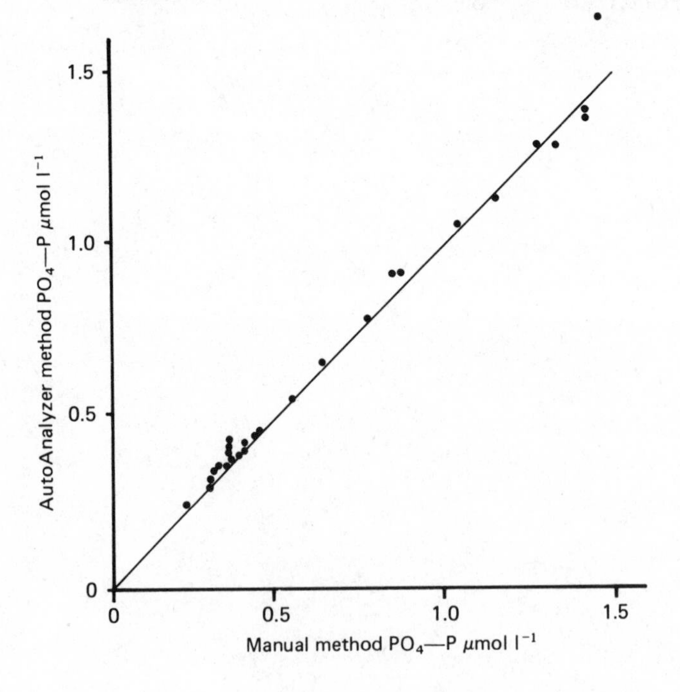

reproducibility of a series of standards containing 1 μmol l^{-1} PO_4—P is illustrated in Fig. 3A.4.

The precision of the method is sufficient for the determination of phosphate in the euphotic zone. However, the precision can only be maintained in waters rich in plankton if the turbidity of each sample is measured.

Comparison with the Manual Method

Various samples of sea water from western Brittany have been measured simultaneously by the method described above and that given by Strickland & Parsons (1972). The results are illustrated in Fig. 3A.5.

The relationship between the two methods can be represented by the equation:

$$(PO_4) \text{ automated} = 0.895 \, (PO_4) \text{ manual} + 0.017$$

with a correlation coefficient of 0.998. There was therefore no significant difference in the results obtained by the two methods for the samples analysed.

Interferences

The Deniges reaction is susceptible to interference from ions such as silicate, arsenate, nitrite, copper(II), iron(III) and also other forms of inorganic phosphorus (pyro- and polyphosphates).

A solution of silicate (5000 μmol l^{-1}) and nitrite (1250 μmol l^{-1}) caused no noticeable interference and 100 μg l^{-1} Cu or 1 μmol l^{-1} Fe do not affect the reaction. A solution of arsenate (0.1 μmol l^{-1}) gave an increase in optical density equivalent to 0.07 μmol l^{-1} PO_4—P.

The influence of pyro- and polyphosphates was also studied. Samples of sea water containing 1 μmol l^{-1} of sodium pyrophosphate and 3μmol l^{-1} of Graham's salt $(NaPO_3)_x$ were analysed along with samples of reference sea water. No differences in optical density could be detected and thus the hydrolysis of pyrophosphate and polyphosphate is not significant at the pH and temperature employed in the automated method.

References

Armstrong, F. A. J. (1965). The determination of phosphorus in sea water. *Oceanography and Marine Biology an Annual Review*, **3**, 79–93.

Armstrong, F. A. J., Stearns, C. R. & Strickland, J. D. H. (1967). The measurement of upwelling and subsequent biological processes by means of a Technicon AutoAnalyzer and associated equipment. *Deep-Sea Research*, **14**, 381–9.

Atkins, W. R. G. (1923). The phosphate content of fresh and salt waters in its relation to the growth of algal plankton. *Journal of the Marine Biological Association of the United Kingdom*, **13**, 119–50.

Atlas, E. L., Hager, S. W., Gordon, L. I. & Park, P. K. (1971). *A Practical Manual for use of the Technicon AutoAnalyzer in Sea-water Nutrient Analyses*. Revised Technical Report, No. 215, Reference 71–22, Oregon State University.

Bernhard, M. & Macchi, G. (1966). Applications and possibilities in automatic chemical analyses in oceanography. *Technicon Fifth International Symposium, Automation in Analytical Chemistry*, pp. 255–9. New York: Mediad.

Brewer, P. G., Chan, K. M. & Riley, J. P. (1966). Automatic determination of certain micro-nutrients in sea water. *Technicon Fifth International Symposium, Automation in Analytical Chemistry*, pp. 308–14. New York: Mediad.

Burton, J. D. & Riley, J. P. (1956). Determination of soluble phosphate and total phosphorus in sea water and of total phosphorus in marine muds. *Mikrochimica Acta*, **9**, 1350–65.

Chan, K. M. & Riley, J. P. (1966). The automatic determination of phosphate in sea water. *Deep-Sea Research*, **13**, 467–71.

Coote, A. R., Duedall, I. M. & Hiltz, R. S. (1970). Automatic analysis at sea. *Technicon International Congress*, pp. 347–51. New York: Mediad.

Coste, B. (1971). Circulation et evolution des composés d'azote et du phosphore (en particulier des sels nutritifs) dans le bassin occidental de la Méditerrané. Influence sur la production organique. Thèse d'état, Université d'Aix-Marseilles, CNRS AO 4495.

Deniges, M. G. (1920). Réaction de coloration extrêment sensible des phosphates et des arséniates. Ses applications. *Comptes Rendues de l'Academie des Sciences, Paris*, 802–4.

Friederich, G. O. & Whitledge, T. E. (1972). AutoAnalyzer Techniques for nutrients. In *Phytoplankton Growth Dynamics*. Special Report No. 52, 38–60. Department of Oceanography, University of Washington.

Grasshoff, K. (1965). Automatische Methoden zur Bestimmung von Fluorid, gelösten Phosphat und Silikat in Meerwasser. In *Automation in der Analytischen Chemie*, pp. 43–52. Frankfurt: Technicon.

Grasshoff, K. (1969). A simultaneous multiple channel system for nutrient analysis in sea water with analog and digital data record. In *Technicon International Congress. Automation in Analytical Chemistry*, pp. 133–45. New York: Mediad.

Hager, S. W., Gordon, L. I. & Park, P. K. (1968). *A Practical Manual for Use of the Technicon AutoAnalyzer in Sea-water Nutrient Analyses*. Report, Department of Oceanography, Oregon State University, Ref. 68–33.

Hager, S. W., Atlas, E. L., Gordon, L. I., Mantyla, A. W. & Park, P. K. (1973). A comparison at sea of manual and AutoAnalyzer analysis of phosphate, nitrate and silicate. *Limnology and Oceanography*, **17**, 931–7.

Harvey, H. W. (1948). The estimation of phosphate and of total phosphorus in sea waters. *Journal of the Marine Biological Association of the United Kingdom*, **46**, 19–32.

Jones, P. G. W. (1966). Comparisons of several methods of determining inorganic phosphate in oceanic sea water. *Journal of the Marine Biological Association of the United Kingdom*, **46**, 19–32.

Kester, D. R. & Pytkowicz, R. M. (1967). Determination of the apparent dissociation constants of phosphoric acid in sea water. *Limnology and Oceanography*, **12**, 243–52.

Mangelsdorf, P. (1972). Methodische Verbesserungen der Phosphat-Bestimmung in Meerwasser mit dem AutoAnalyzer insbesondere für den Bordbetrieb. *Helgoländer wissenschaftliche Meeresuntersuchungen*, **23**, 376–82.

Molof, A. H., Edwards, G. P. & Schneeman, R. W. (1965). An automated analysis for orthophosphates in fresh and saline waters. In *Technicon Symposium, Automation in Analytical Chemistry*, pp. 245–9. New York: Mediad.

Murphy, J. & Riley, J. P. (1958). A single-solution method for the determination of soluble phosphate in sea water. *Journal of the Marine Biological Association of the United Kingdom*, **37**, 9–14.

Murphy, J. & Riley, J. P. (1962). A modified single-solution method for the determination of phosphate in natural waters. *Analytica Chimica Acta*, **27**, 31–6.

Pakalns, P. & McAllister, B. R. (1972). Determination of phosphate in sea water by an isobutyl-acetate extraction procedure. *Journal of Marine Research*, **30**, 305–11.

Piton, B. & Voituriez, B. (1967). L'analyse du phosphate dissous. *Cahiers Oceanographiques*, 761–7.

Proctor, C. M. & Hood, D. W. (1954). Determination of inorganic phosphate in sea water by an iso-butanol extraction procedure. *Journal of Marine Research*, **13**, 122–32.

Revel, J. (1969). Recherches sur l'analyse chimique de l'eau de mer à bord des navire océanographiques. Thèse Faculté d'Pharmacie, Lyon.

Robinson, R. J. & Thompson, T. G. (1948). The determination of phosphates in sea water. *Journal of Marine Research*, **7**, 33–41.

Solórzano, L. & Strickland, J. D. H. (1968). Polyphosphate in sea water. *Limnology and Oceanography*, **13**, 575–8.

Souchay, P. (1969). *Ions Minéraux Condensés*. Paris: Masson.

Stephens, K. (1963). Determination of low phosphate concentrations in lake and marine waters. *Limnology and Oceanography*, **8**, 361–2.

Strickland, J. D. H. & Parsons, T. R. (1972). A practical handbook of sea water analysis. *Bulletin of the Fisheries Research Board of Canada*, No. 167, 2nd edn.

Technicon. (1971 and 1973). Orthophosphate in Water and Sea Water. Industrial Method No. 155–71 W/tentative.

Wooster, W. S. & Rakestraw, N. W. (1951). The estimation of dissolved phosphate in sea water. *Journal of Marine Research*, **10**, 91–100.

4

Trace element analysis

S. R. ASTON

Estuaries are important pathways to the oceans for natural and pollutant trace elements, and are also important in themselves as aquatic environments. The analysis of estuarine waters, sediments and biota for trace elements is usually carried out for one or more of the following reasons:

1. to assess the general levels of trace element concentrations in an estuary (baseline studies);
2. to record data on water quality trends with time (monitoring);
3. to study the fundamental processes in estuaries which control trace element behaviour.

These classes of investigation can give rise to wide variations in the number and type of samples presented to the analyst for trace element determinations. Thus analytical methods must be chosen with care so that the most suitable technique in terms of factors such as precision and accuracy, speed of analysis, and operational facility, is selected.

The sampling and analysis of natural waters, sediments and organisms for their trace element composition are notoriously difficult exercises. In open ocean waters, the very low concentrations of trace elements makes their sampling and determination a most difficult task, and in coastal and estuarine waters where higher concentrations exist there are, however, other problems. These result from the variations in trace element supply into such waters as a result of diurnal and seasonal changes in continental run-off, changes in the chemical forms of natural and pollutant elements, and interactions between the dissolved trace elements and organisms and inorganic particulates.

In any analysis of natural materials for their trace constituents, the question 'What am I really measuring, and how close does this come to something useful for my intended studies?' must always be asked. The problem of element form or speciation is especially relevant when natural waters are involved. In recent years, many advances have been made in the study of the chemical speciation of the dissolved and particulate trace elements in waters. However, we are still very much in the dark when we consider the problem of what the individual methods do actually determine. For a recent review of trace element speciation in natural waters, see Florence (1982).

A knowledge of the chemical forms in which elements exist in natural waters, and hence which form(s) are measured by particular methods of analysis, is important for several reasons. These include: the availability of different forms for biological accumulation and metabolism; the association with dissolved, colloidal, and particulate forms, and hence the behaviour in sedimentary processes; and the division of elements into those associated with inorganic and organic fractions. The question of trace element speciation in sea water has been reviewed several times, see e.g. Stumm & Brauner (1975) and Goldberg (1975). The theme of these publications is, for the most part, directed at *normal* sea water, and only a little attention has been given to speciation in estuarine waters (Dyrssen & Wedborg, 1980). The latter waters are an even more complicated and variable problem than either sea water or fresh water, estuarine waters being derived from the mixing of both water types (Aston, 1978; Dyrssen & Wedborg, 1980).

Discussion of speciation determination in estuarine waters must be limited at present because of the lack of understanding of aquatic trace element chemical speciation itself, and how particular methods of sample treatment and analysis reflect the forms of the elements encountered. Perhaps the simplest practical attempt at measuring different forms of an element is the division into soluble and particulate forms (see below). Even this division is naive and somewhat dangerous, since there are no clear-cut divisions in the dissolved–colloid–particulate system. This problem has for the most part been resolved by choosing an arbitrary convention of particle size, e.g. 0.45 μm for distinguishing 'dissolved' and 'particulate' forms by filtration.

Methods of trace element preconcentration from natural waters for instrumental analysis, e.g. chelating exchange resins, and solvent extraction of metal chelates, will be affected by the speciation of the elements. For example, pre-existing natural chelates of the elements may prevent, or at least, impair, their behaviour with ion-exchange resins which require ionic species. Similarly, natural chelates can hinder the quantitative formation of other chelates for solvent extraction techniques. It has become apparent to analysts that there is no simple, acceptable way of dealing with speciation problems to date, and certainly no one method is completely acceptable for routine use (Florence, 1982). The problem is magnified because the investigator is often unable to decide which fraction(s) of the element present in the water he would like to determine, and would be most useful in his interpretation of chemical, geological or biological processes. Clearly, this is an area in which more attention is required from both the analytical methodology and interpretational view.

The problems of determining trace elements in natural waters, which may often result from different methods 'looking' at different fractions of the total elements present, are exemplified by intercalibration studies. Hume (1975) quotes an example of an intercalibration study between several laboratories for lead in sea water. The reported averages for lead determined by atomic absorption and anodic stripping techniques ranged from 50 to 1300 ng l^{-1}. The accepted value for the water was 14 ± 3 ng l^{-1} as determined by extremely careful mass spectrometry–isotopic dilution analyses. While considering that these horrendous discrepancies will reflect some variations in speciation in relation to the methods used, it must be remembered that contamination probably had an important role. An illuminating study on the analysis of lead in polluted coastal waters is given by Patterson *et al.* (1976).

When the difficulties of speciation and contamination are considered, much of the trace element data reported on estuaries (and other natural waters) must be viewed with extreme caution. These difficulties suggest that considerably more research into methods is required, and that intercalibration and method evaluation programmes will have to form an important role in the future of trace element analysis.

It is the very fact that trace elements occur at such low concentrations ($\ll 1$ mg kg^{-1}) in natural waters that makes their chemical determination an exacting task. These low concentrations necessitate particular care in sampling and handling, and it is perhaps most pertinent to consider these aspects in detail before going on to discuss the analytical techniques. General guidelines about the precautions necessary when attempting to determine trace elements at the levels at which they normally occur in sea water are given by Grasshoff *et al.* (1983).

Sampling and handling methods

The essential feature of a good sampling method for estuarine waters, sediments or organisms is that it provides a representative and uncontaminated specimen for subsequent analysis. The general approaches to representative sampling in estuaries have been dealt with elsewhere (see Chapter 2), and here discussion will be restricted to those problems peculiar to trace elements.

It is unfortunate that many of the materials used in the field, on shipboard and in the ordinary chemical laboratory, are potential sources of contamination, e.g. hydrographic wire (hydrowire), paint, rubber and all metallic implements. Robertson (1968) has evaluated the role of contamination in the trace element analysis of marine waters by conducting high sensitivity neutron activation analysis on a wide variety of solvents reagents and physical materials usually encountered during sampling and analysis. The results of these tests for ten trace elements are presented in Table 4.1, where they are compared with the trace element concentrations for average sea water. It is apparent that certain materials, e.g. rubber products, hydrowire and laboratory tissues are potentially major sources of contamination for some elements. There is no short cut, however, to estimating the potential contamination danger from sampling and handling systems. The most effective means of evaluating the potential of the particular equipment and materials in use by a given laboratory or operator is to conduct exhaustive blank determinations, and trace analyses of the materials actually in use. With this proviso, some general comments on the suitability of sampling techniques can be put forward.

Sampling

For water analysis the sampling containers must obviously be of sufficient size to retain a sample suitable for the subsequent analytical technique and any preconcentration methods required. For many of the more frequently determined elements, a sample of 1 litre is adequate for a multi-element analysis, so that estuarine samples may be taken by using sampling bottles rather than pumping. Polypropylene or polytetrafluoroethylene (PTFE) bottles with polypropylene/PTFE screw caps are suitable for direct sampling of the surface water (see also Chapter 2 for general comments about samping devices). Recent studies on three types of bottles, GO–FLO, Niskin and Hydro-Bios, suggest that modified GO–FLO with silicone 'O' rings are the best (Bewers & Windom, 1982). Metallic flakes stripped from hydrowires and brass messengers are a serious problem during bottle casts, and plastic coated wires are preferable (Bewers & Windom, 1982). Sub-surface samples of estuarine waters should be transferred to polyethylene, polypropylene or PTFE bottles as soon as possible. Such bottles should be cleaned before use by standing overnight full of ~2 M hydrochloric acid, followed by thorough rinsings with redistilled water (see page 152). In a recent off-shore sea-water intercalibration study, systematic differences between the results of analyses of frozen and acidified samples have been noted. For details see Bewers *et al.* (1981).

Table 4.1. *Comparison of trace element concentrations in sea water with those in solvents, reagents, and other materials*[a] *used in the collection and analysis of sea water. After Robertson* (1968).

Sample	Concentration (parts per 10^9)									
	Zn	Fe	Sb	Co	Cr	Sc	Cs	Ag	Cu	Hf
Average sea water[b]	5	2	0.24	0.05	0.3	0.0006	0.4	0.04	0.5	Unknown
Construction materials										
PTFE	9.3	35	0.4	1.7	< 30	< 0.004	< 0.01	< 0.3	22	NM[c]
Plexiglas/Perspex	< 10	< 140	< 0.01	< 0.05	< 10	< 0.002	< 0.06	< 0.03	< 9.5	NM
Polyvinyl chloride (PVC)	7120	2.7×10^5	2690	45	2	4.5	< 1	< 5	630	NM
Surgical rubber tubing	3.08×10^6	UM[d]	< 100	< 30	UM	< 8	< 100	1240	< 6	NM
	4.1×10^7	< 100	360	7500	4.2×10^5	185	580	< 700	NM	NM
	5.35×10^6	UM			UM					
Neoprene rubber	1.82×10^7	UM	290	2300	UM	3090	UM	< 1000	NM	NM
Nylon (block)	UM	UM	UM	1.43×10^6	UM	UM	UM	UM	UM	UM
Polyethylene hose	55	7.4	9000	140	254	11	< 100	< 300	NM	< 100
'Unichrome' organic coating	150	UM	82	42	6.38×10^5	2.2	< 4	< 70	NM	4610
Steel hydrowire	UM	—	54300	57000	UM	< 50	UM	NM	19700	UM
Solvents										
Quartz distilled water	~ 1	~ 1	~ 0.06	~ 0.04	~ 2	0.0022	< 0.01	< 0.02	NM	< 0.005
Quartz distilled water	9.5	< 0.2	~ 0.10	~ 0.20	< 10	0.0025	0.12	< 0.02	NM	< 0.005
Double distilled water	~ 1	< 0.2	< 0.01	< 0.02	~ 2	< 0.0001	< 0.01	< 0.02	NM	< 0.001
Triple distilled water	~ 0.5	~ 1	~ 0.02	< 0.02	12	~ 0.0002	< 0.01	< 0.02	NM	< 0.001
Nitric acid	13	~ 2	~ 0.03	0.018	72	0.0007	< 0.01	~ 0.24	1.3	< 0.005

Double distilled nitric acid	~ 2	~ 1	~ 0.04	~ 0.03	13	~ 0.008	< 0.1	0.29	NM	0.11
Hydrochloric acid	22	~ 1	0.20	0.09	1.1	0.002	< 0.002	< 0.1	82	< 0.005
Double distilled hydrochloric acid	~ 1	~ 1	~ 0.04	~ 0.08	6	~ 0.001	< 0.01	< 0.02	NM	0.010
Purified hydrochloric acid (from HCl gas)	5.4	< 10	0.38	0.04	< 0.8	0.001	< 0.002	< 0.03	NM	< 0.005
Ammonium hydroxide	2.3	< 0.1	< 0.006	~ 0.009	< 0.04	< 0.0003	< 0.002	< 0.1	6.0	
Double distilled CCl_4	1.2	9.8	0.33	~ 0.003	< 50	~ 0.002	< 0.1	< 0.005	0.12	< 0.005
Double distilled $CHCl_3$	2.1	1.6	0.05	~ 0.003	< 100	~ 0.00003	< 0.02	< 0.005	0.29	< 0.005
Container and related materials										
1. Quartz tubing	1.5	395	0.05	0.44	6.5	0.03	1.1	0.05	2.0	< 0.005
2. Quartz tubing	< 1	NM	< 0.01	12	2.5	0.39	< 0.1	< 0.01	0.04	< 0.005
3. Quartz tubing	UM	UM	1940	0.89	UM	UM	1390	< 0.1	0.09	< 0.01
4. Quartz tubing	21	NM	43	0.64	230	0.18	0.30	< 0.1	0.05	27
5. Quartz tubing	33	NM	38	1.1	602	0.16	< 0.1	< 0.1	0.03	26
6. Quartz tubing	20	NM	58	1.7	225	0.10	0.29	< 0.1	0.16	23
Borosilicate glass	730	2.8×10^5	2900	81	UM	106	< 100	< 0.001	NM	597
Vycor glass	UM	UM	1.09×10^6	UM	UM	UM	UM	UM	UM	UM
Polyethylene	28	10400	0.18	0.07	76	0.008	< 0.05	< 0.1	6.6	< 0.5
Polyethylene	25	10600	0.83	0.31	19	0.36	< 0.15	< 0.1	15	< 0.5
Kimwipe tissue	48800	1000	16	24	500	14	< 0.1	~ 0.8	NM	NM
White plastic tape	2.94×10^6	UM	67100	< 1	UM	84	< 50	UM	NM	NM
Scotch tape	1410	5130	33	6.1	< 10	0.48	< 0.3	< 10	NM	NM
Millipore filter	2370	330	39	13	17600	0.79	1.5	< 0.05	NM	< 0.5

Table 4.1 (*continued*)

Sample	Concentration (parts per 10^9)									
	Zn	Fe	Sb	Co	Cr	Sc	Cs	Ag	Cu	Hf
Chelating and chemical reagents										
Dithizone	1150	< 7000	0.8	1.2	< 2000	0.15	10	< 10	420	< 0.1
Cupferron	7000	< 600	0.3	0.68	< 200	0.04	< 1	~ 3	160	< 1
Nitroso-R	1420	3.6×10^5	40	111	UM	0.65	< 100	436	NM	< 100
Thionalide	120	< 300	3.7	5.1	UM	0.10	< 1	4.6	100	< 1
2-Benzimidazolethiol	53000	69000	45	21	UM	0.29	< 10	< 2	0.4	< 100
1-Pyrrolidine carbodithioic acid	1970	~ 5000	1.9	1.3	UM	0.11	< 1	11	4	< 10
Sodium diethyl-dithio-carbamate	40	< 600	42	0.56	UM	0.02	< 1	< 10	NM	< 1
Thenoyltrifluoro-acetone	329	11300	6.2	4.9	192	0.09	< 10	209	16	< 1
8-Hydroxyquinoline	UM	UM	1210	UM	UM	0.14	UM	< 10	290	UM
8-Hydroxyquinoline	370	5700	2.1	1.8	UM	0.06	< 0.1	< 0.8	NM	NM
8-Hydroxyquinoline	< 100	1000	6.0	< 0.4	< 50	< 0.04	< 0.4	< 0.40	NM	NM
8-Hydroxyquinoline	< 40	< 100	< 0.2	< 0.02	< 50	< 0.02	< 0.1	< 1	NM	NM
Sodium hydroxide	< 20	< 900	0.32	5.5	60	0.30	0.69	< 0.2	NM	NM
Potassium hydroxide	1250	2700	1.8	1.7	< 10	0.04	< 0.01	66	NM	NM
Sodium carbonate	74	1400	5.1	1.8	0.76	0.24	< 0.01	< 0.1	NM	NM

[a] Further investigations of the suitability of various materials for the sampling and storage of water samples for trace element analysis are given by IAEA (1970).
[b] Concentrations given by Brewer (1975).
[c] NM = not measured.
[d] UM = unable to measure because of interfering radionuclides.

In most estuaries, concentrations of suspended material are usually high enough to allow sufficient quantities for analysis to be obtained with sampling bottles. If, however, concentrations are low and larger water volumes are required, it may be preferable to use *in situ* pumping techniques such as those described by Perkins & Robertson (1967) and Silker *et al.* (1968). As with bottle sampling, care must be taken to prevent contamination of the sample, particularly by oil leaking from the pump.

Separation of the particulate material may be accomplished by centrifugation or filtration. If large water volumes are involved or substantial quantities of the particulates are required for analysis for size fractionation, continuous centrifugation is to be recommended. The continuous centrifuge allows the collection of up to ~1 g of particulate matter in pre-cleaned plastic liners. Alternatively, filtration, usually through membrane filters, may be employed. By convention, 0.45 μm pore size filters have been adopted to distinguish the 'dissolved' and 'particulate' fractions of trace elements in waters. This size limit has no fundamental significance, but may perhaps be regarded as an approximation to the upper limit of colloidal material in estuaries. The recovery of particulate material from filters for trace analysis is not easy, especially as the amount of material recovered may be less than the weight of the filter itself. Total digestion of the filter, plus its particulate load in acid solution prior to analysis, necessitates careful decontamination of the filters (see Table 4.1). This is achieved by steeping the filters overnight in 2 M hydrochloric acid and multiple rinsings with redistilled water. Details of possible contamination resulting from filtering procedures are given by Robertson (1968) and Robertson *et al.* (1968).

When the proportion of a trace element associated with suspended matter is required, rather than the actual physical recovery of the particulates, a method of differences can be applied. This simply involves the analysis of a water sample, prior to any separation, for its total trace element content, and further analysis of an aliquot of the sample which has been subjected to filtration. The amount of the trace element associated with particulate matter is then calculated from the difference of the 'total' and 'filtrate' analyses. While this procedure removes the need to recover and analyse the particulates, it is not certain that it provides a wholly satisfactory determination of the suspended load of trace elements, and particular care is needed in estuarine waters where flocculation of colloidal material may change the size distribution of particulates over the critical range around 0.45 μm. The 'total' trace element concentration of unfiltered water samples is determined by acidifying the water sample to pH < 1. This acidification brings trace elements held in suspended matter into solution except for those held in silicate lattice positions in suspended detrital minerals. It is unlikely that such lattice-held elements contribute a significant proportion of the total concentration of the element in the estuarine water, the larger proportion of particulate trace elements being located in ion-exchange and surface adsorption sites (Aston & Chester 1975).

A very large variety of devices is available for the collection of estuarine sediments by coring, grabbing or dredging. Holme (1964) and Holme & McIntyre (1984) have reviewed the methods available for sampling marine sediments, and the particular problems of sediment sampling in estuaries have been discussed by Buller & McManus (1979). Mud and sand banks uncovered at low tide in estuaries are often most conveniently sampled by hand, using either a polyethylene scoop or coring tube to avoid metal contamination.

Field observations suggest that coring and grabbing procedures from vessels often lead to disturbance and unrepresentative sampling of the upper few centimetres of the superficial sediments. The disturbances are simply a consequence of the impact of the device with the sediment surface and dislocation of the unconsolidated solids during sampling. Evidence for the unsuitability of gravity coring when collecting sediment profiles for pollution studies, has recently been provided by Baxter *et al.* (1981).

Biological materials recovered from estuarine waters and sediments must be collected with the same careful avoidance of contamination as other materials. Thus macrobiota should be placed in acid-washed polyethylene bags or bottles soon after collection, having been rinsed with redistilled water to remove any adhering clay particles. Microorganisms, e.g. phytoplankton and algae which have been collected by netting or filtration, must be protected from contamination by trace metals by taking precautions similar to those suggested for particulate matter.

Various authors (see, for example, Guillard & Wangersky, 1958; Holmes & Anderson, 1963) have reported that microorganisms are subject to cell-breakage during filtration when the pressure drop across the filter exceeds a critical value. This is obviously unadvisable in any chemical analysis including that for trace elements, and may be reduced by the use of positive pressure filtration. This method employs the use of an inert gas or nitrogen to force the sample through the filter. The technique has been described by Smith & Windom (1972). In estuarine surveys, major difficulties may arise in the trace element analysis of microorganisms due to the inclusion of terrestrial inorganic detritus, e.g. clay minerals, hydrated oxides. These inorganic components are not easily separated from the filtered plankton, but may be assessed by carrying out an analysis for aluminium on the sample. Aluminium is a characteristic indicator of continental weathering debris due to its presence in alumino–silicate minerals. Aluminium analyses provide, however, only a semi-quantitative guide to inorganic contamination of particulate organics.

Trace elements analyses during estuarine research are dominated by samples of waters, suspended matter, biota and sediments; and there is relatively little information on the chemistry of the pore waters of estuarine sediments. These interstitial waters are important in the budget of trace elements in estuaries, providing a sink for natural and pollutant trace constituents (Aston & Chester, 1975). One of the reasons for the relative lack of data on estuarine pore waters has been the difficulty in their sampling and analysis. Manheim & Sayles (1974) have reported the composition of pore waters for deep-sea sediments, and the techniques employed are essentially applicable to the sampling of estuarine deposits. The pore fluids from sediment cores may be recovered in small quantities, e.g. 1 to 10 ml by squeezing sediment samples in small hydraulic squeezers. The pore waters are extruded through micropore filters into polyethylene ampoules. The hydraulic squeezers are preferably constructed of PTFE to avoid metallic contamination. An extensive investigation into the chemistry of pore waters from Narragansett Bay and the Providence River Estuary has been described by Elderfield (1981), Elderfield *et al.* (1981*a*) and Elderfield *et al.* (1981*b*).

Storage and preparation

Problems related to the interaction of samples for trace element analysis with container materials will be most severe during long-term contact in storage, so that the prompt analysis of estuarine samples is advisable. Problems of trace element loss or contamination are most marked for water samples with their low elemental concentrations.

For water containers, the interaction of the water sample with the walls of the bottle can result in either contributions to the trace element concentration in solution or loss from solution to the walls (Robertson *et al.*, 1968). Robertson (1968) has discussed specific instances of losses from solutions, including marine waters, to container walls. In the case of glass bottles, losses from solution are probably due to ion-exchange on the silicate surfaces, and the properties of both glass and plastic containers which lead to trace element adsorption have been discussed by Riley (1975). It is generally considered that plastic or PTFE bottles are preferable to glass (including soda or Pyrex) for the storage of saline samples. Care must, however, be taken in applying the general name 'plastic'. The most suitable materials with low adsorption characteristics and chemical inertness are polyethylene, polypropylene and PTFE. Polypropylene is usually employed, as bottles of suitable form are widely commercially available and more economical than those constructed of PTFE. More exact information is not possible due to the variety of the manufacturers' specifications. Although trace element losses on to container walls are commonly assumed to be the dominant problem in water storage, it should be noted that residues of the catalysts used in polymer acceleration often contain elements such as zinc, cadmium and iron in significant amounts. Blank determinations on the particular bottles used are the only safe guide to their storage characteristics.

Acidification of water samples to pH < 2 to stabilize trace metal ions in solution combined with refrigeration to 3–5 °C are probably the best precautions that can be adopted to preserve samples. Care must be taken in the cleaning of the storage bottles, e.g. washing out with 2 M acid and several rinsings with redistilled water before use. If water samples are to be analysed as filtered samples, or particulates recovered from analysis, the filtration must be carried out before acidification to pH < 2 for storage. In such acid solutions, trace elements will be stripped from clay mineral surfaces and oxides (see page 137).

Biota can be preserved in polyethylene bags or bottles by either short-term refrigeration (2–5 °C) or longer-term deep-freezing (−15 to −20 °C). Biological tissues dried by heating in laboratory ovens at 80–110 °C may be subject to losses of trace elements in volatile constituents. Freeze-drying, although time consuming and expensive, is more acceptable for trace element work when dry tissue information is required.

The trace element contents of marine and estuarine organisms, e.g. fish and molluscs are known to vary considerably from one organ to another. In general, the transition elements are concentrated most strongly in the digestive and renal organs (Riley & Chester, 1971). This trace element heterogeneity of biota must be taken into account in the preparation of samples of estuarine organisms for analysis. On occasions when specific organs are required for trace element analysis, dissection with plastic and glass implements rather than conventional metal scalpels has proved useful in avoiding contamination of tissue by nickel, chromium and other metals. A full discussion of the elemental analysis of biological materials, with special reference to trace elements, has been presented recently (IAEA, 1980).

The suspended matter separated from natural waters by filtering or centrifuging is especially susceptible to the growth of bacteria and fungi during storage. These colonies living on organic debris may change the chemical composition of the particulate sample as a result of their metabolism. Samples of particulate matter or bottom sediments which have been taken in estuaries with high particulate organic loads, e.g. in areas of sewage disposal, should be kept sterile during storage by refrigeration in an atmosphere saturated with formaldehyde. This is easily achieved by storing filter pads, etc. in plastic bags in a desiccator containing a small quantity of 40 per cent formaldehyde solution (formalin).

Estuarine sediments and suspended solids must be brought into aqueous solution for most modern methods of trace element analysis. Total digestion of estuarine sediments requires the use of hydrofluoric acid to dissolve the silicates and silica which are inevitably present in the continentally derived detritus which composes estuarine sediments to a greater or lesser extent. Normally, a small weight of sieved sediment, e.g. $\sim$ 0.5 g of the less than 200 μm size fraction is treated with 10–25 ml of a mixture of hydrofluoric, nitric and perchloric acids in a platinum crucible or PTFE beaker. The acids must be of the highest possible purity, and the vessels pre-cleaned by steeping in hydrochloric acid of equal or greater strength to that to be used for the digestion and rinsed several times with redistilled water before use. Evaporation of the sample to remove the hydrofluoric acid, and then continued to white fumes of perchloric acid, followed by a second treatment with perchloric acid alone is adequate to effect solution of the sediment. The residue is taken up in dilute high purity hydrochloric acid for analysis. Extreme care must be taken when handling either hydrofluoric or perchloric acids.

Various workers have described chemical leaching techniques for the separation of lattice-held and non-lattice-held fractions of trace elements in marine sediments (see Förstner & Wittman, 1979 for a review of these). The fractions of trace elements held in iron and manganese oxides, carbonate shell debris, organic matter and in ion-exchange or adsorbed sites on clay minerals can be leached out by the use of various reagents. The leaching techniques leave those trace elements occupying silicate lattice structural sites in the solid residue. Determination of the lattice-held and non-lattice-held fractions is most conveniently carried out by an analysis of the total sediment and a further analysis of the leached residue. These methods allow a semi-quantitative indication of the sources of the trace elements in estuarine and other marine sediments to be made.

Certain modern methods of trace elements analysis do not require the decomposition and digestion of sediments, e.g. X-ray fluorescence and neutron activation (see page 148). In such circumstances the sediment samples are simply dried and sieved to produce a suitable portion for analysis. The question of the choice of a size-fraction for analysis by either digestion or non-destructive techniques is one which is not easy to define or answer. The particle size distribution in estuarine sediments is very variable, and, to allow comparisons of the trace element composition of samples from different locations, it is advisable to attempt some systematic approach to avoiding the spurious effects associated with comparisons of different size-fractions. Trace elements are mainly associated with the finest size fraction of estuarine sediments because of changes in the ratio of surface area to weight (see, for example, Perhac, 1972). Thus the clay-sized fraction of $< 2\,\mu m$ is extremely important in controlling trace element chemistry of sediments. For most purposes, separation of such a small size-fractions is not practicable or even necessary. Wet or dry sieving of sediments to obtain a size fraction of less than, for example, 60, 100 or $200\,\mu m$ particles is a rapid means of achieving comparable samples. For more detailed studies on the associations of trace elements with sediment size-fractions the arduous task of the separation and recovery of the fractions must be undertaken. The conventional sedimentological technique of sedimentation tubes to separate particles of different diameters falling through a water column at differing rates as a result of Stokes' law is slow, and suffers from possible desorption or adsorption effects during contact with the water. For a review of methods for the collection and separation of suspended sediment in estuaries, see McCave (1979).

As for sediments, estuarine organisms sampled for trace element analysis must usually be dissolved in an appropriate medium. Biological tissue is readily digested in high purity concentrated nitric acid. The oxidation of tissue is promoted by repeated evaporations with aliquots of acid on a hot plate. Suitable containers for biological tissue digestions are small PTFE beakers or conical Pyrex glass flasks. As ever, cleaning with acid prior to use is imperative to avoid contamination. The final soluble residues obtained are brought up into solution in dilute hydrochloric or nitric acid prior to analysis.

Useful general and detailed remarks on analytical quality control in a trace element analytical laboratory that are the responsibility of the analyst in his own laboratory have been presented by Hamilton (1980).

Methods of trace element analysis

There is now available a wide spectrum of techniques for the elemental analysis of natural materials, and to attempt to treat individual methods for selected elements is much beyond the scope of this text. Rather, it is more pertinent to consider some of the modern methods available and discuss their applicability in terms of the problems presented by estuarine samples. The factors which will dominate the final selection of an analytical technique are: availability of the equipment; sensitivity; selectivity; accuracy and precision; practicability and economy.

Given the fact that suitable analytical equipment is available, the first consideration for trace constituents is sensitivity. Minear & Murray (1973) have listed the methods of trace element analysis in common use and their approximate suitable concentration ranges (see Table 4.2). The levels of use are stated in molar concentrations, although the sensitivity of a method for a specific trace element would often be stated as $mg\,l^{-1}$ or $\mu g\,l^{-1}$ concentrations. Considering that trace elements occur in estuarine and other natural waters at concentrations of a few to several hundred $\mu g\,l^{-1}$, the sensitivity of a technique must be around $1\,\mu g\,l^{-1}$ or less. Minear & Murray (1973) have emphasised that it is not realistic to state *exact* sensitivities of individual methods for trace elements, as the sensitivity and ultimate detection limit are functions of the particular instrument in use and its operator.

The selectivity of an analytical method is its ability to determine one constituent without any interference from other substances. Thus selectivity is highly desirable, although preconcentration techniques are usually chosen for their lack of selectivity. Preconcentration of trace elements is often necessary for the analysis of waters, and methods which will enrich several elements efficiently are obviously advisable for multi-element determinations (see page 142).

The accuracy of a method is determined by systematic errors which bias the analysis. The accuracy is best assessed by the analysis of samples of known composition which have been previously analysed by other well-established methods. The systematic errors are often very difficult to find

Table 4.2. *A survey of methods of analysis for trace elements in natural waters, and their approximate useful concentration ranges*

Analytical technique	Approximate useful concentration ranges (molar)
Molecular absorption spectrophotometry	10^{-5} to 10^{-6}
Molecular fluorescence spectrophotometry	10^{-7} to 10^{-8}
Atomic absorption spectrophotometry	10^{-6} to 10^{-7}
Classical polarography	10^{-5} to 10^{-6}
Anodic stripping voltammetry	10^{-8} to 10^{-10}
Neutron activation analysis	10^{-9} to 10^{-10}
Mass spectrometry	$\sim 10^{-6}$

Summarized from Minear & Murray (1973)

and eliminate. Precision of an analytical technique is a function of random errors arising from poor operator technique and/or faulty equipment. Since the errors from these faults obey a normal Gaussian distribution, they can be adequately expressed by one or more standard deviations from the mean. Thus, the precision of a trace element analytical method is fairly easily determined by a series of replicate analyses of separate aliquots of a sample. It must be noted that the precision of the method applies only to the analytical method itself and gives no indication of errors arising from sampling. Sampling errors and changes in sample composition during storage are, however, reflected in the accuracy of the method. For a full discussion of analytical accuracy and precision see Shaw (1969).

The methods of trace element analysis applicable to estuarine samples are considered below in general terms, with reference to some specific problems and individual elements. Some methods, e.g. atomic absorption spectrophotometry have found widespread application in estuarine trace element studies, while others which are, at least, potentially useful have not been adopted by many analysts. In the following sections, the suitability of methods, including their advantages and disadvantages, for estuarine studies is discussed. Other methods of trace element analysis which have either not been used in estuarine research or which have very limited application have been omitted. Priority has also been given to those methods which have already found the most widespread appeal in the various types of estuarine investigations. The criteria for adopting some methods in preference to others, are often a mixture of economics and familiarity of the technique to the users. These users are, as often as not, biologists, ecologists and marine scientists who are not concerned with analytical methodology *per se*. In this context, atomic absorption spectrophotometry and colorimetric methods are dealt with first, while other more financially demanding and less generally available methods are considered later and to a lesser extent than the popular methods.

Atomic absorption spectrophotometry

Atomic absorption spectrophotometry (AAS) has become by far the most commonly used method for the trace element analysis of environmental materials, i.e. waters, soils, sediments and biological tissues (see for example, Angino *et al.*, 1967; Burrell, 1974). The technique involves the absorption of light of a specific wavelength by atomic species of the element as it is excited in a flame or other thermal device. The amount of light absorbed by the atomic species is proportional to the concentration of the element present in its ground state. Thus, if the element can be atomized without excitation, a high sensitivity can be achieved. The light source used is a hollow cathode lamp, with the cathode constructed of the same element as that under analysis or an electrodeless discharge lamp (EDL) which contains a small quantity of the element to be determined. Normally, aspiration of aqueous solutions into a controlled flame is used for the atomization procedure, but various other atomization techniques involving carbon rods, and graphite furnaces are also used. Table 4.3 summarizes the typical sensitivities of the conventional flame and the more recently developed and now widely adopted graphite furnace method, and shows that for many elements the flameless technique offers higher sensitivities than does the flame atomization method (Fernandez & Manning, 1971; Segar & Gonzales, 1972). This advantage of increased sensitivity must, of course, be balanced against the requirements of the estuarine project and the additional cost of the flameless device. For most coastal and estuarine waters which receive any input of industrial or domestic water, the trace element concentrations are such that sufficient sensitivity is obtained by a conventional flame AAS procedure. This assumes that a preconcentration technique has been employed.

Even conventional flame atomization methods are subject to interferences. The interferences themselves result from a variety of physical,

Table 4.3. *Examples of the typical detection limits of the conventional flame and graphite furnace methods of atomic absorption spectrophotometry*

	Detection limits ($\mu g\ ml^{-1}$)	
Element	Flame	Graphite furnace
Ag	0.02	—
Al	0.1	5×10^{-5}
As	0.1	5×10^{-4}
Cd	0.01	1×10^{-6}
Co	0.05	5×10^{-5}
Cr	0.05	5×10^{-5}
Cu	0.05	2×10^{-6}
Fe	0.05	5×10^{-5}
Mn	0.02	5×10^{-6}
Ni	0.05	1×10^{-4}
Pb	0.1	5×10^{-5}
Sn	1.0	—
Ti	1.0	—
V	1.0	1×10^{-4}
Zn	0.01	2×10^{-6}

chemical and matrix effects, e.g. molecular absorptions; line interferences; suppression of ionizations by alkali metals; interelement compound formation; scattering by particulates and atomization suppression by viscosity, surface tension, etc. A useful compilation of interferences and methods of overcoming them has been made by Brezonik (1974). These are summarized in Table 4.4.

Preconcentration techniques

It is apparent from Table 4.3 that the sensitivities of conventional flame and other flameless AAS techniques in which small (ml or μl) quantities of solution are analysed are not adequate for the direct determination of many trace element in marine and estuarine waters. Furthermore, there are serious interelement interferences involving both major and trace elements which require that a preconcentration procedure be adopted for AAS analysis. The problem of interferences from the matrix

Table 4.4. *The interference problems encountered in atomic absorption spectrophotometry methods.*

Elements	Problem	Solution
Al, Ba, Ca, Cr, Mn, Mo, Pb, Sn, Sr, Ti, V	Alkali ionization, especially in N_2O flame	Buffering with Cs ($>$ 100 mg l^{-1}) as ionization suppressor
Al, B, Be, Ge, Mo, Si, Ti, V, W, Zr	Oxide formation in C_2H_2 flame	Use of N_2O–C_2H_2 flame
Ba	Molecular absorption (BaO)	N_2O flame to destroy oxide
Ca, Fe, K, Mg, Mn, Zn	High Na lowers atomization	Preconcentration to isolate from Na
Ca, Mg	Refractory phosphate formation	LaCl buffer ($>$ 100 mg l^{-1})
Cr	Ni interference in C_2H_2 flame	Use of N_2O flame
Fe, Mn	Si interference	Ca buffer (200 mg l^{-1})
Mg	Intermetallic compound with Al in C_2H_2 flame	N_2O flame
Elements with resonance lines $<$ 250 nm	Scattering by particulates	Filtration; high temperature flames; D_2 or H_2 background correction

Details of interferences and their removal in determining trace metals in waters, sediments etc. by AAS are given by Burrell (1974)

of the sample is particularly relevant for estuarine waters which consist of a wide range of mixtures of marine and fresh waters.

The most widely adopted preconcentration techniques now in use for both flame and flameless AAS are chelation solvent extraction and chelating ion-exchange (Morrison & Frieser, 1958; Stary, 1964; Joyner *et al.*, 1967). The most commonly used chelation solvent extraction method is probably the APDC–MIBK (ammonium pyrrolidine dithiocarbamate–methyl isobutyl ketone) procedure (Brooks *et al.*, 1967). In addition to APDC, other derivatives of dithiocarbamic acid have been used to chelate trace metals prior to extraction, principally sodium diethyl dithiocarbamate (NaDDC) and diethylammonium diethyl dithiocarbamate (DDDC). The chelates may be extracted into MIBK, chloroform or Freon TF (1,1,2-trichloro-1,2,2-trifluoroethane). An advantage of the use of chloroform is that it can be recovered and re-used and thus offers some saving in costs over the other solvents, particularly where large numbers of samples are to be analysed. The basic method for the use of NaDDC and chloroform is given by Bryan & Hummerstone (1973) and a detailed procedure for waters of any salinity is described in a technical report of the Applied Geochemistry Research Group, Imperial College, London (AGRG, 1975). Details for using APDC and DDDC with Freon TF are given by Danielsson *et al.* (1978) and with chloroform by Bruland *et al.* (1979).

Chelating ion-exchange offers an alternative method for the preconcentration of trace elements from estuarine waters, and has an advantage in that the resins are easily regenerated and may be used over and over again. The practical details of a suitable procedure are given by Riley & Taylor (1968) and Bruland *et al.* (1979).

Gaseous hydride techniques

Elements such as As, Se, Bi, Sb and Te which form gaseous hydrides can be determined by a special variation of the AAS technique. Two of the above elements, arsenic and selenium, are of potential interest in estuarine waters because of their highly toxic natures and associations with certain effluent discharges. The detection limits for arsenic and selenium by conventional flame atomization techniques are 0.01 and 0.05 μg ml^{-1}, respectively, and they cannot be usually determined in natural waters by this method. Fernandez (1973) has described a method suitable for the determination of arsenic in estuarine waters by AAS, using sodium borohydride to reduce the arsenic to its gaseous hydride, arsine, which is then passed into the flame. By this means the detection limit is a much more useful 1.5×10^{-4} μg ml^{-1}.

A similar high sensitivity technique for selenium determination is also described by Fernandez (1973), in which, following the borohydride reduction of selenium, hydrogen selenide is passed into an argon-entrained hydrogen flame. The detection limit is the same as that for arsenic. Details of hydride generation techniques for the determination of arsenic, antimony and germanium are given by Grasshoff *et al.* (1983). Mercury can be determined by AAS using the monatomic gas of this element, following reduction by stannous chloride (Hatch & Ott, 1968), (see also Chilov, 1975 and Nelson, 1982 for a useful review of methods for the determination of mercury).

Sediments, particulates and biological tissues

Conventional flame atomization AAS methods are suitable for the determination of a wide range of trace elements in estuarine sediments, particulates and biota following the digestion of the materials by appropriate techniques (see above). Relatively small samples, e.g. 0.2 to 0.5 g of the solid samples, digested and brought up to volumes of 25 ml or more, are sufficient for the sensitivities of AAS methods.

Special methods involving hydride formation for elements such as As, Se, Sb etc., can be readily adapted by carrying out borohydride reductions on sample digests and proceeding in the same manner as for water samples. Mercury in estuarine sediments and suspended solids can be determined by a cold vapour method using the rapid technique of Randlesome & Aston (1980).

Colorimetry

Colorimetric methods of trace element analysis have been widely used in the past for natural waters and geological samples. They are applicable to a very wide range of elements, excluding only the alkali metals. Colorimetric methods have enjoyed a long standing popularity with geochemists because of their cheapness, wide applicability and relative ease of operation. More recently, however, there has been a move away from colorimetric techniques in favour of atomic absorption spectrophotometry (see above). The principle of colorimetric methods is the measurement of light of a specific wavelength absorbed by molecules in solution. Some aqueous ions absorb visible and ultraviolet light in solution, but the sensitivity of such direct measurements is poor (> 0.1 g l^{-1}). Most methods therefore involve the conversion of the trace element to a species with a high efficiency for energy absorption, usually by forming an organic complex of the trace element. Such conversions increase the sensitivity to acceptable levels for the direct determination of many elements in natural waters. There are so many colorimetric methods, employing a vast range of complexing reagents, that it is not realistic to compile a summary of detection limits for the trace elements.

Advances in the design of spectrophotometers and the development of countless organic reagents for colorimetric methods have made this technique one of well-established precision and accuracy. It is, however, not a rapid technique for the analysis of large numbers of water or other samples as may often be required in estuarine surveys. Automated colorimetric systems such as those described by Mancy (1971) are applicable to large sample numbers or continuous analysis, and may prove worthwhile in large surveys. Recent developments in the design of compact field instruments for colorimetric analysis have produced the opportunity for on-site monitoring of estuarine waters, and offer an advantage over most other techniques, e.g. atomic absorption spectrophotometry, which are limited to laboratory operation.

The fundamental principles of colorimetric analysis, including a compilation of the applications of this method to the analysis of sea water (and thus to estuarine waters), have been reviewed by Riley (1975). Other reviews dealing with the application of molecular absorption to trace element analysis include Sandell (1965), Charlot (1964), Calder (1969), Stearns (1969), Blotz & Mellon (1970, 1972). The future role of colorimetric analysis in estuarine and other environmental research is not entirely clear, but, for routine rapid analysis of many trace elements, atomic absorption spectrophotometry has much to offer and continues to replace the older colorimetric techniques. Possibly the role of colorimetric methods will depend on the need for field analysis to which it is well suited, and the essential requirement of checking the accuracy of new techniques against old but well-established ones.

Pulse polarography and anodic stripping voltammetry

Until recently electrochemical methods have found little application to the trace element analysis of natural waters. Even the pulse polarograph, with its greater sensitivity than other conventional polarographic methods has little use without preconcentration of the trace elements. For this reason anodic stripping techniques (see below) which can be applied directly to natural water samples have been preferred. The anodic stripping methods are, however, time consuming and of a potentially more restricted application than pulse polarography. However, they do provide an alternative to AAS techniques at a comparable cost.

Abdullah & Royle (1972) have used chelating ion exchange resins to preconcentrate copper, lead, cadmium, nickel, zinc and cobalt from both fresh and marine waters prior to pulse polarographic analysis. By using the chelating resin (Chelex 100) in its calcium form for the preconcentration stage, Abdullah & Royle (1972) have provided a suitable means of eluting the trace elements retained on the resin in a medium for polarography and free from interfering differences in major ion concentrations arising from salinity variations in the original waters. This technique is therefore applicable to estuarine waters. Abdullah & Royle (1972) have concluded that pulse polarography is suitable for the determination of certain trace elements in natural waters by adopting the resin preconcentration procedure, and provide experimental details of the method. Although quite adequate for the analysis of estuarine waters, pulse polarography suffers as an electrochemical technique from the necessity to employ a preconcentration stage. In this respect, the alternative electrochemical trace element method of analysis by the anodic stripping of water samples directly is preferable.

Anodic stripping analysis involves the stripping of metal ions from an electrode with a linear voltage ramp, the metal ions having previously been concentrated on the electrode by electrodeposition. Several types of electrodes may be used for this technique in combination with a standard polarograph having positive and negative scanning facilities. The most widely used electrodes are the hanging mercury drop and graphite–wax types.

Allen *et al.* (1970) have applied the graphite–wax electrode to the anodic stripping analysis of fresh waters for cadmium, copper and lead. These authors have pointed to the problem that unless water samples are acidified to low pH, the presence of natural metal–organic complexes will produce low results by the anodic stripping technique. Acidified samples often gave total metal concentrations of up to 200 per cent greater than those determined without acidification. The use of the hanging mercury drop method of anodic stripping to the analysis saline waters has been reported by Whitnack & Sasselli (1969) who used a single drop electrode for the determination of Zn, Cd, Cu and Pb in sea water. These workers have pointed to two distinct advantages offered by anodic stripping over other methods of determining trace elements in saline waters. Firstly, the method allows direct determination of some elements at very low concentrations (10^{-8}–10^{-9} M), and secondly, the method is relatively free from errors resulting from contamination in concentration procedures and from the glassware and reagents used in most other methods.

Zirino & Healy (1972) have reported an anodic stripping technique for the analysis of Cd, Cu, Pb and Zn, in fresh and saline waters, and which is applicable to the water samples encountered in estuarine sampling programmes. The sensitivity of the conventional single electrode technique has been increased by the use of two hanging drop mercury electrodes in a differential method. The double electrode procedure described in detail by Zirino & Healy (1972) eliminates residual currents and allows high amplification of the signal from the electroplated elements thus making possible slow polarographic voltage scans and improved resolution.

The principles of anodic stripping techniques are discussed by Kemula (1970), Brainina (1971) and Barendrecht (1967). Experimental details of procedures suitable for the anodic stripping analysis of estuarine waters may be found in Allen *et al.* (1970) and Zirino & Healy (1972). A comprehensive account of the electrochemistry of sea water has been recently presented by Whitfield & Jagner (1981).

Molecular fluorescence

Another relatively cheap method of trace element analysis for estuarine waters is molecular fluorescence. However, there are very serious drawbacks as discussed below.

Fluorescence of the molecular species is a much more specific effect than absorption. The term fluorescence refers to the energy released by a molecule in the form of light when the molecular species returns from an excited state to the ground electronic state. The initial excitation is commonly caused by the absorption of radiation, e.g. visible range or shorter, while the fluorescence light is of a longer wavelength than the excitation light.

Aqueous species of trace elements such as simple ions do not fluoresce (except for the lanthanides), and, as for absorption colorimetry, the formation of organic chelates with certain functional groups, e.g. carbonyl and imino groups, is necessary. The effects of fluorescence exhibited by organic chelates are more specific to the individual trace elements and give an improved degree of specificity over molecular absorption. Sensitivities are also higher than most colorimetric techniques. There are, however, serious limitations to the application of fluorescence techniques to estuarine waters, and molecular fluorescence techniques have found little application to the trace element analysis of estuarine waters. This is due to two important drawbacks together with the fact that the emerging field of molecular fluorescence analysis was rapidly overtaken by improvements in atomic absorption spectrophotometry. The first of the problems inherent in applying molecular fluorescence techniques to estuarine waters is background fluorescence. Many natural and pollutant organic substances present in natural waters exhibit strong fluorescences (Ghassemi & Christman, 1968). These effects increase the detection limits in comparison with those which are achievable in pure aqueous solution, and severely lower the precision of fluorescence spectrophotometry. Estuarine waters, which on many occasions carry wide ranging loads of dissolved organic species of natural and pollutant origins, are thus particularly problematical. Destruction of the dissolved organics by oxidation prior to molecular fluorescence is a remedy, but the complete destruction of organics is tedious and leads to severe sample handling limitations.

The second problem presented by the molecular fluorescence analysis of natural waters is that of photo-decomposition and quenching caused by dissolved salts. The wide salinity variations typically found in estuaries make the salt effect an important limitation on the application of this technique to estuarine waters.

In spite of these problems, fluorometric techniques have been satisfactorily applied to the determination of aluminium in estuarine and marine waters (Hydes & Liss, 1976, 1977).

Neutron activation

Unlike colorimetric and atomic absorption spectrophotometric methods, neutron activation analysis (NAA) and mass spectrometry (see page 150) cannot be regarded as general methods for the analysis of trace elements in natural waters. One major reason for this is that these techniques require access to costly equipment, and, in the case of NAA, radio-isotope laboratories and a nuclear reactor. Thus in most estuarine research where the determination of trace elements will probably form only part of a research programme, the use of NAA and mass spectrometry is rarely justified when modern atomic absorption techniques can be used for high production analysis of a wide range of elements at considerably less outlay and running cost. NAA does, however, offer some important advantages over other methods, and for this reason will be given consideration here.

In the NAA technique the sample to be analysed is exposed to irradiation with neutrons so that the elements present in the sample are converted to radioactive species. These neutron-induced radionuclides are then identified and measured either by radiochemical separation of the individual nuclides and counting, or by direct multi-channel spectrometry. Measurement of the radionuclides provides an estimate of the total concentration of the parent element which is calibrated against irradiated standards. The neutron source is normally a high flux nuclear reactor producing thermal neutron fluxes of 10^{11} to 10^{13} n cm^{-2} s^{-1}.

Radiochemical separation of individual nuclides for counting by single channel γ or β spectrometry is a difficult and tedious procedure rarely used for trace element analyses. The rapid multi-element analysis of irradiated samples is typically performed with multi-channel spectrometers involving solid state Ge(Li) detectors or NaI(Tl) scintillation crystals, and a 2000 or 4000 channel pulse-height analyser.

The advantages of NAA multi-element analyses are fourfold:

1. many of the important and widely determined trace elements including those of pollutant origin can be simultaneously analysed;
2. NAA allows the routine analysis of several elements about which little is known geochemically because of the difficulty of determining them by other methods;
3. sample handling and manipulation prior to irradiation are reduced to a minimum, reducing the chances of contamination;
4. while the NAA technique requires expensive and sophisticated equipment which may not be often justified as part of an estuarine research or monitoring programme, NAA is an excellent method by which the accuracy of other routine methods may be checked.

Samples of solid materials, e.g. estuarine biological tissues, sediments and particulates are prepared for neutron activation by weighing a small (< 1 g) freeze-dried portion into a clean polypropylene or silica vial which is heat-sealed. This lack of sample manipulation and handling is highly desirable, although care must be taken to avoid the loss of volatile elements, e.g. Hg, As, Sb during sample drying and vial sealing.

Estuarine waters can be prepared for activation by either evaporation or recovery of the trace elements on ion-exchange resins. Direct neutron irradiation of water samples in sealed vials is dangerous because high pressures, leading to vial explosions, result from the radiolysis of the water to give O_2 and H_2. Water samples of 1 to 10 litres are best evaporated at low temperature or by infra-red radiation in PTFE containers, and again care must be taken to check losses of volatile elements in the method employed. The evaporite sample is transferred to a vial with a small quantity of redistilled water which is then removed by freeze-drying. A major problem arising from the evaporation method is the presence of major elements which when irradiated give high yields of nuclides with prominent gamma ray spectra, e.g. ^{24}Na, ^{42}K, ^{82}Br. The presence of these products requires a suitable period of 'cooling off' before handling of the samples. The preconcentration of water samples for neutron irradiation by employing chelating ion-exchange resins removes the problem of the high yield gamma emitters. Procedures for the use of chelating ion-exchange resins for concentrating trace elements from marine waters have been described by Goya & Lai (1967) and Bruland *et al.* (1979).

Detailed discussions on the application of NAA to the trace element analysis of fresh and marine waters, biological tissues, and sediments are provided by Robertson & Carpenter (1976). The methodology of NAA analysis and its instrumentation are covered by the following reviews and texts – DeVoe & LaFleur (1969), Rahovic (1970), Adams *et al.* (1971), and Kruger (1971). A comparison of the absolute sensitivities of NAA and mass spectrometry for selected trace elements is shown in Table 4.5.

Mass spectrometry

This technique is a multi-element method of high sensitivity, but as with neutron activation analysis it is sophisticated to an extent which may not be compatible with some estuarine research or monitoring exercises. Mass spectrometry is carried out by vaporizing the elements in a vacuum chamber by either arcing or a spark source, and measuring the mass to charge ratio of the vaporized ions. This is achieved by accelerating the ions in an electrical and/or magnetic field and focusing them upon a detector at fixed field strengths which can then be used to quantitatively measure the mass to charge ratios of individual ions. The necessity to excite the sample to produce ions by arcing or sparking requires that water samples are first evaporated or freeze-dried. Low temperature ashing of evaporites and solid samples, e.g. tissues and sediments is also essential to remove organic interferences before excitation. This can be accomplished by, for example, radio-frequency ashing at about 40 °C.

Absolute detection limits for mass spectrometry are given by Morrison (1972), and these compare favourably with those for neutron activation analysis in several instances (see Table 4.5). Crocker & Merritt (1972) have reported the application of spark-source mass spectrometry to fresh-water

Table 4.5. *A comparison of neutron activation analysis and mass spectrometry detection limits for selected trace elements*

	Comparative detection limits (ng)	
Element	Neutron activation analysis	Mass spectrometry
Cr	50	0.02
Fe	5000	0.02
Mo	10	0.5
Ni	5	0.1
Pb	500	0.5
Se	200	0.1
Zn	10	0.1

samples. They employed freeze-drying and low-temperature ashing to river water samples, reporting the detection of 32 trace elements, with detection limits down to 1 part in 10^{11} for some elements.

Wahlgren *et al.* (1971) have also reported the application of the mass spectrometry and its associated problems are discussed in general by Morrison (1972), Smith (1972) and Brown *et al.* (1971).

The high sensitivity and wide elemental application of mass spectrometry to trace determinations will no doubt lead to its greater use in water and environmental analysis. However, at present there are considerable difficulties in its application due to sample preparation, matrix effects and above all cost of installation. The sophisticated focusing geometry, detection systems and computer interfacing put the cost of the mass spectrometry system above that which might be reasonable for trace analysis in estuarine research.

General aspects of trace element analysis

Blanks and standards

The use of representative blank determinations and the application of standard samples to the assessment of accuracy and precision are essential features of trace element analysis.

Blank determinations for estuarine water analyses are best performed on coastal or open ocean water samples which have been stripped of trace metals by chelation and solvent extraction. If, however, these are not available, redistilled water, prepared as described below, may be used. It is essential that the blank sample is treated in exactly the same manner as other samples, whatever the technique in use, if the blank is to be representative.

Blank determinations for sediment, particulate and biological samples may be simulated by carrying out the digestion techniques in use without any sample present. As with water samples, it is essential that the remainder of the analytical procedure is carried out in exactly the same manner as for field samples.

The question of standards for trace element analysis of estuarine samples is problematical. Estuarine waters are subject not only to wide ranges of elemental concentrations, but present highly significant problems as a consequence of their matrix variability, i.e. salinity. Long-term storage of estuarine waters for use as standards or intercalibration samples is not practicable. The results of recent intercalibration studies on the trace element analysis of fresh and marine waters, suggest that either the accuracy of the methods involved was poor or that standard natural water samples are not realistic. The storage of stock standard solutions of trace elements at pH < 2 and concentrations of 100 ppm or greater seems a much more practical step towards producing standards than the preservation of natural samples even though it does not solve the problem of matrix effects. Stock solutions should be diluted to suitable concentrations immediately prior to use, otherwise at the low concentrations they are subject to temporal change due to container contamination and/or adsorption. Stock acid solutions of 100 ppm or more prepared from highly pure elements are stable for several months.

At present, there are few international standards available for modern sediments as is the case for the igneous rocks distributed by the United States Geological Survey. Sediments suffer from several disadvantages as primary analytical standards. The most important of these are:

1. the variability in the particle size distributions of unconsolidated sediments, even between closely neighbouring sites. This can lead to substantial differences in the trace element content of sediments from an apparently identical source;
2. the variability in the organic and shell debris of sediments. This problem is particularly important in near-shore and estuarine sediments, and leads to inhomogeneity of the sediment samples;
3. the formation of concentrations within the sediment leading to highly localized concentration gradients for certain elements, e.g. iron and manganese.

Water and solvents

The availability of high purity water on a large scale is essential to all trace element analysis programmes. Suitable water for the preparation of reagents, dilution operations, and for blank determinations can be prepared by the double distillation of tap water in a silica still. High volume production of redistilled water is achieved by directly passing the distillate of the first condenser into the flask of the second still, using a simple constant head device as a safeguard. The redistilled water should be collected in a large PTFE, quartz or boron nitride container for storage. It is essential to avoid the use of rubber connecting tubes, etc. (see above). All stills 'improve with age' as the silica is gradually leached free of its trace element content, so that initial batches of water may have to be discarded until the ageing process is complete. Hamilton (1980) has described a comprehensive water purification system in some detail.

Many mineral acids and organic solvents commonly used in trace element analysis procedures are now available in very high grades of purity. These commercial products are, however, costly and in some cases suitably pure reagents can be prepared by single or double distillation of cheaper grades using silica apparatus. For any such operations adequate safety precautions must be taken.

Summary

A wide range of methods is now available for the determination of trace elements in estuarine waters and other associated materials, e.g. biological tissues, sediments and suspended matter. The question of suitability of a method for the analytical job to be undertaken will depend on the need for sensitivity, accuracy and precision, selectivity, expense and possible need for field applications. It is not feasible to draw clear-cut conclusions as to which method(s) should be adopted due to two main reasons. Firstly, the scope and nature of estuarine research programmes are wide, and, secondly, the availability of equipment more often than not decides the choice of method. However, some indications of which methods are more or less suitable can be made and should allow the individual to consider his own particular requirements.

1. When a large number of water samples have to be analysed for their range of trace elements, e.g. in a survey programme or routine reconnaissance of metal levels in a number of estuaries, atomic absorption techniques may be the most appropriate. As described above, this method allows multi-element analysis of waters at low cost. Preconcentration with either solvent extraction or ion-exchange procedures is usually necessary, and does slow down productivity, but the method can be used routinely by careful personnel. Sensitivity to enable elements of low concentrations to be determined can be obtained by using non-flame methods. AAS is easily applied to the analysis of biological and geological materials.
2. The need for field determinations and for ease of operation with relatively unsophisticated instrumentation may be fulfilled by colorimetric methods. Spectrophotometers are quite robust for use in shipboard or field laboratories. A wide range of methods for many elements is available, with varying degrees of specificity and sensitivity, so that individual methods must be assessed for the task at hand. Colorimetric methods have been automated for trace element analysis of natural waters.
3. Molecular fluorescence techniques, except in the case of aluminium, are unlikely to be the first choice in any estuarine chemistry project due to the high background fluorescence exhibited by natural waters, and secondly the problem of quenching by dissolved salts. The presence of natural or pollutant organic

matter and the wide variations in salinity in estuaries cause particular problems with this technique.

4. Electrochemical methods of trace element analysis, e.g. pulse polarography and particularly anodic stripping have enjoyed an increasing popularity in recent years. Anodic stripping with its high sensitivity and lack of a need for preconcentration procedures makes it an attractive method. However, there is some doubt as to the effects of metal–organic complexes on this method, and the question of speciation must be taken into account. Anodic stripping techniques offer an alternative to atomic absorption spectrophotometry, and do have one possible advantage in that contamination from preconcentration methods is avoided.
5. Sophisticated instrumental techniques including neutron activation analysis, mass spectrometry and X-ray fluorescence analysis have applications to the analysis of estuarine waters. The use of such costly facilities will usually depend on their availability to the individual researcher. Thus, when available, neutron activation analysis has the advantages of a wide range of elements, routine analysis of elements not easily determined by other methods, and a reduction of contamination to a minimum. Apart from the presently unsolved problems of sample preparation to avoid matrix effects, mass spectrometry could have an important future in water analysis. In the absence of existing facilities, the installation of these techniques for most estuarine research programmes would be hard to justify.

References

Abdullah, M. I. & Royle, L. (1972). The determination of copper, lead, nickel, zinc and cobalt in natural waters by pulse polarography. *Analytica Chimica Acta*, **58**, 283–8.

Adams, F., Van den Winkel, P., Gijbels, R., De Soete, D., Hoste, J. & Op de Beeck, J. P. (1971). Activation analysis. *Critical Reviews in Analytical Chemistry*, **1**, 455–586.

AGRG (1975). Applied Geochemistry Research Group Technical Report No. 62, Imperial College, London.

Allen, H., Matson, W. & Nancy, K. (1970). Trace metal characterization in aquatic environments by anodic stripping voltammetry. *Journal of the Water Pollution Control Federation*, **42**, 573–81.

Angino, E. E. & Billings, G. K. (1967). *Atomic Absorption Spectrometry in Geology*. Amsterdam: Elsevier.

Aston, S. R. & Chester, R. (1975). Estuarine sedimentary processes. In *Estuarine Chemistry*, ed. J. D. Burton & P. S. Liss, pp. 37–52. London: Academic Press.

Barendrecht, E. (1967). Stripping voltammetry. In *Electroanalytical Chemistry*, vol. 2, ed. A. J. Bard, pp. 53–109. New York: Marcel Dekker.

Baxter, M. S., Farmer, J. G., McKinley, I. G., Swan, D. S. & Jack, W. (1981). Evidence of the unsuitability of gravity coring for collecting sediment in pollution and sedimentation rate studies. *Environmental Science and Technology*, **15**, 843–6.

Bewers, J. M., Dalziel, J., Yeats, P. A. & Barron, J. L. (1981). An intercalibration for trace metals in sea water. *Marine Chemistry*, **10**, 173–93.

Bewers, J. M. & Windom, H. L. (1982). Comparison of sampling devices for trace metal determinations in sea water. *Marine Chemistry*, **11**, 71–86.

Blotz, D. F. & Mellon, M. G. (1970). Light absorption spectrometry. *Analytical Chemistry*, **42**, 152 R.

Blotz, D. F. & Mellon, M. G. (1972). Light absorption spectrometry. *Analytical Chemistry*, **44**, 300 R.

Brainina, K. Z. (1971). Film stripping voltammetry. *Talanta*, **18**, 513–39.

Brewer, P. G. (1975). Minor elements in sea water. In *Chemical Oceanography*, ed. J. P. Riley & G. Skirrow, 2nd edn., vol. 1, pp. 415–96. London: Academic Press.

Brezonik, P. L. (1974). Analysis and speciation of trace metals in water supplies. In *Aqueous-Environmental Chemistry of Metals*, ed. A. J. Rubin, pp. 167–91. Ann Arbor: Ann Arbor Science.

Brooks, R. R., Presley, B. J. & Kaplan, I. R. (1967). APDC–MIBK extraction system for the determination of trace metals in saline waters by atomic absorption spectrophotometry. *Talanta*, **14**, 809–16.

Brown, R., Powers, P. & Wolstenholm, W. A. (1971). Computerized recording and interpretation of spark-source mass spectra. *Analytical Chemistry*, **43**, 1079–85.

Bruland, K. W., Franks, R. P., Knauer, G. A. & Martin, J. H. (1979). Sampling and analytical methods for the determination of copper, cadmium, zinc and nickel at the nanogram per liter level in sea water. *Analytica Chimica Acta*, **103**, 233–45.

Bryan, G. W. & Hummerstone, L. G. (1973). Brown seaweed as an indicator of heavy metals in estuaries in south-west England. *Journal of the Marine Biological Association of the United Kingdom*, **53**, 705–20.

Buller, A. T. & McManus, J. (1979). Sediment sampling and analysis. In *Estuarine Hydrography and Sedimentation*, ed. K. R. Dyer, pp. 87–130. Cambridge: Cambridge University Press.

Burrell, D. C. (1974). *Atomic Spectrometric Analysis of Heavy Metal Pollutants in Water*. Ann Arbor: Ann Arbor Science.

Calder, A. B. (1969). *Photometric Methods of Analysis*. Amsterdam: Elsevier.

Charlot, G. (1964). *Colorimetric Determination of Elements, Principles and Methods*. Amsterdam: Elsevier.

Chilov, S. (1975). Determination of small amounts of mercury. *Talanta*, **22**, 205–22.

Crocker, I. H. & Merritt, W. F. (1972). Analysis of environmental samples by spark-source mass spectrometry – 1. Trace elements in water. *Water Research*, **6**, 285–95.

Danielsson, L.-G., Magnusson, B. & Westerlund, S. (1978). An improved metal extraction procedure for the determination of trace metals in sea water by atomic absorption spectrometry with electrothermal atomization. *Analytica Chimica Acta*, **98**, 47–57.

DeVoe, J. R. & LaFleur, P. D. (1969). *Modern Trends in Neutron Activation Analysis*. National Bureau of Standards. Special Publication No. 312. Washington D.C.: National Bureau of Standards.

Dyrssen, D. & Wedborg, M. (1980). Major and minor elements, chemical speciation in estuarine waters. In *Chemistry and Biogeochemistry of Estuaries*, ed. E. Olausson & I. Cato, pp. 71–119. Chichester: John Wiley & Sons.

Elderfield, H. (1981). Metal–organic associations in interstitial waters of Narragansett Bay sediments. *American Journal of Science*, **281**, 1184–96.

Elderfield, H., Bender, M., McCaffrey, R. J. & Luedtke, N. (1981*a*). Benthic flux studies in Narragansett Bay. *American Journal of Science*, **281**, 768–87.

Elderfield, H., McCaffrey, R. J., Luedtke, N., Bender, M. & Truesdale, V. W. (1981*b*). Chemical diagenesis in Narragansett Bay sediments. *American Journal of Science*, **281**, 1021–55.

Fernandez, F. J. (1973). Atomic absorption determination of gaseous/hydrides utilizing sodium borohydride reduction. *Atomic Absorption Newsletter*, **12**, 93–7.

Fernandez, F. J. & Manning, D. (1971). Atomic absorption analyses of metal pollutants in water using a heated graphite atomizer. *Atomic Absorption Newsletter*, **10**, 65–9.

Florence, T. M. (1982). The speciation of trace elements in waters. *Talanta*, **29**, 345–64.

Förstner, U. & Wittmann, G. T. W. (1979). *Metal Pollution in the Aquatic Environment*. Berlin: Springer-Verlag.

Ghassemi, M. & Christman, R. F. (1968). Properties of the yellow organic acids of natural waters. *Limnology and Oceanography*, **13**, 583–97.

Goldberg, E. D. (1975). *The Nature of Sea Water*. Berlin: Dahlem Konferensen.

Goya, H. A. & Lai, M. G. (1967). *Adsorption of Trace Elements from Sea Water by Chelex*-100. Report USNRDL-TR-67-129. San Francisco: Naval Radiological Defense Laboratory.

Grasshoff, K., Ehrhardt, M. & Kremling, K. (eds) (1983). *Methods of Seawater Analysis*, 2nd edn. Weinheim: Verlag Chemie.

Guillard, R. R. & Wangersky, P. J. (1958). The production of extra cellular carbohydrates by some marine flagellates. *Limnology and Oceanography*, **3**, 449–54.

Hamilton, E. I. (1980). The chemical laboratory and trace element analysis. In *Elemental Analysis of Biological Materials*, pp. 303–15. Vienna: International Atomic Energy Agency.

Hatch, W. R. & Ott, W. L. (1968). Determination of sub-microgram quantities of mercury by atomic absorption spectrophotometry. *Analytical Chemistry*, **40**, 2085–7.

Holme, N. A. (1964). Methods of sampling the benthos. *Advances in Marine Biology*, **2**, 171–260.

Holme, N. A. & McIntyre, A. D. (eds) (1984). *Methods for the Study of Marine Benthos.* International Biological Programme Handbook No. 16, 2nd edn. Oxford: Blackwell Scientific Publications.

Holmes, R. W. & Anderson, G. C. (1963). In *Symposium on Marine Microbiology*, ed. C. H. Oppenheimer. Springfield: C. C. Thomas.

Hume, D. N. (1975). Fundamental problems in oceanographic analysis. In *Analytical Methods in Oceanography*, ed. T. R. P. Gibbs, pp. 1–8. Washington, D.C.: American Chemical Society.

Hydes, D. J. & Liss, P. S. (1976). Fluorimetric method for the determination of low concentrations of dissolved aluminium in natural waters. *Analyst, London*, **101**, 922–31.

Hydes, D. J. & Liss, P. S. (1977). The behaviour of dissolved aluminium in estuarine and coastal waters. *Estuarine and Coastal Marine Science*, **5**, 755–69.

IAEA (1970). *Reference Methods for Marine Radioactivity Studies.* International Atomic Energy Agency Technical Report No. 118. Vienna: International Atomic Energy Agency.

IAEA (1980). *Elemental Analysis of Biological Materials.* International Atomic Energy Agency Technical Report No. 197. Vienna: International Atomic Energy Agency.

Joyner, T., Healy, M. L., Chakravarti, D. & Koyanagi, T. (1967). Preconcentration for trace analysis of sea water. *Environmental Science and Technology*, **1**, 417–24.

Kemula, W. (1970). The application of stripping processes in voltammetry. *Pure and Applied Chemistry*, **21**, 449–60.

Kruger, P. (1971). *Principles of Activation Analysis.* New York: Wiley Interscience.

Mancy, K. H. (1971). *Instrumental Analysis for Pollution Control.* Ann Arbor: Ann Arbor Science.

Manheim, F. T. & Sayles, F. L. (1974). Composition and origin of interstitial waters of marine sediments, based on deep sea drill cores. In *The Sea*, ed. E. D. Goldberg, vol. 5, pp. 527–68. New York: Wiley Interscience.

McCave, I. N. (1979). Suspended sediment. In *Estuarine Hydrography and Sedimentation*, ed. K. R. Dyer, pp. 131–85. Cambridge: Cambridge University Press.

Minear, R. A. & Murray, B. B. (1973). Methods of trace analysis in aquatic systems. In *Trace Metals and Metal–Organic Interactions in Natural Waters*, ed. P. C. Singer. Ann Arbor: Ann Arbor Science.

Morrison, G. H. (1972). Spark-source mass spectrometry for the study of the geochemical environment. *Annals of the New York Academy of Science*, **199**, 162–72.

Morrison, G. H. & Frieser, H. (1958). *Solvent Extraction in Analytical Chemistry.* New York: John Wiley & Sons.

Nelson, L. A. (1982). The measurement of mercury in water. *Marine Pollution Bulletin*, **13**, 149–50.

Patterson, C., Settle, D. & Glover, B. (1976). Analysis of lead in polluted coastal sea water. *Marine Chemistry*, **4**, 305–19.

Perhac, R. M. (1972). Distribution of Cd, Co, Cu, Fe, Mn, Ni, Pb and Zn in dissolved and particulate solids from two streams in Tennessee. *Journal of Hydrology*, **15**, 177–86.

Perkins, R. W. & Robertson, D. E. (1967). The application of nuclear techniques to seawater analysis. *American Chemical Society Meeting Report* No. BNWL-SA-1007.

Rahovic, N. (1970). *Activation Analysis*. Cleveland: C.R.C. Press.

Randlesome, J. E. & Aston, S. R. (1980). A rapid method for the determination of mercury in sediments, suspended solids and soils. *Environmental Technology Letters*, **1**, 3–8.

Riley, J. P. (1975). Analytical chemistry of sea water. In *Chemical Oceanography*, ed. J. P. Riley & G. Skirrow, 2nd end., vol. 3, pp. 193–514. London: Academic Press.

Riley, J. P. & Chester, R. (1971). *Introduction to Marine Chemistry*. London: Academic Press.

Riley, J. P. & Taylor, D. (1968). Chelating resins for the concentration of trace elements from sea water and their analytical use in conjunction with atomic absorption spectrophotometry. *Analytica Chimica Acta*, **40**, 479–85.

Robertson, D. E. (1968). Role of contamination in trace element analysis of sea water. *Analytical Chemistry*, **40**, 1067–72.

Robertson, D. E. & Carpenter, R. (1976). Activation analysis. In *Strategies for Marine Pollution Monitoring*, ed. E. D. Goldberg, pp. 93–159. New York: John Wiley & Sons.

Robertson, D. E., Rancitelli, L. A. & Perkins, R. W. (1968). Multielement analysis of sea water, marine organisms and sediments by neutron activation without chemical separation. *Proceedings of an International Symposium on the Application of Neutron Activation Analysis in Oceanography*. Brussels: Institut Royal des Sciences Naturelles de Belgique.

Sandell, E. B. (1965). *Colorimetric Metal Analysis*, 3rd edn., vol. 3, New York: Wiley Interscience.

Segar, D. & Gonzales, J. (1972). Evaluation of atomic absorption with a heated graphite atomizer for the direct determination of trace transition metals in sea water. *Analytical Chimica Acta*, **58**, 7–14.

Shaw, D. M. (1969). In *Handbook of Geochemistry*, ed. K. H. Wedepohl. Berlin: Springer-Verlag.

Silker, W. B., Robertson, D. E., Rieck, H. G., Perkins, R. W. & Prospero, J. M. (1968). Beryllium-7 in ocean water. *Science, New York*, **161**, 879.

Smith, D. S. (1972). Mass spectrometry. In *Guide to Modern Methods of Instrumental Analysis*, ed. T. H. Gouw, pp. 357–92. New York: Wiley Interscience.

Smith, R. G. & Windom, H. L. (1972). Georgia Marine Science Center – Technical Report. No. 72–6.

Stary, J. (1964). *The Solvent Extraction of Metal Chelates.* London: Macmillan.

Stearns, E. I. (1969). *The Practice of Absorption Spectrophotometry.* New York: Wiley Interscience.

Stumm, W. & Brauner, P. A. (1975). Chemical speciation. In *Chemical Oceanography*, ed. J. P. Riley & G. Skirrow, 2nd edn., vol. 1, pp. 173–239. London: Academic Press.

Wahlgren, M. A., Edgington, D. N. & Rawlings, F. F. (1971). Radiological Physics Division Report No. ANL-7860-IV. Argonne National Laboratory.

Whitfield. M. & Jagner, D. (1981). *Marine Electrochemistry: A Practical Introduction.* New York: Wiley Interscience.

Whitnack, G. C. & Sasselli, R. (1969). Application of anodic stripping voltammetry to the determination of some trace elements in sea water. *Analytica Chimica Acta*, **47**, 267–74.

Zirino, A. & Healy, M. (1972). pH-controlled differential voltammetry of certain trace transition elements in natural waters. *Environmental Science and Technology*, **6**, 243–9.

5

Analysis: organic matter

P. J. leB WILLIAMS

Estuaries are the interfaces between rivers and the sea. Sea water generally has a lower organic content than river water and thus the organic chemistry of estuaries may be regarded as that of fresh water diluted to varying degrees with sea water. In general, methods developed for sea water, where they exist, will have the required sensitivity for the estuarine situation and in the past the application of such methods to estuarine samples has often been successful. Thus it is worth giving first consideration to sea-water methods. However, certain factors, such as the presence of large amounts of 'humic' material, may modify the situation.

Estuaries are poorly understood areas. Their hydrology is complex and their chemistry is further complicated by a variable ionic content and by a multiplicity of sources of organic material. Sewage and industrial effluents, rivers, planktonic and benthic processes, as well as mixing processes with the sea – all may introduce organic material into estuaries. In some instances, for example certain industrial effluents, the additions will consist of compounds unique to the source; in such situations, given a suitable analytical method, it is possible to set up a programme to monitor the distribution and dispersion of the effluent material. In practice such situations are uncommon; more often a single compound may originate from two or more quite distinct sources. For example, many of the compounds present in sewage will also be intermediates in the normal cycle of the estuarine biota. Even in an urbanized estuary, the inputs from planktonic processes may be comparable with, or even many times greater than, that from sewage. This means that, in contrast to rivers, simple indices of pollution such as biochemical oxygen demand or ammonia concentration have to be used with caution. This situation complicates the seemingly simple problem of determining the extent of gross organic pollution of an estuary, for its organic content may be controlled as much by internal biological processes as by external effluent loading. There appears to be no ready solution to this problem, other than to recognize it and make due allowance in planning survey programmes and interpreting their results. Some resolution may be achieved by exploiting the fact that many biological processes within an estuary, especially planktonic ones, will show a somewhat predictable seasonal fluctuation, in contrast to many 'pollution' additions which may remain effectively constant throughout the year (see e.g. de Souza Lima & Williams, 1978).

The present account does not aim to provide a set of methods for the organic analysis of estuarine materials. The subject is a continuous and active state of development and it is common experience in such situations that to set down recommended methods only serves to inhibit progress. The intention of the present account is to indicate or explore the existing possibilities and to comment, where possible and appropriate, on the merits and weaknesses of the various alternative approaches. Useful reviews for the organic analysis of sea water have been produced by Wangersky & Zika (1977) and are included in Duursma & Dawson (1981).

Preliminary aspects

There are certain steps which precede the analysis and these need special consideration, for they affect the way the sample is obtained and processed. Three topics will be considered: contamination, filtration and storage.

Contamination

When undertaking the organic analysis of estuarine water it is prudent to regard it as clean from the outset. Except in highly polluted situations, the concentration of individual organic compounds in natural waters is low, typically less than 10 μg l^{-1} (i.e. about 10^{-7} M). As a consequence, contamination problems can be acute. Perhaps the simplest illustration of the potential seriousness of contamination is gained by noting that the quantities of compounds such as amino acids, urea and ammonia in a litre of estuarine water may be comparable with those reported for a fingerprint (Eastoe, 1966).

Instances of contamination at various stages in the overall analysis are given below to illustrate the types of problem that might occur. Most of the examples are taken from sea-water studies and, although they might exaggerate the problem somewhat, the need for care cannot be overstressed.

Sampling

If a boat is used to obtain water samples then it is a potential source of both organic and inorganic contamination. An oil slick soon surrounds a stationary ship. Sampling equipment can cause problems: Williams (1965) reported the release of up to 164 μ l^{-1} of lauric acid from an NIO polypropylene water sampler (see Chapter 2 for information about NIO and other water samplers).

Filters

Williams (1965) reported the release of up to 8 μg of individual fatty acids from Millipore membrane filters.

Reagents

These are a common source of contamination which must be monitored by the frequent running of blanks. In many cases distilled water may be used for the preparation of blanks, but it is often wise to reduce its organic content, which may be about 0·5 mg C l^{-1}, by refluxing before distillation from a strong oxidizing agent. Acid peroxodisulphate (persulphate) is a very suitable reagent for this purpose. If organic-free sea water is required, this may be conveniently prepared by photo-oxidation (see Armstrong *et al.*, 1966; Armstrong & Tibbits, 1968).

Environment

The atmosphere can be an important source of contamination which is too easily overlooked. A striking example of serious atmospheric contamination was reported by Blumer (1965). Work in his laboratory on the trace analysis of hydrocarbons was hindered for over a year by the contamination of organic solvents by phthalates, which were eventually found to originate from the air-conditioning system. He reported that 200 ml of pentane exposed to the laboratory atmosphere overnight picked up 694 μg of non-volatile material, principally phthalate esters.

Filtration

At some stage during the design of the analytical programme it will be necessary to consider whether or not to filter the sample. There are advantages and disadvantages in filtration. The obvious disadvantages are that time, equipment and facilities are needed and that the filtration step provides an opportunity for contamination. Furthermore, the separation achieved by filtration is poorly understood (see page 164); thus, after dividing the sample into so-called 'particulate' and 'dissolved' fractions, they turn out to have little fundamental significance. The papers by Sharp (1973*a*, 1975) should be read for a discussion of these problems.

There are, nonetheless, advantages to filtering. Filtration through submicrometer pore diameter filters removes a substantial part of the biological population and this will retard the rate of change of composition in samples during storage. This is a very real consideration, for, although the samples may well be stored in a frozen state eventually, the time between sampling and freezing may be several hours. A second practical consideration is that if the particulate fraction is not removed it may well settle out on standing, giving rise to a variety of analytical problems. But probably the most important reason for filtering is that there is information associated with the particulate fraction which would be lost if only the whole sample were analysed. The particulate fraction, for example, contains the biological part of the ecosystem. In the sea and coastal waters the dissolved organic fraction is usually by far the major organic pool, although it shows much smaller temporal and spatial variations than the particulate fraction. Unless a separate analysis is made of the particulate fraction, its fluctuations and the processes giving rise to them could be overlooked.

A variety of filters is now available. There is no universally acceptable filter and careful selection is necessary. This usually means some compromise has to be made. It is conventional to divide filters into two general types: depth filters and barrier filters.

Depth filters

These consist of a finely divided and randomly distributed matrix of small particles or fibres. Two common examples are the traditional filter paper and the glass-fibre mat filter. Precipitates of metal oxides or carbonates have also been used in the past. The depth filter has no precise pore size and relies upon contact and entrapment during filtration. The greater the depth of the filter the finer the particle that will be retained.

The Whatman range of glass-fibre mat filters (GF/A, GF/B, GF/C, and GF/F) is used extensively in water analysis. They can cope with quite high particulate loads and may be freed from organic contamination by heating overnight at 500 °C. The finest in the series (the GF/F) and its equivalents (e.g. Reeve-Angel 984H), are probably the most useful all-purpose filters for the organic analysis of water. Glass-fibre filters do seem to adsorb some material from solution (see Menzel, 1966; Banoub & Williams, 1972; Marvin *et al.*, 1972).

Barrier filters

There is now a variety of filters prepared with more or less defined pore sizes. The first type to become available was the cellulose ester membrane, which has been used extensively in microbiological and inorganic analysis. They are still used for chlorophyll analyses, but are less satisfactory than glass-fibre filters for most organic work because of the problem of contamination. Various alternatives to the original cellulose ester filters are now available. The best from the point of view of size fractionation are Nucleopore perforated polycarbonate membranes. Metallic barrier filters, prepared by compacting silver precipitates, are also available. They seem less prone to adsorb material from solution than glass-fibre filters (see Gordon & Sutcliffe, 1974), but on the other hand they are expensive and difficult to clean.

Studies on the performance of some commonly used filters have been made by Sheldon & Sutcliffe (1969) and Sheldon (1972). Figure 5.1 taken from the latter paper, shows their retention characteristics and Fig. 5.2 reproduces scanning electron micrographs of four commonly used types of filter. The performance of some filters, especially depth filters, will change as the particulate load on the filter increases. Wangersky & Hincks (1978)

Fig. 5.1. Retention curves for different filtration media: *a* Millipore, *b* Whatman GF/C and GF/A and Reeve Angel 984H, *c* Nucleopore, *d* Flotronics. The numbers over the curves are the pore sizes (μm) quoted by the makers (after Sheldon, 1972).

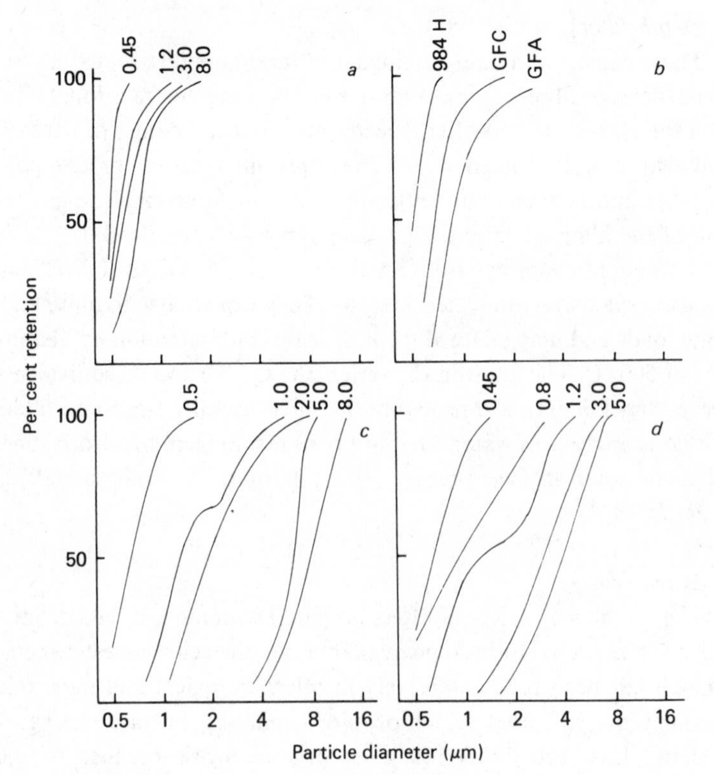

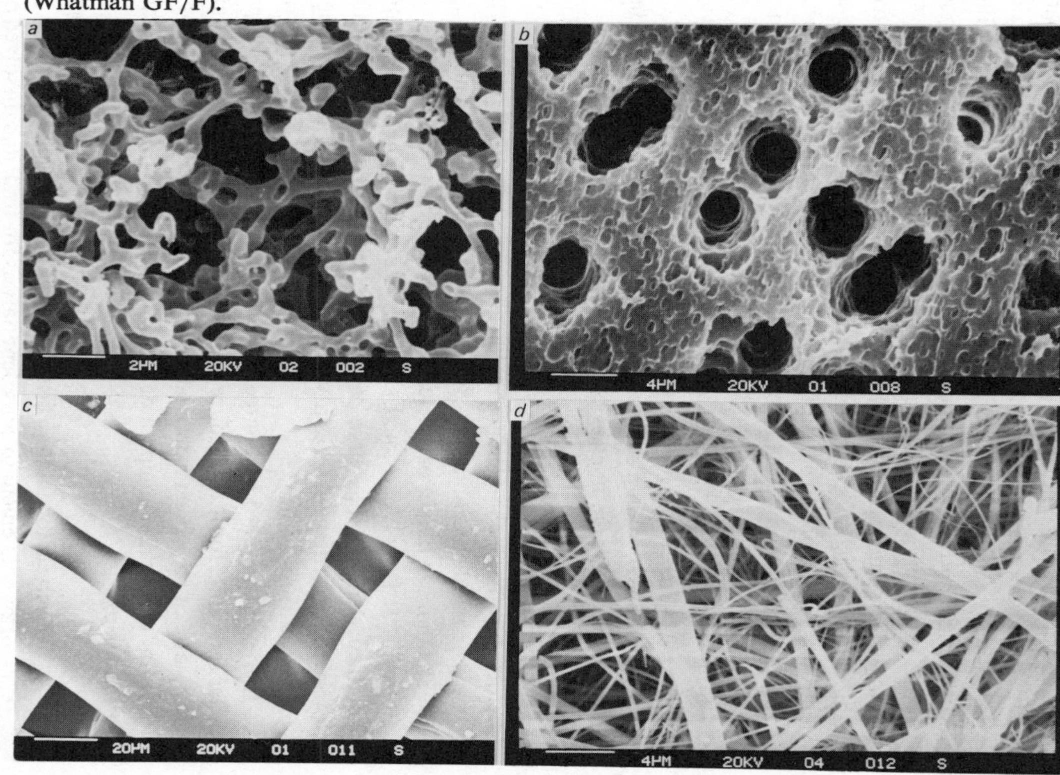

Fig. 5.2. Scanning electron micrographs of four commonly used filtration media: *a* 3 μm Sartorius membrane, *b* 5 μm Nucleopore membrane, *c* 20 μm plankton net, *d* 0.7 μm glass-fibre mat (Whatman GF/F).

have discussed the use of various filters for particulate organic carbon analysis.

Storage

The manner of storage between sampling and analysis needs careful consideration. The turnover time of some organic compounds in estuarine water may be only a few hours, so unless analysis is carried out immediately some form of preservation is essential. Although filtration will reduce the rate of biological processes it cannot be relied upon as a means of preservation. Samples should be stored in tightly sealed vessels, ideally glass, although polyethylene or polypropylene containers have been found to be satisfactory for storing samples for dissolved organic carbon analysis. They must be carefully cleaned and, if the sample is to be stored frozen, then it is advisable to put the bottles through a couple of freezing and thawing cycles beforehand.

Some means of preventing microbial changes is invariably necessary. In many cases freezing is the most satisfactory method; solid carbon dioxide is a very convenient and rapid means of freezing water samples in the field. If samples are frozen in glass bottles they should be half-filled and frozen at an angle of 45° to allow a large surface for expansion. When many samples are placed in a deep-freeze there may be a delay of several hours before they freeze completely. Acidification appears to be quite an effective way to inhibit microbial activity in water samples (Williams & Askew, 1968), but it can induce or accelerate some chemical reactions. For example, Brezonik & Lee (1966) noted a greater decrease in the nitrite content of acidified water samples than of untreated ones. They suggested that at a low pH the nitrite undergoes a Van Slyke reaction with amino compounds present in the sample.

If the sample is to be filtered, this should be done as soon as possible after sampling, for it will reduce the rate of microbial activity. Filtration of samples that have been frozen is probably unwise because of the possibility that not all organic compounds will return to their original physical state on thawing.

Analysis for total or dissolved organic material

It is convenient to consider analysis for total and dissolved organic material at the same time. The reason for this is that the two analyses have much in common, in some instances the only difference being whether or not the sample has been filtered.

It should be stressed again that the terms dissolved and particulate have no fundamental meaning in the present context, but are only to be regarded as operational terms defined by the type of the filter and the conditions used. In nature there is often, if not invariably, a continuous spectrum of material from the truly dissolved to the unambiguously particulate. The difficulty in interpreting the terms particulate and dissolved (generally preferred to suspended and soluble) has been discussed by Olsen (1967), Sharp (1973*a*, 1975) and others. The analytical and other advantages of filtering water samples have already been mentioned and whether or not samples are filtered will depend upon the requirements and aims of the analytical programme.

There is no direct method for measuring a water sample's organic content as such, nor is there any known way of separating all the organic material from the water in a quantifiable form. As a consequence, analyses for total or dissolved organic material must treat the water sample as a whole, although they are, in a chemical sense, indirect. Briggs *et al.* (1976) have reviewed some of the methods available for measuring these parameters in pollution monitoring. Here they will be considered under three headings:

1. physical methods;
2. measurement of biologically or chemically oxidizable material;
3. elemental analysis.

Physical methods

Light extinction and, to a lesser extent, surface tension have been used as a means of estimating or monitoring the organic content of natural and polluted waters. These methods have the obvious appeal that, once set up, they are often simple to operate and give immediate and continuous readings. The underlying problem with physical methods is that there is no readily measurable universal yet unique physical property of organic matter. Light attenuation, either in the visible or ultraviolet (UV) region, is highly dependent upon molecular structure. The paper by Foster & Morris (1971) illustrates the generally poor relationship between light extinction (in their case integrated between 250 and 350 nm) and the organic content of water. The relationship is strongly dependent upon composition and thus the technique will function best in areas where the inputs are limited in this respect: an estuary receiving a single trade waste might be such a case (see e.g. Briggs *et al.*, 1976). The normal estuarine situation, in which there is a multiplicity of sources of organic material, probably represents the worst situation for an optical method.

Physical methods are most likely to be of value as alarm monitors for spills. In such situations an accurate measurement is unnecessary; all that is required is an indication of any significant increase in organic content. Surface tension indicators (e.g. McMullen *et al.*, 1975) would be most appropriate for oil spills; fluorescence and UV light extinction have more general applications.

Measurement of biologically or chemically oxidizable material

Biochemical oxygen demand (BOD) and chemical oxygen demand (COD) have been used extensively in fresh-water quality studies. Details of the methods are given in various standard works (e.g. HMSO, 1983; American Public Health Association, 1976; Golterman *et al.*, 1978).

The BOD test, in which a sample of water is incubated, usually for five days at 20 °C, measures the quantity of dissolved oxygen removed by microbial respiration. It provides an estimate of the organic burden on the oxygen resources of a water body and is used extensively to monitor effluents and in calculations of the potential oxygen status of a receiving water. Beyond this use, its value in estuarine chemistry is open to question. As a measure or indication of the total organic content of a body of water, its only advantage over chemical methods based on elemental analysis is that the BOD test is simple and inexpensive to set up and carry out. On the other hand, it is tedious, slow and generally imprecise; furthermore, its interpretation is often complex. As discussed in the introduction, its value as a 'pollution' indicator in an estuary depends very much upon the magnitude of the organic flux resulting from autochthonous photosynthesis.

The conventional COD methods similarly seem to have little to offer in estuarine work. They rely upon determining the amount of oxidizing agent reduced by the organic material present in a water sample. If a strong oxidizing agent (e.g. acid dichromate) is used, then chloride will also be oxidized. Alternatively, weak oxidizing agents (e.g. alkaline permanganate) may be used, but these will effect only a partial oxidation and the resultant determination normally has little meaning.

Elemental analysis

If some suitable means of oxidizing the organic material in the water sample can be devised, then the products of oxidation (e.g. CO_2, NH_3) may be used as a measure of the organic content. These methods have the general appeal that their interpretation is comparatively straightforward. However, although simple in concept, they are not often simple in practice. In some instances the equipment is expensive to set up and requires a high degree of skill to operate.

The choice of which of the several major elements to determine is governed by a variety of considerations. Carbon has many advantages: it is the major constituent of organic matter and it can easily be arranged for its only oxidation product to be carbon dioxide (cf. nitrogen) which, being a gas, may readily be removed from the sample. Sensitive and continuous methods for the determination of carbon dioxide are now available. Phosphorus and nitrogen are also commonly determined, but they are lesser and of course more variable constituents of organic matter. The forms in which they are liberated by oxidation are usually not volatile (it is much more difficult to remove ammonia than carbon dioxide from water) and consequently the normal approach is to measure the organic fraction as the difference between determination of the inorganic fraction before and after oxidation. In the case of nitrogen this can entail two analyses for nitrate, nitrite and ammonia which, on a routine basis, is quite an undertaking. Phosphorus is more straightforward, since only phosphate need be determined.

Methods for the determination of some products of oxidation, including phosphate, nitrate, nitrite and ammonia, are considered in Chapter 3. Carbon dioxide may be determined by a variety of chemical procedures (e.g. gravimetry, titrimetry), but modern instrumental techniques are generally preferred. Two methods have been used extensively: non-dispersive infrared (IR) gas analysis and flame ionization detection (FID). The former is simple, highly sensitive and very reliable. In the latter method the carbon dioxide is first reduced to methane and then determined as such with a modified gas-chromatograph detector. This technique is relatively complex, but has certain advantages over the IR analysis in that the response of the FID is both rapid and linear over several orders of magnitude and, being a mass detector, gas flow rates are not critical.

It is usually necessary to decarbonate the sample prior to organic carbon analysis. This is easily achieved by acidification. Extreme conditions are not necessary, since inorganic carbon is present almost exclusively as carbonic acid and carbon dioxide below about pH 4.5. However, it involves a possibility of losing volatile material: organic acids are slowly removed from acid solution and low molecular weight hydrocarbons (C_8 and below) are readily lost. This can be largely overcome by stirring the acidified sample in a small, stoppered container in the presence of sodium hydroxide held in a cup attached to the stopper (see Van Hall *et al.*, 1965).

The various approaches open to the water chemist are best considered and classified in terms of the oxidation procedure employed. Three general categories can be recognized: high temperature combustion, photo-oxidation and wet chemical oxidation.

High temperature combustion

This type of method has an advantage over photo-oxidation and wet chemical oxidation in that its completeness of oxidation is widely accepted. It should be clearly recognized that chemical methods will only be accurate if they completely decompose all the organic material in the sample. This cannot be proved, although one can prove that a method is *in*accurate by showing that some material is incompletely oxidized. The problem involved in developing accurate high temperature combustion methods for dissolved organic carbon are well described by Wangersky (1975).

The simplest dry combustion method entails evaporating the water sample and burning the residue in a combustion tube. This approach is tedious and it is a common experience that the blank in such approaches is equivalent to 1–10 mg C l^{-1}, comparable to the levels found in most estuarine waters. Successful methods have been described by Montgomery & Thom (1962), Gordon & Sutcliffe (1973) and MacKinnon (1978). Commercial analysers have been developed based on this approach, but general experience is that such instruments are usually only suitable for the analysis of effluents and highly polluted waters.

A variety of methods has been described which involve the direct injection of the sample into a combustion tube (Van Hall *et al.*, 1963; Croll, 1972; Sharp, 1973*b*). The first two methods are now available in the form of commercial instruments. The method of Van Hall *et al.* suffers from a relatively high blank when used for unpolluted waters. Croll's method is continuous and this gives it certain advantages: its blank for instance is lower. Presently the most suitably dry combustion method for the analysis of unpolluted saline waters appears to be that described by MacKinnon (1978), although its use in estuarine work will be limited to higher salinity samples; conversely Croll's method, which works very satisfactorily with fresh-water samples, will be limited to lower salinities. Salonen (1979) has described a dry combustion method that can be used for both fresh and saline samples. It is simple in principle; however, it does seem to give suspiciously high results with sea water.

High temperature combustion is not normally used in the analysis of water for organic nitrogen or phosphorus. If the need exists, it should be possible to develop such a method for total nitrogen.

Photo-oxidation

The photochemical combustion of organic material in fresh- or sea-water samples using high wattage medium-pressure mercury lamps is now well established (Armstrong *et al.*, 1966; Armstrong & Tibbits, 1968; Grasshoff, 1966; Erhardt, 1969; Henriksen, 1970; Grasshoff *et al.*, 1983). The technique may be used batchwise (e.g. Armstrong & Tibbitts, 1968; Henriksen, 1970) or in a continuous mode (e.g. Grasshoff, 1966; Afghan *et al.*, 1971; Collins & Williams, 1977; Schreurs, 1978; Mantoura & Woodward, 1983; Statham & Williams, 1983). The appeal of these methods lies in their simplicity and versatility: few if any reagents need be added to the water sample and many aspects of its inorganic chemistry will be unchanged after oxidation. This approach has a variety of uses to the application discussed here, for example the preparation of blanks in organic analysis, liberating dissolved organically bound metals (P. M. Williams, 1969*a*) and purifying tracer solutions (e.g. Williams *et al.*, 1972). At the same time one should sound a note of caution: the mechanism of photo-oxidation has been inadequately studied and the selection of UV sources and irradiation conditions is at best empirical and in some cases arbitrary.

Some general points about the photo-oxidation process should be noted in passing. Medium-pressure mercury sources have almost invariably been used (the classification of lamps is based on that adopted by Calvert & Pitts, 1966), although photo-oxidation can be achieved by low-pressure lamps with less heating of the sample (see Collins & Williams, 1977; Schreurs, 1978). Lamp performance varies from batch to batch as well as from manufacturer to manufacturer and the useful output of UV radiation decreases with use. Presently the performance of lamps can only be determined empirically. When setting up a method this must be taken into consideration by allowing a safety margin and making frequent checks on performance. In most cases the oxidation rate is faster in the presence of compounds such as hydrogen peroxide or potassium persulphate at concentrations in the region of 10^{-2} M. It is presumed that these compounds provide free radicals in the oxidation. Hydrogen peroxide is most suitable when the irradiated solution is to undergo chemical analysis (e.g. for inorganic nitrogen or phosphorus), because unlike persulphate its decomposition products do not interfere. Persulphate has the advantage that it is a more stable compound and solutions containing exact quantities may easily be prepared; it is usually suitable if a gaseous oxidation product (e.g. carbon dioxide) is to be measured.

It is known that pH can affect both the rate and extent of photo-oxidation. The matter gives signs of being complex and the papers by Afghan *et al.* (1971), Henriksen (1970), Manny *et al.* (1971) and Collins & Williams (1977) should be studied. Whenever possible the irradiation should be carried out at a pH between 5 and 9. The diameter of the vessel containing the sample, not surprisingly, has a pronounced effect on the rate of photo-oxidation. In some cases rates almost a hundred times greater are found in millimetre sized containers than in centimetre sized ones, leading to the tentative conclusions that much of the reaction proceeds near the surface of the vessel, and that in large containers convection currents are relied upon to expose the bulk of the solution to effective radiation. In view of this possibility, as well as considerations such as cooling, the simple batchwise irradiation of litre volumes of water is inadvisable, unless a careful study of the kinetics has first been made.

The major end-product of the photo-oxidation of the carbon in organic material is carbon dioxide; for phosphorus it is orthophosphate. Organic nitrogen may end up as nitrate, nitrite and ammonia in varying proportions. Under acid and perhaps other conditions, nitrogenous end-products other than these three appear to be produced, so that the normal analytical programme will give incomplete results. Photo-oxidation will oxidize C–P bonds but, at natural pH values, it does not affect the hydrolysis of

Fig. 5.3. Diagram of batchwise photo-oxidation apparatus.

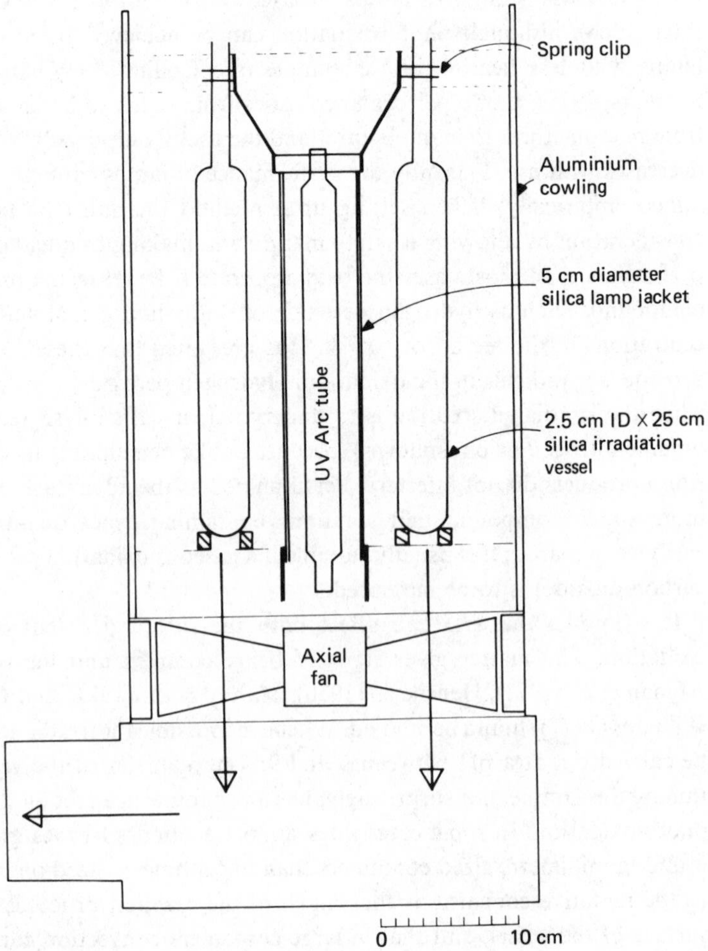

inorganic polyphosphate. This fact has been exploited to distinguish polyphosphates from other forms of dissolved phosphorus (see Chapter 3 and Solórzano & Strickland, 1968).

Adequate descriptions of the batchwise irradiation of water for dissolved organic nitrogen and phosphorus analysis are given by Armstrong & Tibbitts (1968), Henriksen (1970), Strickland & Parsons (1972), and Stainton *et al.* (1977). A suitable apparatus developed over the years in the author's laboratory is shown in Fig. 5.3. Batchwise determination of dissolved organic carbon has been used, but it is slow and there are reports of high blanks. Continuous methods have been described for the analysis of dissolved organic phosphorus and nitrogen (see Chapter 3), and carbon (Ehrhardt, 1969; Baker *et al.*, 1974; Goulden & Brooksbank, 1975; Collins & Williams, 1977; Stainton *et al.*, 1977; Mantoura & Woodward, 1983; Statham & Willlams, 1983). Most of these methods have not evolved from any detailed study of the photo-oxidation step and some of them, for example those of Baker *et al.* and Goulden & Brooksbank, employ short irradiation times which could give incomplete results under certain conditions. Continuous photo-oxidation methods have certain advantages over batchwise procedures: an important one from the analytical point of view is that they tend to be self-cleansing and as a result their blanks are often low and constant. They can also be incorporated into wholly automatic procedures. Schreurs (1978) has described a very promising system for dissolved organic carbon analysis based on continuous photo-oxidation with a low-pressure mercury lamp.

Wet chemical oxidation

These may be regarded as an extension of the COD method, with the important difference that it is the product of oxidation that is measured, rather than the extent of reduction of the oxidizing agent. They key to these methods lies in choosing a convenient and effective oxidant. Sulphuric, dichromic, persulphuric and perchloric acids have all been used, singly and in combination, for the analysis of dissolved organic carbon, nitrogen or phosphorus. The use of concentrated acids such as perchloric and sulphuric has obvious disadvantages. Persulphate has been widely used as an oxidizing agent in water analysis and it appears to be very effective at concentrations as low as 10^{-2} M at pH 2 (Menzel & Vaccaro, 1964; P. J. leB. Williams, 1969; Sharp, 1973*b*; MacKinnon, 1978).

Methods based on wet oxidation with persulphate offer some of the simplest approaches to the measurement of dissolved organic carbon in estuarine waters. Menzel & Vaccaro (1964) devised a method for sea water in which the sample was heated with persulphuric acid in a sealed glass ampoule at 130 °C; the ampoule was subsequently opened and the carbon dioxide produced was flushed through an IR gas analyser. The method has found extensive application in marine work and, apart from the IR analyser, it requires a minimum of equipment. Its disadvantages are that it is tedious and spurious values are often encountered. Goulden & Brooksbank (1975) have described a continuous method for fresh water using persulphate with silver ions (which would be precipitated by the chloride in sea water) as a catalyst. They found that, although the persulphate method gave results about 3 per cent lower than dry combustion and photo-oxidation methods, it was more convenient and precise. A similar difference was found by P. M. Williams (1968*b*), who compared results obtained from sea water by the persulphate and photo-oxidation methods.

Persulphate has also been used as an oxidant for the analysis of dissolved organic phosphorus (Menzel & Corwin, 1965; Solórzano & Sharp, 1980*a*; Grasshoff *et al.*, 1983) and dissolved organic nitrogen (D'Elia *et al.*, 1977; Nydahl, 1978; Solórzano & Sharp, 1980*b*; Grasshoff *et al.*, 1983). Wet oxidation has some advantages over photochemical combustion in these analyses: it lends itself to the processing of relatively large numbers of samples and the cost of apparatus is low. Finally, in the case of nitrogen, nitrate is apparently the sole oxidation product in alkaline solution (Koroleff, 1973), which simplifies the procedure considerably. Persulphate interferes with many inorganic nutrient analyses and it is necessary to ensure that none remains after the oxidation.

Descriptions of earlier wet combustion methods based on acid dichromate are given by Barnes (1959), Szekielda (1967) and Golterman *et al.*, (1978). These methods have probably been made redundant by the simpler persulphate techniques.

General points concerning elemental analysis

The present position regarding the various approaches for the determination of the total or dissolved organic content of waters of varying salinity appears to be as follows. In the case of dissolved organic nitrogen and phosphorus, batchwise analysis using photo-oxidation would appear to be most satisfactory. Continuous versions of the photo-oxidation procedure exist for these elements but they are technically much more demanding. The situation is more complex in the case of dissolved organic carbon. In general, current batchwise methods are limited to samples of high (> 10 mg C l^{-1}) organic content; below this concentration continuous methods are usually more satisfactory. The method of Menzel & Vaccaro (1964) is perhaps an exception here. Continuous high temperature oxidation methods for fresh water have been developed and are now being

Table 5.1. *Summary of methods for the analysis of saline water for dissolved organic carbon*

Oxidation procedure	Comments	CO_2 detection	Range	Precision	Reference
Acid dichromate	Technically difficult and time-consuming		0.1–8 mg C l^{-1}	±0.03 mg C l^{-1}	Duursma (1961)
Persulphate	Simple, but precision can be poor	Infrared	0.1–10 mg C l^{-1}	±0.1 mg C l^{-1}	Menzel & Vaccaro (1964)
Continuous photo-oxidation	Equipment comparatively complex and expensive. Good precision and low blanks	Conductimetric	0.1–10 mg C l^{-1}	SD 0.06 mg C l^{-1}	Ehrhardt (1969)
		Infrared	Up to 25 mg C l^{-1}*	Better than ±2 %	Baker *et al.* (1974)
		Infrared	Down to 0.01 mg C l^{-1}*	SD 1.2 % at 1 mg C l^{-1}*	Goulden & Brooksbank (1975)
		Infrared	Up to 40 mg C l^{-1}	Approaching 0.03 mg C l^{-1} Below 1 mg C l^{-1}	Collins & Williams (1977)
		Colorimetric	0.1–10 mg C l^{-1}	SD 5.2 % at 1 mg C l^{-1}	Schreurs (1978)
Batchwise high temperature combustion	Technically difficult. Some systems give high blanks, others are not suitable for low salinity	Infrared	Down to 2 mg C l^{-1}	SD 1.0 % at 100 mg C l^{-1}	Van Hall *et al.* (1963)
		CHN analyser	—†	SD 4.1 % at 2 mg C l^{-1}	Gordon & Sutcliffe (1973)
		Infrared	—	SD 4.2 % at 1 mg C l^{-1}	Sharp (1973*b*)
		Infrared	—	SD 2.5 % at 1 mg C l^{-1}	MacKinnon (1978)
		Infrared	—	SD 5.0 % at 2 mg C l^{-1}	Salonen (1979)
Continuous high temperature combustion	Only suitable for samples of low salinity	Flame ionization	0.1–10 mg C l^{-1}*	SD 7 % at 0.5 mg C l^{-1}*	Croll (1972)

* In fresh water. † Range not given.

adapted to saline waters. When available, they are undoubtedly the methods of preference. The equipment, however, is sophisticated, expensive and wholly dedicated. The continuous persulphate and photo-oxidation methods are less expensive to set up and the apparatus more versatile.

In conclusion, one should note that this is an area of analysis in which the skill and competence of the analyst is as important, if not more so, than the chemistry of the method involved. A detailed intercomparison of three dissolved organic carbon methods undertaken by Gershey *et al.* (1979) found that dry combustion and photo-oxidation gave, for all intents and purposes, identical results, whilst the persulphate method of Menzel & Vaccaro (1964) gave values about 15 per cent lower. This paper also contains a detailed discussion of the various approaches to the analysis of dissolved organic carbon in saline water, which are summarized in Table 5.1.

Analysis of particulate organic material

The separate analysis of the particulate fraction may be undertaken for a variety of reasons as part of general studies on the organic load, balance or sedimentation of material in estuaries. It is commonly an integral part of biological programmes. The methods fall into two broad categories: gross analysis of particulate organic material and analyses for specific biochemical constituents associated with the living fraction. In the former category, analyses for total organic matter and for the major elements carbon, nitrogen and phosphorus will be considered. In the latter category, attention will be restricted to analyses for chlorophyll and adenosine triphosphate (ATP).

Analysis for total particulate organic material

In common with analysis for the dissolved fraction, a variety of alternative approaches exists, each having particular advantages and disadvantages. The various methods will be reviewed briefly and a summary is given in Table 5.2. Physical methods, based on turbidity, are used for the estimation of total particulate material; they cannot, however, distinguish between inorganic and organic and therefore have little general application in chemical studies. Three other approaches will be considered:

1. loss of weight on ignition;
2. chemically oxidizable material;
3. elemental analysis.

All these methods are normally preceded by collecting the particulate material on inorganic filters. Glass-fibre filters are extensively used in this type of work. It is advisable, if not essential, and certainly

Table 5.2. *Summary of methods for the measurement of particulate organic matter and particulate organic carbon*

Determinand	Combustion procedure	Detection procedure	Range	Sensitivity or precision	Reference
Organic matter	Dry combustion at 500 °C in a muffle furnace	Loss of weight	—[a]	Relatively low	Department of the Environment (1972)
Organic matter	Wet oxidation with dichromate at 100 °C	Titration	100–2000 μg C	60 μg as C	Strickland & Parsons (1972)
Carbon	Dry combustion at 800 °C in a muffle tube	Collection of a constant volume of gas and measurement of CO_2 concentration with an IR analyser	5–500 μg C 5–10000 μg C	10 μg C 2 μg C (SD)	Menzel & Vaccaro (1964) Banoub & Williams (1972) Baker *et al.* (1974)
Carbon	Dry combustion at 700 °C in a muffle tube	Collection of CO_2 in alkali and measurement of conductivity change	50–700 μg C	25 μg C (SD)	Dal Pont & Newell (1963)
Carbon	Dry combustion in a glass ampoule at 500 °C	Ampoule opened and the CO_2 flushed through an IR analyser and the CO_2 content determined by integration of pulse	10–200 μg C	2 μg C (SD)	Holm-Hansen *et al.* (1967)
Carbon	Dry combustion at 750–800 °C in muffle tube	Measurement of CO_2 pulse produced during combustion with an IR analyser	—[a]	7.4 μg C (SD)	Wangersky & Gordon (1965)

[a] Range not given.

good practice to free them from organic contamination prior to use by heating them overnight at 500 °C. This of course will not remove any phosphate present, for which acid leaching may be necessary. Some workers have preferred to accept the contamination associated with the filters as purchased, making a correction by subtracting the blank value obtained by analysing unused filters. This may save a small amount of time and avoid the risk of changes in porosity due to sintering, but in the long run it would seem to generate more problems than it solves, particularly since an unknown quantity of contaminating material may be leached from the filters during use. As mentioned above, glass-fibre filters adsorb small amounts of organic matter from solution, but the effect on determinations of particulate organic material is negligible in most estuarine situations.

Loss of weight on ignition

This procedure is commonly used in biological studies. The filter with its organic load is dried at a temperature in the region of 105 °C and weighed, then further heated at a temperature in the region of 500 °C and finally reweighed. The organic content is taken as the difference between the two weights. The method is simple, it gives a measure of total particulate organic material, rather than some constituent, and it can also provide information on the organic content. It has two disadvantages: first, it is insensitive; the organic fraction is determined by difference and thus the precision of the estimate is low, generally no better than a milligram or so; secondly, particulate carbonates can affect the results. The selection of the temperature and duration of the drying and combustion steps needs careful consideration. In the former it is important to drive off all water, but to avoid the loss of volatile organic compounds. During the latter it is necessary to destroy the organic material without decomposing carbonates. Hirota & Szyper (1975) resolved between inorganic and organic carbon by heating in air at 500 °C for four hours. Under these conditions the organic fraction is decomposed, whereas calcium carbonate does not dissociate. Carbonates can be removed before analysis by washing, or preferably fuming, with acid. Procedures for minimizing carbonate interference will be considered in more detail below (p. 181).

Chemically oxidizable material

Another method frequently used to measure total organic content is to heat the filter with acid dichromate at 100 °C and determine the residual dichromate spectrophotometrically or, more simply and conveniently, titrimetrically. This is analogous to the chemical oxygen demand method for water samples. It appears in various manuals (e.g. Strickland & Parsons, 1972) and need not be described in detail here. Manuals of freshwater analysis cover the overall procedure and particularly the titration in some detail as part of the COD method (Golterman *et al.*, 1978; Mackereth *et al.*, 1978).

It would appear (el Wakeel & Riley, 1957; Copin-Montegut & Copin-Montegut, 1973) that acid dichromate, even under the relatively mild conditions used, effects the complete combustion of organic material. Inorganic carbonate does not interfere, but of course there will be severe interference from halides, if present. They are most readily removed by washing the filter with isotonic sodium sulphate, taking care that the periphery, as well as the centre, is treated. An alternative procedure is to heat the filter with phosphoric acid before adding the dichromate, but complete removal is less readily achieved in this manner.

One minor disadvantage of the method is that the results are expressed in chemical equivalents rather than as mass. These units are inconvenient for many studies and some arbitrary or empirical conversion factor is needed to transform the results into mass units (see Golterman *et al.* (1978) for further information about the use of chemical equivalents). There is, however, one situation in which units of equivalence have an advantage over mass units and that is nutritional and kindred studies, where energy units are often preferred. The relationship between 'oxygen demand' and calorific value is closer than that between organic mass and energy. A figure of 4.83 cal ml^{-1} O_2 is often taken as the oxycalorific ratio, which corresponds to 0.863 cal $meq.^{-1}$. For a discussion of this relationship, the paper by Elliott & Davidson (1975) should be consulted. It is possible, with care, to determine the calorific value of the organic material on a filter directly by bomb calorimetry. This requires a fair degree of skill and practice and is only suitable for specialized applications, but it does have its use in certain types of biological study concerned with nutrition and energy flow.

Elemental analysis

The third approach is to oxidize the organic material and determine its inorganic oxidation products. These methods can be very sensitive, although with estuarine samples this is not necessarily an advantage. The reason is that the particulate fraction is heterogeneous in its distribution (see e.g. Wangersky, 1974) and it is usually desirable to overcome this by filtering as large a volume of water as is consistent with the sampling programme or the capacity of the filter. Volumes of 250–1000 ml are commonly used; small volumes, i.e. less than 100 ml should be avoided if possible. However, if the analytical technique is too sensitive, it will be necessary to dilute the sample at some stage and in such situations a more satisfactory solution is to turn to a less sensitive method. In doing so, contamination problems may also be reduced. Analysis is typically limited to the carbonaceous and nitrogenous parts of the organic material and their analyses are conveniently considered more or less separately.

The determination of particulate carbon can be made with commercial CNH analysers or 'home-made' carbon analysers. The latter commonly consist of a combustion system, some means of collecting the combustion gases, and a carbon dioxide detector such as a non-dispersive IR analyser. They are comparatively easy to construct and Fig. 5.4 is a diagram of one developed by the author which is capable of the routine analysis of samples of particulate material containing from 10 to 10000 μg of organic carbon. The analysis time is less than ten minutes per sample. Gordon (1969) found no significant difference between the results obtained by instruments of

Fig. 5.4. Diagram of a particulate organic-carbon analyser suitable for analysing particulate material containing between 10 and 10000 μg of organic carbon.

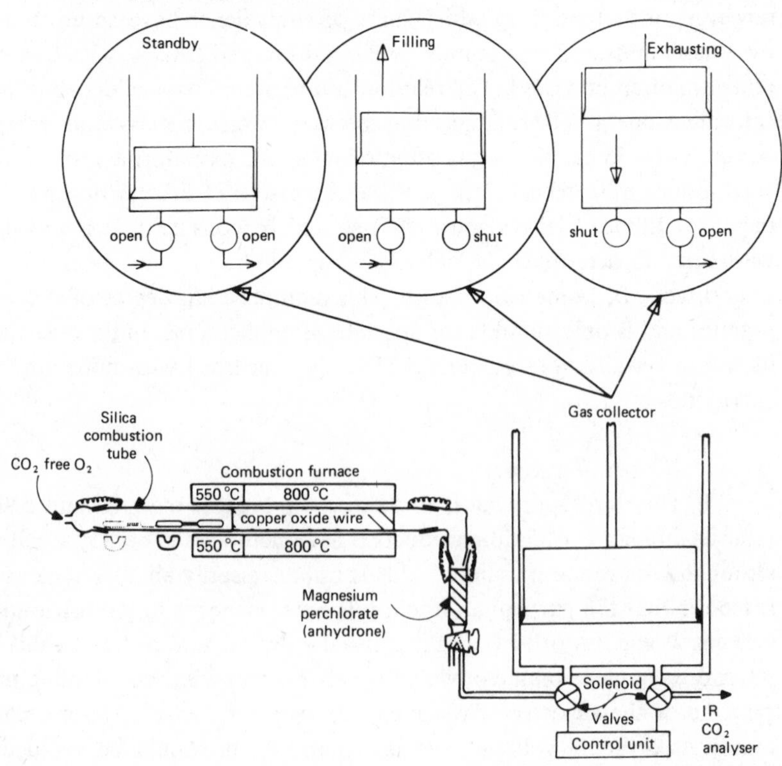

these two types. He also described a procedure for calculating the confidence limits of individual particulate organic carbon determinations.

Carbonates are a potential source of interference in this type of analysis. This may be overcome by acidification, by careful selection of the temperature of combustion, or by determining the organic fraction by a differential method. Washing with acid is open to the immediate criticism that it can cause the lysis of delicate organisms and consequent loss of organic material; fuming in acid vapour is a better alternative. In some circumstances it is possible to devise combustion conditions in which the rate of carbonate decomposition is slow compared with that of organic matter. Statham (1972), for example, found that in oxygen at 550 °C essentially complete oxidation of the organic material was achieved in about ten minutes, whilst finely divided calcium carbonate decomposed at a rate of only 0.35 per cent per minute. This approach is satisfactory for samples which are low in carbonate, but substantial errors will occur if the inorganic carbon content greatly exceeds that of organic carbon.

An alternative approach to determining organic carbon in the presence of carbonates is to filter two replicate samples (Hirota & Szyper, 1975). One is heated in air (not oxygen) for four hours at 500 °C. Under these conditions the organic fraction will be decomposed, but the temperature is below that at which calcium carbonate dissociates. Subsequently both samples are heated at 1100 °C, when both the inorganic and organic fractions are decomposed. The organic content is calculated from the difference between the quantities of carbon dioxide liberated from the two filters during the second combustion. This procedure is more lengthy than those described above, but it should give a more exact resolution between the two fractions.

Particulate organic nitrogen may be determined by dry or wet combustion procedures. Commercial instruments are available, either dedicated for nitrogen analysis or as CHN analysers. They are relatively expensive items and wet oxidation offers a cheaper alternative. The latter is generally based on a simplified version of the Kjeldahl procedure. The sample is digested in aqueous sulphuric acid and the ammonia produced is determined in the neutralized digest by a colorimetric technique. The methods are simple to set up and carry out. The original procedure, to distil the ammonia and titrate it with acid, can incur high blanks, requires considerable skill and has largely been abandoned in favour of the simpler colorimetric determination of ammonia.

The persulphate oxidation methods already mentioned for dissolved organic nitrogen and phosphorus can readily be adapted to particulate material. In the case of nitrogen, the filter is digested with alkaline

Table 5.3. *Summary of methods for the determination of particulate organic nitrogen*

Determinand	Combustion procedure	Detection procedure	Range	Sensitivity or precision	Reference
Nitrogen	Dumas dry combustion	Automatic measurement of N_2 gas (Model 29 Coleman Nitrogen Analyser)	50–1000 μg N	10 μg N (SD)	Strickland & Parsons (1972) Stainton *et al.* (1977)
Nitrogen	Kjeldahl combustion	NH_3 determination by ninhydrin method	0.5–50 μg N	0.24 μg N (SD)	Holm-Hansen (1968)
Nitrogen	Kjeldahl combustion	NH_3 determination by ninhydrin method	—[a]	—[a]	Dal Pont & Newell (1963)
Nitrogen	Kjeldahl combustion	NH_3 determination as indophenol blue	0–100 μg N	1 μg N (SD)	Banoub (1972)

[a] Range not given.

persulphate and the resulting nitrate determined as nitrite after reduction with cadmium. D'Elia *et al.* (1977), Nydahl (1978) and Grasshoff *et al.* (1983) have all described the procedure for water samples. This approach originates from work by Koroleff (1973). Although in principle the particulate nitrogen fraction may be determined as the difference between the total and the dissolved fraction, this procedure is not recommended because of its low precision (D'Elia *et al.*, 1977). Table 5.3 contains details of a selection of particulate organic nitrogen methods.

Several methods are available for the determination of particulate phosphorus (e.g. Strickland & Parsons, 1972). Grasshoff *et al.* (1983) have described a method for the measurement of total phosphorus based on oxidation with acid persulphate which could be adapted for particulate phosphorus determinations. This analysis is not common in water science and its interpretation is more difficult than those of the corresponding carbon and nitrogen analyses.

Biochemical analysis of particulate material for ATP and chlorophyll

A full coverage of the biochemical analysis of particulate material is not merited in the present account. There is, unfortunately, no comprehensive review of the subject, but details may be found in the articles by Strickland (1965) and Giese (1967) and practical details of some methods will be found in Strickland & Parsons (1972) and Grasshoff *et al.* (1983). The book by Jones (1979) discusses methodology and is a valuable source of practical details.

Only two analyses will be considered here: ATP and chlorophyll. The former is used as an estimate of total biomass; the latter as a measure of plant biomass and may form an important part of programmes concerned with the study of productivity and of eutrophication of estuaries.

ATP analysis

There is often a need to determine the amount of living material in a sample. Direct biological methods based on counting the number of organisms are tedious and imprecise. In the case of bacteria this is very difficult in environments such as estuaries, where large amounts of suspended material are also present. Various approaches have been examined, but the most successful to date is the determination of ATP (Holm-Hansen & Booth, 1966). This method relies on the assumption that ATP is universal in living organisms and absent from non-living material. In order to interpret the analyses it is necessary to assume some mean ATP content for living material. This is a major weakness in the method, but one should note that a similar situation in fact exists with chlorophyll and in this case it is customary not to convert the data to plant biomass but to interpret it as chlorophyll. Hamilton & Holm-Hansen (1967) and Holm-Hansen (1970) have examined the ATP content of a variety of algae and bacteria, and although the extreme range was 50-fold the majority of values lay within a factor of five of the median value, which was found to be 0.4 per cent of the cell carbon (i.e. about 0.16 per cent of dry weight).

The ATP determination is based upon the well-established luciferin/luciferase method, in which the ATP is used by the enzyme–substrate complex as a source of energy for a reaction that generates light. The photon output may be related to the ATP content of an extract. Once set up, the method is simple and very sensitive. It is capable of detecting the ATP present in less than 0.1 μg of cell material (dry weight). The measurement of light emission may be made in purpose-built instruments, but many fluorometers and liquid scintillation counters may be adapted for ATP analyses. Details of the original method are given in the paper of Holm-Hansen & Booth (1966). Samples with high particulate loads can give erroneously low results (Sutcliffe *et al.*, 1976). This effect can be reduced by filtering the minimum volume of water. Jones & Simon (1977) avoided this interference in fresh water by carrying out the determination directly on unfiltered samples. Their paper contains a valuable review of the method and describes modifications which improve upon the sensitivity of the original method. Details will also be found in Jones (1979).

Chlorophyll analysis

Chlorophyll analyses feature in many estuarine programmes and several alternative approaches exist. The details of the methods will be found in various texts (Strickland & Parsons, 1972; Vollenweider, 1974; Stainton *et al.*, 1977; Jones, 1979) and need not be repeated here. The purpose of the present account is to consider the various steps and alternatives, to give some guidance and to indicate where care should be taken. The routine analysis of sea-water samples for chlorophyll has been considered by a group of experts and their report (UNESCO, 1966) is worth studying, although it is now somewhat out of date.

The overall aim of the usual chlorophyll analysis is to extract the chlorophyll from the particulate material quantitatively and without degradation and then to determine the chlorophyll(s) in the extract without interference from related pigments, notably the phaeophytins. Phaeophytins are degradation products of the chlorophylls, produced by the irreversible loss of the magnesium atom from the porphyrin ring. Although their chemical and spectral properties are similar to those of the chlorophylls, they cannot participate in photosynthesis. They are sometimes loosely referred to as 'dead chlorophyll'. On occasions, phaeophytin can be present in particulate material at a level comparable to that of chlorophyll and, if no attempt is made to resolve between chlorophyll and phaeophytin, then wholly erroneous values for chlorophyll can result. The article by Yentsch (1967) contains a very useful discussion of chlorophyll degradation and the problem of its analysis.

The determination of chlorophyll may be conveniently considered in three steps: filtration, extraction and analysis.

Filtration. Glass-fibre mat, membrane and paper filters have all been used for chlorophyll analysis. The type of filter used will be mainly governed by the extraction procedure adopted. Glass-fibre filters have several advantages: they are inexpensive and can act as an excellent abrasive during the extraction procedure. They can also be used for particulate carbon and nitrogen analysis, thus reducing the variety of filters needed for field work. The finest grade (Whatman GF/F or equivalent) is generally recommended.

In early work it was conventional to filter onto a bed of magnesium carbonate which was intended to prevent chlorophyll degradation to phaeophytin. However it appears that magnesium carbonate need not be present at any stage (Holm-Hansen & Rieman, 1978). The filters bearing the particulate material for chlorophyll analysis must be stored dry in the dark at 0 °C or preferably below.

Extraction. The chlorophyll can be determined directly on the filter by absorbance or reflectance measurements, although the methods are unlikely to be particularly accurate. The conventional analysis is usually commenced by extracting the chlorophyll into an organic solvent: either methanol or 90 per cent (v/v) aqueous acetone. Chlorophyll is not very readily extracted from algae by either solvent and merely allowing the filters to soak in the solvent may not be satisfactory (although see Holm-Hansen & Rieman, 1978). Ultrasonics have been used to accelerate extraction and although they do indeed achieve this, some more vigorous method is generally adopted. Grinding in an all-glass homogenizer has been recommended and is very effective. Glass fibres act as a very good abrasive and thus a very simple and convenient procedure is to collect the samples on GF/C or preferably GF/F filters and to homogenize the filters either with a power or hand homogenizer. Heating may also be used to extract the chlorophyll when methanol is used as the solvent. Short periods of boiling in organic solvents apparently do not cause the degradation of chlorophyll to phaeophytin. Holm-Hansen & Rieman (1978) in their examination of the method of chlorophyll determination recommended methanol as an extracting solvent. They concluded that with methanol there appears to be little need to grind or homogenize the filter. For sediment samples the extraction procedure described by Tett *et al.* (1975) would appear to be satisfactory. After extraction it is usually necessary to centrifuge the sample to remove any added magnesium carbonate and the residue of the glass-fibre filters. Cellulose acetate filters, if used, dissolve in both 90 per cent acetone and methanol, but they apparently can impart a slight haze to the extract which must be allowed for in the analysis (see Strickland & Parsons, 1972).

Analysis. At least three quite distinct types of analysis may be made on the extract:

1. thin-layer chromatographic (TLC) separation of the pigments;
2. measurement of chlorophyll *a* and phaeophytin *a*;
3. resolution of the three chlorophylls, *a*, *b* and *c*.

The first of the three approaches listed above, TLC separation of the pigments, would not normally be used in routine work and will not be considered in any detail in the present text. A variety of methods are available in the literature (see e.g. Riley & Wilson, 1965; Madgwick, 1966; Yentsch, 1967); general accounts of TLC techniques will be found in Stahl (1969). The second two approaches listed above are extensively used in estuarine studies and details of procedure will be found in various texts (e.g. UNESCO, 1966; Strickland & Parsons, 1972; Vollenweider, 1974; Golterman *et al.* 1978; Jones, 1979).

The resolution of chlorophyll and phaeophytin is based on the observation that chlorophyll rapidly degrades to the equivalent phaeopigment in acid solution, resulting in a fall in absorption. Thus, by measuring the extinction of a pigment extract before and after acidification, it is possible to determine both the chlorophyll and phaeophytin in the extract. In acetone there is an approximately 40 per cent decrease in extinction at 665 nm on conversion of chlorophyll to phaeophytin (Loftus & Carpenter, 1971); in methanol the decrease is about 75 per cent (Tett *et al.*, 1975). It would appear that a more precise determination of chlorophyll may be made in methanol than acetone, but the spectra suggest that the wavelength would be more critical in methanol, Furthermore, it would appear (Holm-Hansen & Rieman, 1978), that in methanol the absorbance of phaeophytin is pH-dependent, so that it is necessary to readjust the pH after acidification.

The resolution of chlorophylls *a*, *b* and *c* may be achieved by measurement of the absorption of the pigment extract at three wavelengths. The method was originally devised by Richards with Thompson (1952) and minor modification of the method has been described subsequently (UNESCO, 1966; Strickland & Parsons, 1972). The equations are not to be regarded as universal and their accuracy depends to some extent upon the floristic composition of the population. For this analysis the accuracy of the wavelength settings of the spectrophotometer is important; Strickland & Parsons recommend that it should be accurate to 2 nm and describe how this may be achieved. Recently high-performance liquid chromatography (HPLC) methods have been developed to resolve both chlorophylls and other phytoplankton pigments (see e.g. Abaychi & Riley, 1979).

Fluorescent techniques may be used in place of absorptiometric ones. The main advantage of fluorescence methods in chlorophyll determination is their greater sensitivity; a tenfold or more gain is obtained. The disadvantages are that the methods are not absolute and, because pure chlorophyll preparations are not readily available, the technique must be standardized against absorptiometric methods. Secondly, the interpretation of fluorescence data can be more complex than absorption data (see Loftus & Carpenter, 1971). In estuarine work, where chlorophyll levels are generally high, there would seem to be little gained by using fluorescence to determine the chlorophyll content of extracts.

Often sufficient time is available to allow only one of the alternative chlorophyll analyses to be made. The one selected will depend upon the specific needs of the programme. In general the 'acid-ratio' resolution of chlorophyll and phaeophytin will be found to be more useful than the trichromatic resolution of the three chlorophylls. The reason for this is twofold, first, the proportion of phaeophytin in extracts can be quite high and as a consequence uncorrected chlorophyll data can be misleading; secondly, it is rare that full use can be made of data for chlorophyll *b* and *c*.

Finally, some mention should be made of the determination of chlorophyll by its *in situ* fluorescence. Modern fluorometers permit the direct determination of algal chlorophyll in water samples at concentrations down to 0.1 $\mu g\ l^{-1}$ (Lorenzen, 1966). It is quite feasible to make continuous measurements of chlorophyll using this approach, by pumping the water through the fluorometer cell or, more recently, by *in situ* fluorometers. Thus detailed studies of the regional and temporal distribution of chlorophyll may be undertaken. The main drawback of the method is that the measured *in situ* fluorescence of a given amount of chlorophyll depends on a variety of factors including species composition, cell physiology, cell morphology and ambient light intensity (Kiefer, 1973). There may be, for example, fourfold diurnal variations in the fluorescence yield of chlorophyll. Thus, although *in situ* fluorescence is unquestionably a valuable technique, the data it yields need to be interpreted with care. Slovacek &

Hannan (1977) appear to have successfully overcome this problem by adding the photosynthesis inhibitor 3-(3, 4 dichlorophenyl)-1,1-dimethylurea (DCMU) to the sample prior to making the measurement of fluorescence. In the presence of 10 μmolar DCMU, there was a very close relationship between chlorophyll *a* and fluorescence with both cultures and natural populations.

Analysis for anthropogenic compounds

The analysis of estuarine samples for individual biochemical compounds, such as lipids, amino acids or sugars, is normally undertaken as part of some specialized programme and usually those contemplating such work will be sufficiently familiar with the literature and available instrumentation and methodology that there is little to be gained, in the present account, by reviewing this extensive field. Reviews of the marine literature have been given by Williams (1975), Morris & Culkin (1975) and Dawson & Liebezeit (1981).

It was thought, on the other hand, that because water quality control and kindred laboratories were called upon on occasions to undertake analyses of samples for traces of anthropogenic compounds, a brief review of the subject would be helpful. This has been undertaken with reservations, for it is a complex topic, still undergoing method development: for example, in the case of oil, fundamental differences of opinion exist over whether one should analyse for a single target compound or for broad molecular classes. The present account will restrict itself to petroleum-associated hydrocarbons, chlorinated pesticides (the DDT group) and polychlorinated biphenyls (PCBs). No attempt will be made to discuss methods in detail, rather the subject will be considered in outline and the reader directed, where possible, to authoritative reports and papers. The following publications will give some guide to the overall strategies and the required laboratory facilities in addition to providing, in some cases, details of methodology: FAO (1971); National Academy of Sciences (1975); Keith (1976); Duursma & Dawson (1981) and Grasshoff *et al.* (1983).

At the outset it must be stressed that the problems in undertaking these analyses are formidable and must be recognized. Considerable experience of trace organic analysis is needed if meaningful results are expected. Instruments often need to be operated at or near their limit of sensitivity; chemicals, especially solvents, may need thorough purification before use. Finally, it is essential to have control over the laboratory environment, for contamination problems are acute (see e.g. Blumer, 1965). This type of work is most satisfactorily undertaken in a dedicated laboratory and an analyst considering starting up this type of work is strongly advised to contact a laboratory with experience in the type of analysis. The FAO (1971) report listed the then existing centres for this type of work.

Separation procedures

Presently there are no satisfactory direct methods for the analysis of these three classes of compound in natural samples. It is possible that the fluorescence method (Levy, 1971) may achieve sufficient sensitivity to permit the direct detection of oil in water, but because fluorescence is not a universally constant property of the constituents of oil and also because of the interference by the fluorescence of non-hydrocarbon compounds in natural waters, fluorescence will probably find its main use in alarm monitors for spills, like the measurement of surface tension (McMullen *et al.*, 1975), or in forensic work. Thus, presently, all commonly used methods involve an initial separation step. Solvent extraction is widely used for water and solid samples; however, two additional approaches should be noted. Volatile hydrocarbons are not conveniently determined by solvent extraction procedures because they are lost during the evaporation of the solvent. For such compounds, some sort of volatilization procedure, for example stripping with a gas-stream or head-space analysis must be used. Giger's work (Giger, 1977; Giger *et al.*, 1976) demonstrates the potential of such methodology. As is illustrated by this work, these types of compounds lend themselves to the capillary gas chromatography – mass spectroscopy (GC–MS) approach.

Alternative separation procedures usually involve removal onto a surface or into an immiscible solvent. Hydrocarbons and a variety of non-polar and semipolar compounds may be extracted from water by the polystyrene microbead matrix Amberlite XAD-1 (Riley & Taylor, 1969). This has the advantage that large volumes of water may be processed by passing them through comparatively small columns of the resin. This technique reduces the opportunity for contamination, which otherwise can be a problem when there are large samples of water to be processed. The adsorbed material may be eluted from the column directly with an appropriate solvent or extracted from the column in a Soxhlet extraction apparatus.

Whereas solvent extraction procedures are often satisfactory and comparatively straightforward for water samples, solid material, especially tissue and sediment samples, needs careful attention in order to obtain complete extraction. Soxhlet extraction provides one approach. Prior to extraction the sample may need homogenizing. Drying, especially freeze-drying, will improve the extraction but may bring along with it contamination problems. Saponification in 1 N KOH provides a powerful way of disrupting and opening up the sample matrix and it is especially useful for sediment and tissue samples. It will also hydrolyse triglycerides which facilitate clean-up, which is useful for biological material or material of recent biological origin. It does, however, provide an opportunity for chemical reactions to occur and accordingly may introduce undesirable changes. Wong & Williams (1980) have examined the performance of three extraction procedures for the separation of hydrocarbons from estuarine sediments.

Analysis

Once the compound(s) of interest has been extracted by the solvent, the subsequent analysis may take on varying levels of sophistication. Crude estimates of petroleum-hydrocarbons may be made at this stage by IR analysis of carbon tetrachloride extracts. The method (e.g. American Petroleum Institute, 1957) which is widely used as a form of effluent monitoring relies upon the absorption at a wavenumber of 2930 cm^{-1} (wavelength 3.41 μm) associated with the bond stretching of aliphatic structures. Obviously the results obtained by this method will be biased by the ratio of aliphatics to aromatics in the sample. For monitoring purposes this shortcoming is probably outweighed by the extreme simplicity of the method. A different, but equally simple, approach to determining total hydrocarbons is to weigh the extract, subsequent to a saponification step. Although simple, neither of these methods would be suitable for other than grossly contaminated samples.

More sophisticated methods usually incorporate a clean-up step between solvent extraction and quantification. Typically, silica–alumina columns are used to separate polar material such as lipids, from the hydrocarbons of interest. The details of the column preparation and sample processing for petroleum and chlorinated hydrocarbons may be found in Ehrhard, (1972), Holden & Marsden (1969), Duinker & Hillebrand (1978) and Grasshoff *et al.* (1983).

Subsequent to the clean-up, the methods for the PCBs and the chlorinated hydrocarbon pesticides typically involve separation and quantification with a gas–liquid chromatograph, using an electron capture detector. For both classes of compounds, complex chromatograms result. In the case of the chlorinated pesticides, standards may be run and identification made. In the case of the PCBs, which are manufactured as complex mixtures quantification is less straightforward. It is conventional to match the resultant chromatogram with commercial preparations and to undertake quantification in this manner. Duinker & Hillebrand (1983) give a very thorough discussion of the preparation of standards and analysis of GLC traces for PCBs and chlorinated hydrocarbon pesticides.

Crude oil and the effluent from oil refineries are complex mixtures of an exceedingly large number of compounds and it is in no way feasible to resolve this mixture completely. Indeed, even if it were possible, it is difficult to see what would be achieved in doing so. Presently and in the foreseeable future all analytical approaches will provide some form of partial resolution or analysis of the sample. Detailed analyses of certain oil fractions may be undertaken, the simplest being the *n*-alkane series. These may be readily separated using GLC techniques and easily identified. With other oil fractions, the branched chain alkanes, olefins and aromatics, the number of isomers becomes so great that resolution is either difficult or impossible. Even if resolution techniques can be devised the problem of data logging, processing and interpretation would be formidable. High resolution capillary column GC–MS can provide very detailed analysis of natural samples (e.g. Giger, 1977; Giger *et al.*, 1976). Often in toxicological studies, only certain target compounds (e.g. benzpyrene) need be analysed. In such situations the mass spectrograph can be used as a highly specific detector for the gas chromatograph.

Alternative approaches to the problem usually entail making some partial resolution of the hydrocarbons into gross molecular categories: alkanes, olefins, cyclic paraffins, one-, two-, etc. ring aromatics. Thin-layer chromatography on silica gel plates effects partial separation of hydrocarbon mixtures (Ehrhardt, 1972). The fractions may be visualized under a UV lamp and then eluted off the thin-layer plate and quantified either by weighing or by combustion and carbon dioxide determination (see e.g. DiSalvo *et al.*, 1973). The latter method is simple and sensitive. Alternatively, approximate quantification of the fractions may be achieved by UV or IR absorption or fluorescence of the eluates.

HPLC may be used for hydrocarbon work: Dawson & Ehrhardt (1976) used an improved version of Zsolnay's method (Zsolnay, 1973) to determine total aromatics by HPLC. The method which used a UV detection at 254 nm, makes no attempt to resolve individual aromatics. Dawson & Ehrhardt point out that because the extinction coefficients of individual compounds vary considerably the choice of a standard is critical and the method cannot be expected to give an accurate estimate of the mass of hydrocarbons. HPLC, as yet, does not seem to be able to provide the same resolution as GLC; its main drawback would seem to be the detection part of the system. Fluorescence, UV absorption and refractive index are commonly used in HPLC detectors, but none of them gives a true mass response unless the system can be made to resolve individual compounds.

If this can be achieved, identification will become a problem because interfacing a HPLC with a mass spectrograph is technically a great deal more difficult than it is with a GLC system. Figure 5.5 illustrates the various strategies available for hydrocarbon and chlorinated hydrocarbon analysis.

Fig. 5.5. Strategies available for the analysis of hydrocarbons and chlorinated hydrocarbons. Identification methods include infra-red absorption (IR absorption), gas chromatography (GC), mass spectrometry (MS), high-performance liquid-chromatography (HPLC) and thin-layer chromatography (TLC).

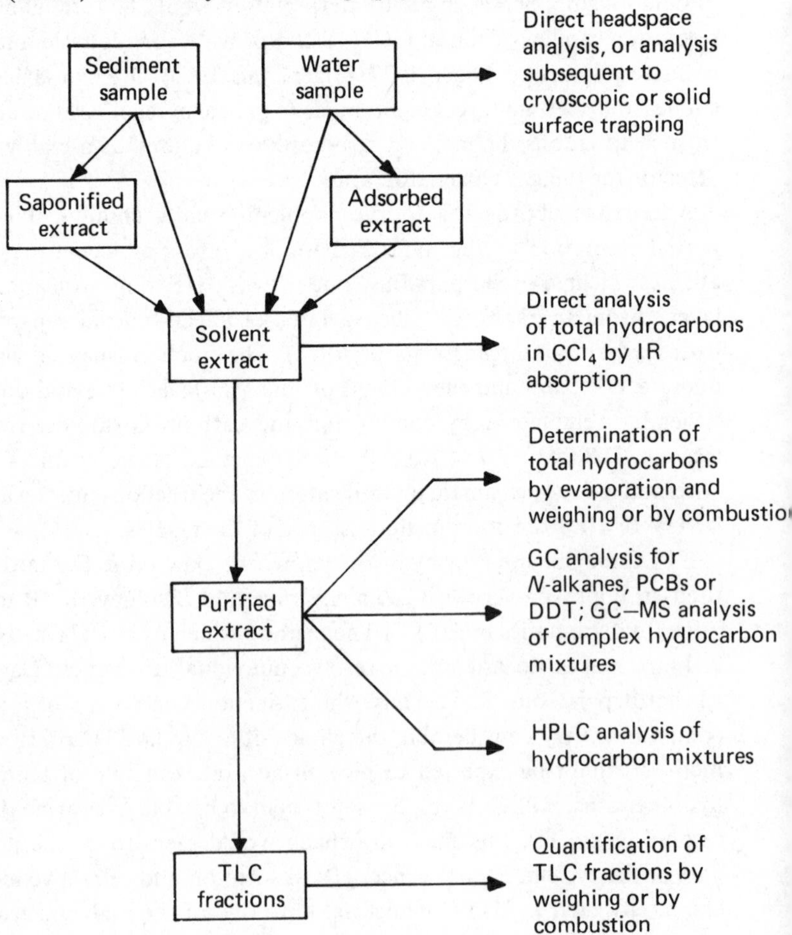

References

Abaychi, J. K. & Riley, J. P. (1979). The determination of phytoplankton pigments by high-performance liquid chromatography. *Analytica Chimica Acta*, **107**, 1–11.

Afghan, B. J., Goulden, P. D. & Ryan, J. F. (1971). An automated method for the determination of soluble nitrogen in natural waters. In *Advances in Automated Analysis*. vol. II, *Industrial Analysis*, pp. 291–7. Miami: Thurman Associates.

American Petroleum Institute (1957). *Manual on Disposal of Refinery Wastes*, vol. IV, *Sampling and Analysis of Waste Water*. Washington, D.C.: American Petroleum Institute.

American Public Health Association (1976). *Standard Methods for the Examination of Water and Waste Water*. 14th edn. Washington, D.C.: American Public Health Association – American Water Works Association – Water Pollution Control Federation.

Armstrong, F. A. J. & Tibbitts, S. (1968). Photochemical combustion of organic matter in sea water, for nitrogen, phosphorus and carbon determination. *Journal of the Marine Biological Association of the United Kingdom*, **48**, 143–52.

Armstrong, F. A. J., Williams, P. M. & Strickland, J. D. H. (1966). Photo-oxidation of organic matter in sea water by ultra-violet radiation, analytical and other applications. *Nature, London*, **211**, 481–3.

Baker, C. D., Bartlett, P. D., Farr, I. S. & Williams, G. I. (1974). Improved methods for the measurement of dissolved and particulate organic carbon in fresh water and their application to chalk streams. *Freshwater Biology*, **4**, 467–81.

Banoub, M. W. (1972). A method for the determination of particulate organic nitrogen in natural waters. *International Journal of Environmental Analytical Chemistry*, **2**, 107–11.

Banoub, M. W. & Williams, P. J. leB. (1972). Measurements of microbial activity and organic material in the western Mediterranean Sea. *Deep-Sea Research*, **19**, 433–43.

Barnes, H. (1959). *Apparatus and Methods of Oceanography*. Part One: *Chemical*. London: George Allen & Unwin.

Blumer, M. (1965). Contamination of a laboratory building by air filters. *Journal of the American Association for Contamination Control*, **4**, 13–15.

Brezonik, P. L. & Lee, G. F. (1966). Preservation of water samples for inorganic nitrogen analyses with mercuric chloride. *Air and Water Pollution*, **10**, 549–53.

Briggs, R., Schofield, J. W. & Gorton, P. A. (1976). Instrumental methods of monitoring organic pollution. *Water Pollution Control*, **75**, 47–57.

Calvert, J. G. & Pitts, J. N. (1966). *Photochemistry*. London: John Wiley & Sons.

Collins, K. J. & Williams, P. J. leB. (1977). An automated photochemical method for the determination of dissolved organic carbon in sea and estuarine waters. *Marine Chemistry*, **5**, 123–41.

Copin-Montegut, C. & Copin-Montegut, G. (1973). Comparison between two processes of determination of particulate organic carbon in sea water. *Marine Chemistry*, **1**, 151–6.

Croll, B. T. (1972). Determination of organic carbon in water. *Chemistry and Industry*, 386.

Dal Pont, G. & Newell, B. (1963). Suspended organic matter in the Tasman Sea. *Australian Journal of Marine and Freshwater Research*, **14**, 155–65.

Dawson, R. & Ehrhardt, M. (1976). Determination of aromatic hydrocarbons in seawater. In *Methods of Seawater Analysis*, ed. K. Grasshoff, pp. 227–34. Weinheim: Verlag Chemie.

Dawson, R. & Liebezeit, G. (1981). The analytical methods for the characterization of organics in sea water. In *Marine Organic Chemistry*, ed. E. K. Duursma & R. Dawson, pp. 445–96. Amsterdam: Elsevier.

D'Elia, C. F., Steudler, P. A. & Corwin, N. (1977). Determination of total nitrogen in aqueous samples using persulfate digestion. *Limnology and Oceanography*, **22**, 760–4.

Department of the Environment (1972). *Analysis of Raw, Potable and Waste Waters*. London: Her Majesty's Stationery Office.

de Souza Lima, H. & Williams, P. J. leB. (1978). Oxygen consumption by the planktonic population of an estuary – Southampton Water. *Estuarine and Marine Science*, **6**, 515–21.

DiSalvo, L. H., Guard, H. E., Hunter, L. & Cobet, A. B. (1973). Hydrocarbons of suspected pollutant origin in aquatic organisms of San Francisco Bay: methods and preliminary results. In *The Microbial Degradation of Oil Pollutants*, ed. D. G. Ahearn & S. D. Meyers. Center for Wetland Resources, Report No. LSU-SG-73-01. Baton Rouge: Louisiana State University.

Duinker, J. C. & Hillebrand, M. Th. J. (1978). Minimizing blank values in chlorinated hydrocarbon analyses. *Journal of Chromatography*, **150**, 195–9.

Duinker, J. C. & Hillebrand, M. Th. J. (1983). Determination of selected organochlorines in sea water. In *Methods of Seawater Analysis*, ed. K. Grasshoff, M. Ehrhardt & K. Kremling, 2nd edn., pp. 290–309. Weinheim: Verlag Chemie.

Duursma, E. K. (1961). Dissolved organic carbon, nitrogen and phosphorus in the sea. *Netherlands Journal of Sea Research*, **1**, 1–147.

Duursma, E. K. & Dawson, R. (eds) (1981). *Marine Organic Chemistry*. Amsterdam: Elsevier.

Eastoe, J. E. (1966). Separation of amino acids. *British Medical Bulletin*, **22**, 174–9.

Ehrhardt, M. (1969). A new method for the automatic measurement of dissolved organic carbon in sea water. *Deep-Sea Research*, **16**, 393–7.

Ehrhardt, M. (1972). Petroleum hydrocarbons in oysters from Galveston Bay. *Environmental Pollution*, **3**, 257–71.

Elliott, J. M. & Davison, W. (1975). Energy equivalents of oxygen consumption in animal energetics. *Oecologia*, **19**, 195–201.

el Wakeel, S. K. & Riley, J. P. (1957). The determination of organic carbon in marine muds. *Journal du Conseil Permanent International pour l'Exploration de la Mer*, **22**, 180–3.

FAO (1971). *Report on the seminar on methods of detection, measurement and monitoring of pollutants in the marine environment. Fisheries Report No.* 99, *Supplement* 1. Rome: FAO.

Foster, P. & Morris, A. W. (1971). The use of ultraviolet absorption measurements for the estimation of organic pollution in inshore sea waters. *Water Research*, **5**, 19–27.

Gershey, R. M., MacKinnon, M. D., Williams, P. J. leB. & Moore, R. M. (1979). Comparison of three oxidation methods used for the analysis of the dissolved organic carbon in seawater. *Marine Chemistry*, **7**, 289–306.

Giese, A. C. (1967). Some methods for study of the biochemical constitution of marine invertebrates. In *Oceanography and Marine Biology, An Annual Review*, ed. H. Barnes, vol. 5, pp. 159–86. London: George Allen & Unwin.

Giger, W. (1977). Inventory of organic gases and volatiles in the marine environment. *Marine Chemistry*, **5**, 429–42.

Giger, W., Reinhard, M., Schaffner, C. & Zürcher, F. (1976). Analysis of organic constituents in water by high-resolution gas chromatography in combination with specific detection and computer assisted mass spectrometry. In *Identification and Analysis of Organic Pollutants in Water*, ed. L. H. Keith, pp. 433–52. Ann Arbor: Ann Arbor Science.

Golterman, H. L., Clymo, R. S. & Ohnstad, M. A. M. (1978). *Methods for Physical and Chemical Analysis of Fresh Waters*, International Biological Programme Handbook No. 8, 2nd edn. Oxford: Blackwell Scientific Publications.

Gordon, D. C. (1969). Examination of methods of particulate organic carbon analysis. *Deep-Sea Research*, **16**, 661–5.

Gordon, D. C. & Sutcliffe, W. H. (1973). A new dry combustion method for the simultaneous determination of total organic carbon and nitrogen in seawater. *Marine Chemistry*, **1**, 231–44.

Gordon, D. C. & Sutcliffe, W. H. (1974). Filtration of seawater using silver filters for particulate nitrogen and carbon analysis. *Limnology and Oceanography*, **19**, 989–93.

Goulden, P. D. & Brooksbank, P. (1975). Automated determinations of dissolved organic carbon in lake water. *Analytical Chemistry*, **47**, 1943–6.

Grasshoff, K. (1966). Uber eine Methode zur automatischen Bestimmung von Gesamtphosphat in Meerwasser durch Aufschluss mit ultraviolettem Licht. *Zeitschrift für Analytische Chemie*, **220**, 89–95.

Grasshoff, K., Ehrhardt, M. & Kremling, K. (eds) (1983). *Methods of Seawater Analysis*, 2nd edn. Weinheim: Verlag Chemie.

Hamilton, R. D. & Holm-Hansen, O. (1967). Adenosine triphosphate content or marine bacteria. *Limnology and Oceanography*, **12**, 319–24.

Henriksen, A. (1970). Determination of total nitrogen, phosphorus and iron in fresh water by photo-oxidation with ultraviolet radiation. *Analyst, London*, **95**, 601–8.

Hirota, J. & Szyper, J. P. (1975). Separation of total particulate carbon into inorganic and organic components. *Limnology and Oceanography*, **20**, 896–900.

HMSO (1983). *Biochemical Oxygen Demand* 1981. Methods for the Examination of Waters and Associated Materials. London: Her Majesty's Stationery Office.

Holden, A. V. & Marsden, K. (1969). Single-stage clean-up of animal tissue extracts for organochlorine residue analysis. *Journal of Chromatography*, **44**, 481–92.

Holm-Hansen, O. (1968). Determination of particulate organic nitrogen. *Limnology and Oceanography*, **13**, 175–8.

Holm-Hansen, O. (1970). ATP levels in algal cells as influenced by environmental conditions. *Plant and Cell Physiology*, **11**, 689–700.

Holm-Hansen, O. & Booth, C. R. (1966). The measurement of adenosine triphosphate in the ocean and its ecological significance. *Limnology and Oceanography*, **11**, 510–19.

Holm-Hansen, O., Coombs, J., Volcani, B. E. & Williams, P. M. (1967). Quantitative micro-determination of lipid carbon in microorganisms. *Analytical Biochemistry*, **19**, 561–8.

Holm-Hansen, O. & Riemann, B. (1978). Chlorophyll *a* determination: improvements in methodology. *Oikos*, **30**, 438–47.

Jones, J. G. (1979). *A Guide to Methods for Estimating Microbiol Numbers and Biomass in Fresh Water*. Freshwater Biological Association, Scientific Publication No. 39. Ambleside: Freshwater Biological Association.

Jones, J. G. & Simon, B. M. (1977). Increased sensitivity in the measurement of ATP in fresh-water samples with a comment on the adverse effect of membrane filtration. *Freshwater Biology*, **7**, 253–60.

Keith, L. H. (ed.) (1976). *Identification and Analysis of Organic Pollutants in Water*. Ann Arbor: Ann Arbor Science.

Kiefer, D. A. (1973). Fluorescence properties of natural phytoplankton populations. *Marine Biology*, **22**, 263–9.

Koroleff, F. (1973). Bestämning av total nitrogen i vatten. In *Interkalibrering av Metoder för Bestämning a Nitrat och Totalnitrogen*, pp. 39–40. Helsinki: Nordforsk Miljövårdssekretariatet.

Levy, E. M. (1971). The presence of petroleum residues off the east coast of Nova Scotia, in the Gulf of St Lawrence, and the St Lawrence River. *Water Research*, **5**, 723–33.

Loftus, M. E. & Carpenter, J. H. (1971). A fluorometric method for determining chlorophyll *a*, *b* and *c*. *Journal of Marine Research*, **29**, 319–38.

Lorenzen, C. J. (1966). A method for the continuous measurement of *in vivo* chlorophyll concentration. *Deep-Sea Research*, **13**, 223–7.

Mackereth, F. J. H., Heron, J. & Talling, J. F. (1978). *Water Analysis: Some Revised Methods for Limnologists*. Freshwater Biological Association, Scientific Publication No. 36. Ambleside: Freshwater Biological Association.

MacKinnon, M. D. (1978). A dry oxidation method for the analysis of the TOC in seawater. *Marine Chemistry*, **7**, 17–37.

McMullen, A. I., Monk, J. F. & Stuart, M. J. (1975). Detection and measurement of pollutants of water surfaces. *American Laboratory*, **7** (2), 87–92.

Madgwick, J. C. (1966). Chromatographic determination of chlorophylls in algal cultures and phytoplankton. *Deep-Sea Research*, **13**, 459–66.

Manny, B. A., Miller, M. C. & Wetzel, R. G. (1971). Ultraviolet combustion of dissolved organic nitrogen compounds in lake waters. *Limnology and Oceanography*, **16**, 71–85.

Mantoura, R. F. C. & Woodward, E. M. S. (1983). Conservative behavior of riverine dissolved organic carbon in the Severn Estuary: chemical and geochemical implications. *Geochimica et Cosmochimica Acta*, **47**, 1293–309.

Marvin, K. T., Proctor, R. R. & Neal, R. A. (1972). Some effects of filtration on the determination of nutrients in fresh and salt water. *Limnology and Oceanology*, **17**, 777–84.

Menzel, D. W. (1966). Bubbling of sea water and the production of organic particles: a re-evaluation. *Deep-Sea Research*, **13**, 963–6.

Menzel, D. W. & Corwin, N. (1965). The measurement of total phosphorus in seawater based on the liberation of organically bound fractions by persulfate oxidation. *Limnology and Oceanography*, **10**, 280–2.

Menzel, D. W. & Vaccaro, R. F. (1964). The measurement of dissolved organic and particulate carbon in seawater. *Limnology and Oceanography*, **9**, 138–42.

Montgomery, H. A. C. & Thom, N. S. (1962). The determination of low concentrations of organic carbon in water. *Analyst*, **87**, 689–97.

Morris, R. J. & Culkin, F. (1975). Environmental organic chemistry of oceans, fjords and anoxic basins. In *Environmental Chemistry*, vol. 1, *Specialist Periodical Report*, pp. 81–108. London: The Chemical Society.

National Academy of Sciences (1975). *Petroleum in the Marine Environment.* Washington, DC: National Academy of Sciences.

Nydahl, F. (1978). On the peroxodisulphate oxidation of total nitrogen in waters to nitrate. *Water Research*, **12**, 1123–30.

Olsen, S. (1967). Recent trends in the determination of orthophosphate in water. In *Chemical Environment in the Aquatic Habitat*. ed. H. L. Golterman & R. S. Clymo, pp. 63–105. Amsterdam: North Holland.

Richards, F. A. with Thompson, T. G. (1952). The estimation and characterization of plankton populations by pigment analyses. II. A spectrophotometric method for the estimation of plankton pigments. *Journal of Marine Research*, **11**, 156–72.

Riley, J. P. & Taylor, D. (1969). The analytical concentration of traces of dissolved organic materials from sea water with Amberlite XAD-1 resin. *Analytica Chimica Acta*, **46**, 307–9.

Riley, J. P. & Wilson, T. R. S. (1965). The use of thin-layer chromatography for the separation and identification of phytoplankton pigments. *Journal of the Marine Biological Association of the United Kingdom*, **45**, 583–91.

Salonen, K. (1979). A versatile method for the rapid and accurate determination of carbon by high temperature combustion. *Limnology and Oceanography*, **24**, 177–83.

Schreurs, W. (1978). An automated colorimetric method for the determination of dissolved organic carbon in sea water by UV destruction. *Hydrobiological Bulletin*, **12**, 137–42.

Sharp, J. H. (1973*a*). Size classes of organic carbon in seawater. *Limnology and Oceanography*, **18**, 441–7.

Sharp, J. H. (1973*b*). Total organic carbon in seawater – comparison of measurements using persulfate oxidation and high temperature combustion. *Marine Chemistry*, **1**, 211–29.

Sharp, J. H. (1975). Gross analyses of organic matter in seawater: why, how and from where. In *Marine Chemistry in the Coastal Environment*, ed. T. M. Church, ACS Symposium Series, No. 18, pp. 682–96. Washington, D.C.: American Chemical Society.

Sheldon, R. W. (1972). Size separation of marine seston by membrane and glass-fiber filters. *Limnology and Oceanography*, **17**, 494–8.

Sheldon, R. W. & Sutcliffe, W. H. (1969). Retention of marine particles by screens and filters. *Limnology and Oceanology*, **14**, 441–4.

Slovacek, R. E. & Hannan, P. J. (1977). *In vivo* fluorescence determinations of phytoplankton chlorophyll *a*. *Limnology and Oceanography*, **22**, 919–25.

Solórzano, L. & Sharp, J. H. (1980*a*). Determination of total dissolved phosphorus and particulate phosphorus in natural waters. *Limnology and Oceanography*, **25**, 754–8.

Solórzano, L. & Sharp, J. H. (1980*b*). Determination of total dissolved nitrogen in natural waters. *Limnology and Oceanography*, **25**, 751–4.

Solórzano, L. & Strickland, J. D. H. (1968). Polyphosphate in seawater. *Limnology and Oceanography*, **13**, 515–8.

Stahl, E. (ed.) (1969). *Thin-Layer Chromatography, A Laboratory Handbook*, 2nd edn. London: George Allen & Unwin.

Stainton, M. P., Capel, M. J. & Armstrong, F. A. J. (1977). *The Chemical Analysis of Fresh Water. Miscellaneous Special Publication* No. 25, 2nd edn. Ottawa: Fisheries and Environment, Canada.

Statham, P. J. (1972). *A Preliminary Survey of Some Aspects of the Geochemistry of Zn and Mn in Southampton Water and Solent Sediments*. M.Sc. dissertation, University of Southampton.

Statham, P. J. & Williams, P. J. leB. (1983). The automatic determination of dissolved organic carbon. In *Methods of Seawater Analysis*, ed. K. Grasshoff, M. Ehrhardt & K. Kremling, 2nd edn., pp. 380–95. Weinheim: Verlag Chemie.

Strickland, J. D. H. (1965). Production of organic matter in the primary stages of the marine food chain. In *Chemical Oceanography*, ed. J. P. Riley & G. Skirrow, vol. 1, pp. 477–610. London: Academic Press.

Strickland, J. D. H. & Parsons, T. R. (1972). A practical handbook of seawater analysis. *Bulletin of the Fisheries Research Board of Canada*, No. 187, 2nd edn. Ottawa: Fisheries Research Board of Canada.

Sutcliffe, W. H., Orr, E. A. & Holm-Hansen, O. (1976). Difficulties with ATP measurement in inshore waters. *Limnology and Oceanography*, **21**, 145–9.

Szekielda, K.-H. (1967). Methods for the determination of particulate and dissolved carbon in aqueous solutions (literature review). In *Chemical Environment in the Aquatic Habitat*, ed. H. L. Golterman & R. S. Clymo, pp. 150–7. Amsterdam: North Holland.

Tett, P., Kelly, M. G. & Hornberger, G. M. (1975). A method for the spectrophotometric measurement of chlorophyll *a* and phaeophytin *a* in benthic microalgae. *Limnology and Oceanography*, **20**, 887–96.

UNESCO (1966). *Determination of Photosynthetic Pigments in Sea Water.* Monographs on oceanographic methodology No. 1. Paris: UNESCO.

Van Hall, C. E., Barth, D. & Stenger, V. A. (1965). Elimination of carbonates from aqueous solutions prior to organic carbon determination. *Analytical Chemistry*, **37**, 769–71.

Van Hall, C. E., Safranko, J. & Stenger, V. A. (1963). Rapid combustion method for the determination of organic substances in aqueous solutions. *Analytical Chemistry*, **35**, 315–19.

Vollenweider, R. A. (ed.) (1974). *A Manual on Methods for Measuring Primary Production in Aquatic Environments.* International Biological Programme Handbook No. 12, 2nd edn. Oxford: Blackwell Scientific Publications.

Wangersky, P. J. (1974). Particulate organic carbon: sampling variability. *Limnology and Oceanography*, **19**, 980–4.

Wangersky, P. J. (1975). Measurement of organic carbon in seawater. In *Analytical Methods in Oceanography*, ed. T. R. P. Gibb, pp. 148–62. Advances in Chemistry Series, No. 147. Washington, D.C.: American Chemical Society.

Wangersky, P. J. & Gordon, D. C. (1965). Particulate carbonate, organic carbon and Mn^{2+} in the open ocean. *Limnology and Oceanography*, **10**, 544–50.

Wangersky, P. J. & Hincks, A. V. (1978). *Shipboard Intercalibration of Filters Used to Measure Particulate Organic Carbon.* Publication 16767. Ottawa: National Research Council of Canada.

Wangersky, P. J. & Zika, R. G. (1977). *The Analysis of Organic Compounds in Sea Water.* Publication 16566. Ottawa: National Research Council of Canada.

Williams, P. J. leB. (1969). The wet oxidation of organic matter in seawater. *Limnology and Oceanography*, **14**, 292–7.

Williams, P. J. leB. (1975). Analytical chemistry of sea water: determination of organic components. In *Chemical Oceanography*, ed. J. P. Riley & G. Skirrow, 2nd edn., vol. 3, pp. 443–77. London: Academic Press.

Williams, P. J. leB. & Askew, C. (1968). A method of measuring the mineralization by micro-organisms of organic compounds in sea water. *Deep-Sea Research*, **15**, 365–75.

Williams, P. J. leB., Berman, T. & Holm-Hansen, O. (1972). Potential sources of error in the measurement of low rates of planktonic photosynthesis and excretion. *Nature New Biology*, **236**, 91–2.

Williams, P. M. (1965). Fatty acids derived from lipids of marine origin. *Journal of the Fisheries Research Board of Canada*, **22**, 1107–22.

Williams, P. M. (1969*a*). The association of copper with dissolved organic matter in seawater. *Limnology and Oceanography*, **14**, 156–8.

Williams, P. M. (1969*b*). The determination of dissolved organic carbon in seawater: a comparison of two methods. *Limnology and Oceanography*, **14**, 297–8.

Wong, M. K. & Williams, P. J. leB. (1980). A study of three extraction methods for hydrocarbons in marine sediments. *Marine Chemistry*, **9**, 183–90.

Yentsch, C. S. (1967). The measurement of chloroplastic pigments – thirty years of progress? In *Chemical Environment in the Aquatic Habitat*, ed. H. L. Golterman & R. S. Clymo, pp. 255–70. Amsterdam: North Holland.

Zsolnay, A. (1973). The relative distribution of non-aromatic hydrocarbons in the Baltic in September 1971. *Marine Chemistry*, **1**, 127–36.

6

Ion-selective electrodes in estuarine analysis

M. WHITFIELD

Estuaries exhibit extreme variability in chemical properties in both space and time. On the passage from the river to the sea, the background ionic strength rises from 1 mmol l^{-1} to 0.73 mol l^{-1} and the pH can range from 6 to 9 with the greatest variability being exhibited at lower salinities. Particle loads may vary from a few mg l^{-1} to several g l^{-1} and mixing processes and chemical reactions can cause the concentrations of some dissolved components to alter by several orders of magnitude. The properties of the solution may shift dramatically from one sample to the next depending on the location, the state of the tide and the river flow rate.

Against such a background of variability, consider the seductive appeal of a technique that boasts the following characteristics:

1. The sensor package gives a direct readout in millivolts which can be selectively related to the concentration of a particular dissolved component.
2. The reading is not influenced by sample colour nor, in saline waters, by particle load.
3. The response is logarithmic in concentration so that the sensitivity is the same over the accessible concentration range (usually 1 μmol l^{-1} to 1 mol l^{-1}) and the possibility of 'off-scale' readings is reduced.
4. The response time of the sensors is rapid, usually of the order of seconds, so that concentration data are available in real time.
5. The instrumentation is cheap and robust and digital recording techniques, even for *in situ* measurements, are well developed.
6. Interchangeable sensors are available for more than two dozen dissolved components: most of which are important in natural or polluted environments.

7. *In situ* instruments with as many as six sensors have been described in the literature and individual sensors have been used in a wide variety of on-stream assemblies including autoanalysers and continuous flow cells.

The purpose of this chapter is to look critically at potentiometric methods in the estuarine context in order to establish which components can presently be measured using ion-selective electrodes (ISEs) and to examine the advantages and disadvantages of the various methods available. A clear definition of the problems presently encountered should enable techniques which *are* reliable to gain wider acceptance and speed up the development of new procedures and sensors.

Principles and use of ion-selective electrodes

Direct potentiometry

If an electrode selective to an ion X is immersed in a solution containing that ion, a potential is developed across the electrode–solution interface that is related to the activity of the ion (a_X) by the Nernst equation

$$\epsilon_X = \epsilon^o{}_X + kT \log_{10} a_X \qquad (6.1)$$

where $\epsilon^o{}_X$ is the standard potential (in millivolts) of the electrode, T is the absolute temperature and ideally, $k = 0.1984/n$, with n equal to the stoichiometric charge on the selected ion (negative for an anion, positive for a cation). The cell response is restricted to about 60 mV for each ten-fold change in the concentration of monovalent ions although the theoretical (or Nernstian) slope factor kT is not necessarily achieved in practical cells. The response for divalent ions is only one-half of this value. As the precision of direct potentiometry is inherently limited by the size of kT, it is rarely possible for analysis in the field to approach a precision better than about ± 5 per cent even for monovalent ions. Greater precision can be obtained by using ISEs to follow the course of a titration and both standard addition of the selected ion and removal of the selected ion by precipitation or complex formation have been used. By using mathematical techniques to locate the endpoint, these procedures are capable of a precision better than 0.1 per cent (see pp. 205–210). The sensitivity of direct potentiometry is restricted by the range of validity of equation 6.1. In unbuffered solutions the lower limit of response is generally 1 μmol l^{-1} (Midgley, 1981) but if the free ion concentration is effectively buffered by a complexing system that contains a high *total* concentration of the selected ion then both anion and cation selective electrodes are capable of working down to very low free ion concentrations indeed (Moody & Thomas, 1981).

Although the activity of the ion is the fundamentally important parameter linking the solution composition to the energetics of chemical processes, the estimation and interpretation of activity values in natural multi-component systems is rather complicated at the present time (Whitfield, 1975*a*, 1979). Consequently most ISE measurements are expressed in terms of the concentration of the ion X (m_X) which is related to the activity by

$$a_X = m_X \gamma_X \tag{6.2}$$

where γ_X is the activity coefficient. Equation 6.1 then becomes

$$\begin{aligned} \epsilon_X &= (\epsilon^o{}_X + kT\log_{10}\gamma_X) + kT\log_{10}m_X \\ &= \epsilon^{o\prime}{}_X + kT\log_{10}m_X \end{aligned} \tag{6.3}$$

The new 'standard' potential $\epsilon^{o\prime}{}_X$ varies with solution composition so that there is usually no simple linear relationship between ϵ_X and m_X.

To measure the potential across the ISE/sample interface it is necessary to establish a complete cell containing the ISE or working electrode (WE) and a reference electrode (RE). In doing this, additional sources of potential are introduced (see Fig. 6.1) so that the overall cell potential E_X is given by

$$E_X = \epsilon_{WE} + \epsilon_X - \epsilon_{RE} \tag{6.4}$$

where ϵ_{WE} is the sum of potentials within the working electrode. ϵ_{RE} is the potential of the reference half-cell and comprises ϵ_R the reference electrode potential and ϵ_J the liquid junction potential. By combining equations 6.3 and 6.4. E_X is related to m_X by the equation

$$E_X = (\epsilon_{WE} + \epsilon^o{}_X - \epsilon_R) + (kT\log_{10}\gamma_X - \epsilon_J) + kT\log_{10}m_X \tag{6.5}$$

To measure m_X it is necessary to ensure that all contributions to E_X except $kT\log_{10}m_X$ remain almost constant in the face of variations in the composition and the temperature of the sample. Electrode calibration should make adequate corrections for any variability that cannot be eliminated by electrode design and should provide an unequivocal relationship between the cell potential and the concentration of the selected ion.

A complicating factor arises because ISEs rarely respond exclusively to a single ionic species. In general, for the case of an interfering ion J, equation 6.1 must be replaced by

$$\epsilon_X = \epsilon^o{}_X + kT\log_{10}(a_X + \Sigma_J K_{XJ} a_J{}^{Z_X/Z_J}) \tag{6.6}$$

where Z_X and Z_J are the absolute values of the charges on X and J respectively. K_{XJ} is the selectivity coefficient of the electrode for ion J relative to ion X. The empirical nature of equation (6.6) and the irreproducible behaviour of many ISEs in the presence of an excess of interfering ions implies that K_{XJ} values can only be used to define the useful range of application of the electrode and not to provide corrections for interference effects when they occur.

Additional problems may occur at low concentrations as the response time of the electrode increases and causes the cell potential to drift for some time after a change in m_X has been experienced. If slow response time is a problem, it is sometimes possible to estimate the equilibrium emf by either:

1. plotting $1/t$ versus 10 exp (E_X) and extrapolating to $1/t = 0$; or
2. plotting t/E_X versus t and equating the reciprocal of the slope of the straight line to E_X at equilibrium.

Fig. 6.1. Sources of potential in the complete cell. The potential of the ISE is the sum of the inner reference potential (ϵ_1), the inner membrane potential (ϵ_2), the asymmetry potential (ϵ_3) and the outer membrane (or ion-selective) potential (ϵ_X, equation 6.1). The potential of the reference half-cell is the sum of the reference electrode potential (ϵ_R) and the liquid junction potential (ϵ_J).

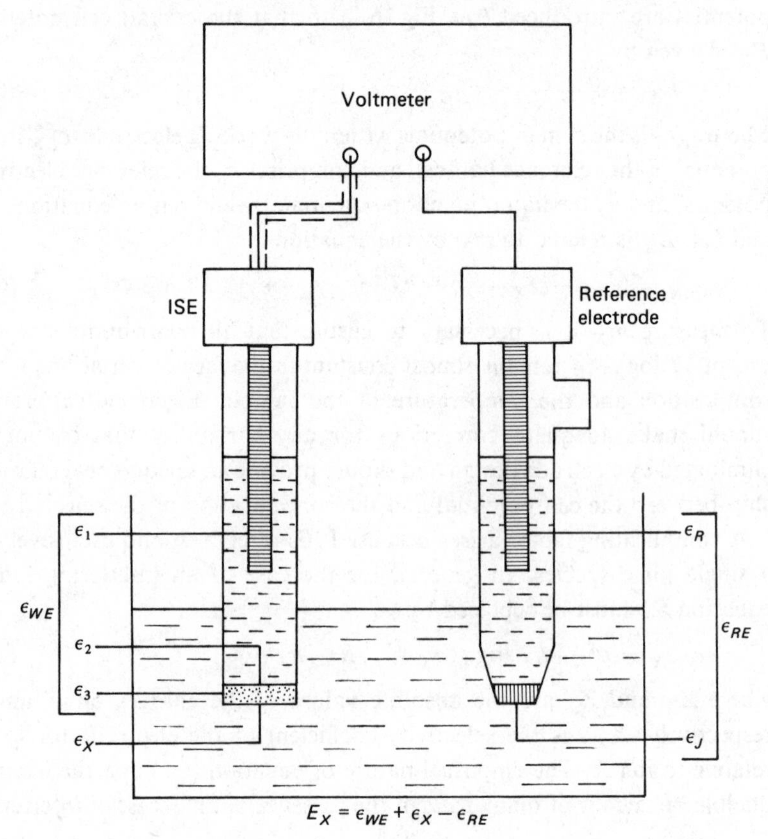

Calibration

Calibration is achieved by transferring the cell containing the working and reference electrodes from the sample to a standard, S, at the same temperature and pressure. The change in potential observed (ΔE) will be given by

$$\Delta E = E_X - E_{X,S} = kT \log_{10}(a_X/a_{X,S}) - \Delta\epsilon_J \tag{6.7}$$

where $\Delta\epsilon_J = \epsilon_J - \epsilon_{J,S}$. Rearranging this equation we find that

$$a_X = a_{X,S} 10 \exp(\Delta E + \Delta\epsilon_J)/kT \tag{6.8}$$

or in concentration terms

$$m_X = m_{X,S}(\gamma_{X,S}/\gamma_X) 10 \exp(\Delta E + \Delta\epsilon_J)/kT \tag{6.9}$$

If the ionic strength and composition of the sample are closely matched then

$$m_X \approx m_{X,S} 10 \exp(\Delta E/kT) \tag{6.10}$$

The change in potential is then directly related to the difference in concentration of the selected ion in the sample and the standard solution. In later sections we will look at the difficulties of realizing this condition in practice. The general problems surrounding the preparation of calibration standards for ISEs have been discussed by Covington (1979).

If the electrodes are stable and a number of standard solutions are available, a calibration curve can be prepared. If the electrode drift is appreciable, one standard can be used to fix the slope of the calibration curve every hour or so. Since drifts in the response slope are usually slow, it is sufficient to calibrate the electrode with two or three standards at the beginning of a day's work and to check its performance in the same way a few times during the day. If the electrode is subjected to rough handling, if samples are laden with particulate matter or if large excesses of interfering ions are encountered, more frequent 'two standard' calibration is advisable. Typically ISE cells exhibit a cumulative drift of about 2 mV per day.

Potentiometric titrations

Much greater precision can be attained, at the expense of greater experimental complexity and longer analysis times, by using ISEs as end-point indicators in potentiometric titrations (Jagner, 1981). In addition, the scope of potentiometry is widened since the electrode can also be used indirectly to monitor the change in concentration of the titrant, or of an added indicator species. The introduction of a titrant also provides an additional variable that can be used to improve the selectivity of the analysis. Titrimetric procedures make less rigorous demands on the performance of the electrodes themselves. The precision of the method is not limited by the Nernstian slope since very large potential changes (usually several hundred millivolts) are generated by altering the concentrations of the selected ion by several orders of magnitude. The accuracy of the endpoint location is not seriously affected by errors of a few tenths of a millivolt in the individual potential readings. The electrode slope (kT) must be determined experimentally since the electrodes in practice do not always exhibit a theoretical slope and uncertainties here can result in systematic errors in the estimation of the endpoint (Jagner, 1981).

In the simplest case, the selected ion X is removed quantitatively by titration with a complexing or precipitating agent N according to the reaction

$$N+X \rightleftharpoons NX$$

where NX might be either an insoluble salt, a metal complex or an electrically neutral compound (Meites, 1963; Wagner & Hull, 1971). At constant solution composition and ionic strength, the stoichiometric stability constant of the reaction may be written as

$$K^* = m_{NX}/m_N m_X \tag{6.11}$$

For precipitation reactions K^* is equivalent to the reciprocal of the solubility product. During the course of the titration the cell potential (E_X) measured in the presence of an electrode perfectly selective to X is given by

$$E_X = E'_X + kT \log_{10} m_X \tag{6.12}$$

where E'_X, which represents the first two terms in equation (6.5), is assumed to remain constant throughout the titration.

Thus $$m_X = 10 \exp [(E_X - E'_X)/kT] \tag{6.13}$$

As the titrant is added, the concentration of X remaining unreacted is given by

$$m_X = (m_o V_o - m_g v_g)/(v_o + v_g) \tag{6.14}$$

where the subscript o indicates the properties of the original solution containing X and the subscript g indicates the properties of the titrant solution containing N. The equivalence volume (v_e) is related to the concentrations of the two solutions by

$$m_o v_o = m_g v_e \tag{6.15}$$

Combining equations (6.13), (6.14) and (6.15) gives

$$(v_o + v_g) . 10 \exp [(E_X - E'_X)/kT] = m_g(v_e - v_g) \tag{6.16}$$

The left-hand side of equation 6.16 contains the experimental parameters characterizing the titration and is known as the first Gran function (Gran,

Fig. 6.2. Simulated titration curves using equations 6.15, 6.16 and 6.17 with the following conditions (see text for notation):

$v_o = 100$ ml; $m_o = 10$ mmol l^{-1}; $E_X = 0$ mV;
$kT = 59.16$ mV/decade; $K^* = 10^{11}$ l mol^{-1}.

It is assumed that $\overline{m}_X \ll K_{XJ}m_J$. The values of $K_{XJ}m_J$ used are

$A = 10^{-4}$, $B = 10^{-6}$, $C = 10^{-8}$, $D = 10^{-10}$.

The dashed curve shows the first derivative of curve D with a sharp maximum at $f = 1.0$, the titration endpoint. As $K_{XJ}m_J$ increases, this curve will become more asymmetrical and the maximum more difficult to locate. The Gran functions are identical for all titration curves. The open circles indicate deviations due to dilution effects if $m_g = 0.1$ mol l^{-1}.

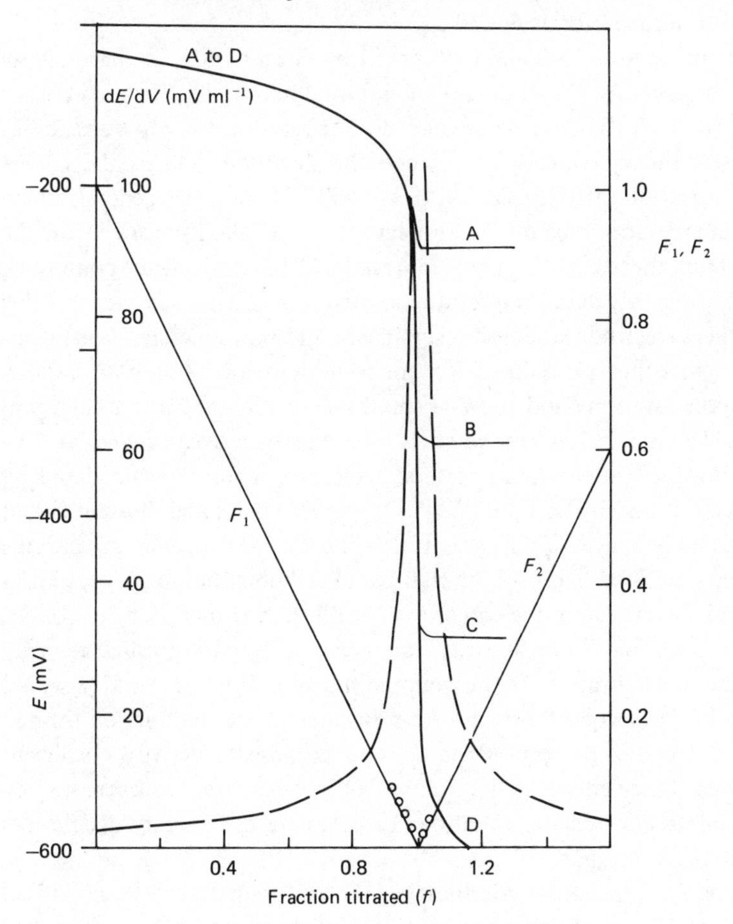

1952; usually designated F_1). When this term is plotted against v_g it extrapolates to zero when $v_g = v_e$ (Fig. 6.2).

A similar derivation can be used to define the second Gran function (F_2) after the endpoint for an electrode selective to the titrant, N. Here we find that

$$(v_o+v_g).10 \exp[(E'_N-E_N)/kT] = m_g(v_g-v_e) \tag{6.17}$$

where the left-hand side of the equation defines F_2. If an insoluble-salt electrode (page 219) is used to monitor the progress of a precipitation titration (e.g. the $AgCl/Ag_2S$ electrode to follow the titration of chloride ions by silver nitrate, Jagner & Årén, 1970), the same electrode can often be used to measure both F_1 and F_2.

These procedures, originally proposed by Gran (1952), are particularly useful for locating the endpoint of potentiometric titrations since they make use of all the data points and do not rely too heavily on readings taken near the equivalence point where the electrode itself will be subject to the greatest interference. Also as ISEs do not, in general, show Nernstian response below total concentrations of 1 μmol l^{-1} for the selected ion, the use of Gran procedures allows the equivalence point to be located using only data from solutions with concentrations in excess of this limit where electrode response is significant and reproducible. When compared with other procedures for endpoint location (Anfält & Jagner, 1971*b*), the Gran method shows a small systematic error and a high precision. The straight line can be fitted by eye or by a least squares fit. The errors involved in the Gran method have been considered in detail by Buffle (1972) and by Buffle *et al.* (1972). Jagner (1981) and Mascini (1980) provide detailed reviews of its practical application. An ingenious apparatus has been described that uses the changes in cell potential to plot the Gran functions directly with due correction for dilution (Johannson, 1974).

Some deviation from linearity can occur if activity coefficients vary throughout the titration. In the example shown in Fig. 6.2, dilution would cause a deviation of 0.5 per cent from linearity when the fraction titrated (f) = 0.99 and 2 per cent when $f = 1.5$ because of activity coefficient variations. These effects, if significant, can be adequately compensated by using simple electrostatic equations to calculate the activity coefficients (McCallum & Midgley, 1973).

Alternative procedures for locating the endpoint (Isbell *et al.*, 1973) include estimation of the maximum slope of the titration curve (Fig. 6.2, first derivative) or titration to a fixed potential (dead-stop titrations). Dead-stop titrations are extremely rapid, since it is not necessary to record the whole titration curve, and they are easily automated. The electrode assembly is calibrated by titrating a standard of known concentration to the equivalence point and noting the potential recorded. Provided the same electrodes are used and there are no significant electrode drifts and changes in solution composition, ionic strength or temperature, this potential can be used to locate the endpoint in subsequent titrations. For estuarine work it will usually be necessary to add a swamping electrolyte (e.g. 3 M $Mg(NO_3)_2$ for chlorinity titrations) to the sample if dead-stop titrations are to be used.

Effects of interfering ions

In practice the response of the electrode near the endpoint and beyond will be restricted by the presence of interfering ions in the solution. For a single interfering ion of the same charge we can write, by analogy with equation 6.6,

$$E_X = E'_X + kT \log_{10}(m_X + K_{XJ} m_J + \bar{m}_X) \tag{6.18}$$

where $\bar{m}_X$ is the concentration limit set by the solubility of the membrane material (Midgley, 1981). This equation has been used with some success to predict the shape of potentiometric titration curves (Whitfield & Leyendekkers, 1969; Whitfield *et al.*, 1969; Van der Meer *et al.*, 1975). The solubility limits of the membrane material are seldom responsible for poor electrode response in potentiometric titrations since the free concentration of the selected ion is effectively buffered by the presence of NX (Carr, 1971). When interference levels are high, the term $K_{XJ} m_J$ effectively sets a lower limit on the electrode response so that the change in potential for each addition of titrant is reduced and the titration curve flattens out immediately after the endpoint is reached (Curve B, Fig. 6.2). Procedures that depend on the determination of the point of maximum slope of the titration curve to locate the endpoint will, therefore, be susceptible to error if electrode interferences are significant. The endpoint estimation by the Gran method is largely unaffected provided that $K_{XJ} m_J$ remains constant throughout the titration. Even where selectivity effects vary, the Gran plot is probably less susceptible to error than other methods (Anfält & Jagner, 1971*b*, 1973).

If the selected ion X is involved in a number of side reactions the Gran function must be modified so that

$$\begin{aligned} F_3 &= (v_o + v_g)\, m_X \\ &= (v_o + v_g)\, m_X{}^F (1 + \Sigma_L \beta_{LX_n} m_L{}^n) \end{aligned} \tag{6.19}$$

Here $m_X{}^F$ is the concentration of X *not* involved in side reactions, m_L is the concentration of the competing ion L, and β_{LX_n} is the stability constant for the side reaction

$$L + nX \rightleftharpoons LX_n \tag{6.20}$$

If the competing ions are present in great excess, F_3 is equivalent to F_1 (equation 6.16). These competing reactions usually cause some curvature in the Gran plot. If the solution chemistry is sufficiently well characterized, it is possible to unravel the various competing side reactions using a generalized computer program (Anfält & Jagner, 1969; Hansson & Jagner, 1973). In other cases it is necessary to accept the curvature and use the coincidence of Gran plots before and after the endpoint to locate v_e.

Standard addition methods

The standard addition method is a special form of potentiometric titration in which the selected ion is added to the sample in a series of known increments rather than being quantitatively removed by a complexing or precipitating reagent. It is particularly useful for the determination of ions in samples with unknown ionic strengths and/or undetermined levels of interfering ions.

For a cell with a potential E_X and a working electrode perfectly selective to X, equation 6.12 applies and a small increment in the concentration of X (Δm_X) will result in small shift in potential (ΔE_X) such that

$$E_X + \Delta E_X = E'_X + kT \log_{10} (m_X + \Delta m_X) \tag{6.21}$$

Assuming that the volume of the increment is negligible and that neither the activity coefficient of X nor the liquid junction potential are affected by the increment of X, equation 6.12 can be subtracted from equation 6.21 to give

$$\Delta E_X = kT \log_{10} [(m_X + \Delta m_X)/m_X] \tag{6.22}$$

which on rearrangement gives

$$m_X = -\Delta m_X/[(1 - 10 \exp (\Delta E_X/kT)] \tag{6.23}$$

Thus m_X can be calculated directly from ΔE_X providing that the response slope of the cell is accurately known. If the electrodes are not behaving ideally the theoretical slope (kT) must be replaced by an empirical electrode slope (S).

If the electrode is sensitive to more than one ion in solution then an equation analogous to equation 6.23 can be derived (Whitfield, 1971*a*) from equation 6.6, viz.

$$m_X = -[\Delta m_X/(1 - 10 \exp (\Delta E_X/kT)] - \Sigma_J (K_{XJ} a_J^{Z_X/Z_J})/\gamma_X \tag{6.24}$$

At moderate salinities (5‰ and more) in estuarine waters, the natural salt content is sufficient to ensure a constant ionic background throughout the measurements. At lower salinities a swamping electrolyte must be added before measuring E_X. The empirical electrode slope (S) must be determined at least once or twice a day. This can be conveniently done with a series of solutions prepared by dilution of a well-characterized sample. After the standard addition has been made and the potential recorded, the solution is diluted, usually to twice its original volume by the addition of distilled water (or the swamping electrolyte where appropriate) and the slope is related to the change in potential observed on dilution ($\Delta E_{X\ \mathrm{dil.}}$) by

$$S = \Delta E_{X\ \mathrm{dil.}} . \log_{10} (v_o/v_f) \tag{6.25}$$

where v_o and v_f are the volumes before and after dilution. If the electrode is working near the lower limit of its concentration range, or there is any reason to believe that its slope is variable, then S can be determined *in each sample* by the dilution method.

Tables of $m_X/\Delta m_X$ can be prepared (Table 6.1) so that the concentration values can be calculated directly from the potential shift observed. The same values can be used to provide a graduated standard addition scale on

Table 6.1. *Values of $m_X/\Delta m_X$ (equation 6.23) for the standard addition method[a] for monovalent ions assuming a theoretical slope at 25 °C[b]*

(mV)	0.0	0.1	0.2	0.3	0.4	0.5	0.6	0.7	0.8	0.9
1.0	25.19	22.86	20.91	19.27	17.85	16.63	15.56	14.62	13.78	13.03
2.0	12.35	11.74	11.18	10.68	10.21	9.78	9.39	9.02	8.68	8.37
3.0	8.073	7.797	7.539	7.296	7.067	6.851	6.648	6.455	6.273	6.100
4.0	5.935	5.779	5.630	5.488	5.353	5.223	5.100	4.891	4.868	4.759
5.0	4.654	4.554	4.457	4.364	4.275	4.189	4.106	4.025	3.948	3.873
6.0	3.801	3.731	3.664	3.598	3.535	3.473	3.414	3.356	3.300	3.246
7.0	3.193	3.141	3.091	3.043	3.996	2.950	2.905	2.861	2.819	2.777
8.0	2.737	2.698	2.659	2.622	2.586	2.550	2.515	2.481	2.448	2.415
9.0	2.386	2.353	2.322	2.292	2.263	2.235	2.207	2.180	2.153	2.127
10.0	2.101	2.076	2.052	2.027	2.004	1.981	1.958	1.936	1.914	1.892
11.0	1.871	1.850	1.830	1.810	1.790	1.771	1.752	1.734	1.715	1.697
12.0	1.680	1.662	1.645	1.628	1.612	1.596	1.580	1.564	1.548	1.533
13.0	1.518	1.503	1.489	1.475	1.460	1.447	1.433	1.419	1.406	1.393
14.0	1.380	1.367	1.355	1.343	1.330	1.319	1.307	1.295	1.284	1.272
15.0	1.261	1.250	1.239	1.228	1.218	1.207	1.197	1.187	1.177	1.167
16.0	1.157	1.148	1.138	1.129	1.119	1.110	1.101	1.092	1.083	1.075
17.0	1.066	1.057	1.049	1.041	1.032	1.024	1.016	1.008	1.001	0.993
18.0	0.985	0.978	0.970	0.963	0.955	0.948	0.941	0.934	0.927	0.920
19.0	0.913	0.906	0.900	0.893	0.887	0.880	0.874	0.867	0.861	0.855
20.0	0.849	0.843	0.837	0.831	0.825	0.819	0.813	0.808	0.802	0.796
21.0	0.791	0.785	0.780	0.774	0.769	0.764	0.759	0.753	0.748	0.743
22.0	0.738	0.733	0.728	0.723	0.719	0.714	0.709	0.704	0.700	0.695
23.0	0.691	0.686	0.682	0.677	0.673	0.668	0.664	0.660	0.656	0.651

a A summary of this table in nomograph form has been prepared by Karlberg (1971).

b For other slopes multiply ΔE_X (observed) by $59.157/S$ (observed) and use the corrected ΔE_X value to estimate $m_X/\Delta m_X$.

a voltmeter to give direct concentration readings, provided that the electrode slope (S) is close to the theoretical value (kT). The standard addition procedure also lends itself to full automation and a number of computer-controlled systems have been described (see for example Jagner, 1981; Mascini, 1980).

The standard addition method offers a number of practical advantages in estuarine waters. The experimental procedure itself is very simple, only one concentrated standard solution is required to cover the whole salinity range, a detailed knowledge of the sample composition is not required and the electrodes remain undisturbed throughout, ensuring stable performance. If the concentration estimated from a second standard addition agrees with the first, this confirms that the assumptions made in deriving equation 6.23 are valid in that particular sample. However, the method depends on the estimation of a relatively small potential shift and is approximately twice

Fig. 6.3. Errors in the determination of monovalent ions resulting from uncertainties in the measured cell potential. - - - direct potentiometry. —— single standard addition potentiometry with the ΔE value shown. Errors for divalent ions will be twice as large (see also Ratzlaff, 1979).

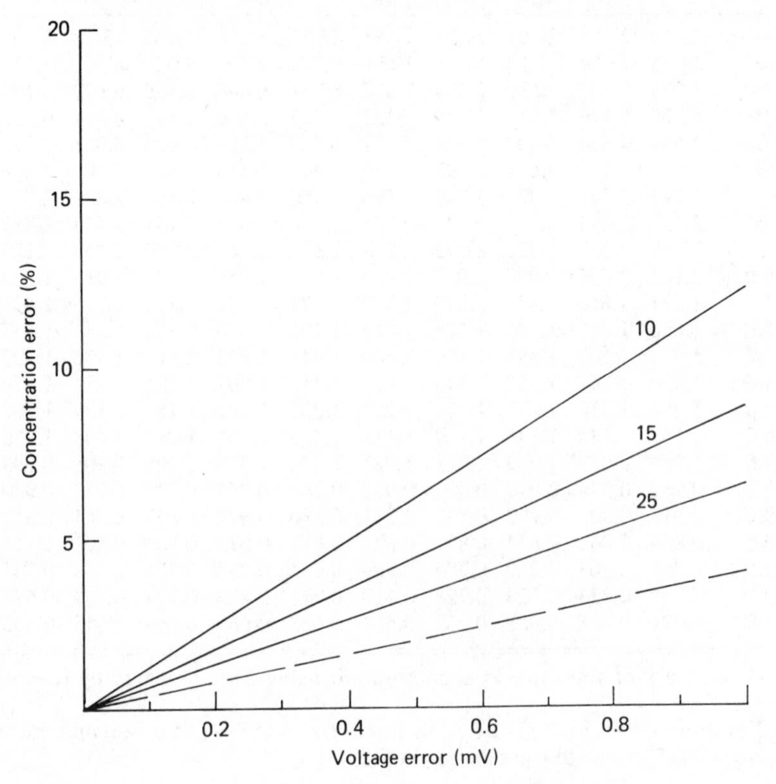

as sensitive as direct potentiometry to errors in the measured potential (Fig. 6.3). In practice, therefore, some compromise must be made between the desirability of adding small solution increments to ensure the validity of equation 6.23 and large increments to provide a large value of ΔE_X (Ratzlaff, 1979). As a general rule of thumb, the volume of solution added should be less than 1 per cent of sample volume and the concentration of the selected ion in the standard solution should be approximately one hundred times greater than in the sample. The potential shift should be not less than 15 mV. The initial potential when the electrodes are placed in the sample can be used to give a rough estimate of m_X which can help to optimise ΔE_X and Δm_X. An excellent review of the scope and limitations of the standard addition method is provided by Mascini (1980).

The sensitivity of the standard addition technique can be increased by using a multiple standard addition titrimetric procedure. With the notation adopted earlier in the discussion of titrimetric techniques, equation 6.21 may be written in the form (see equation 6.14 for notation),

$$\Delta E_X = kT \log_{10} [(m_o v_o + m_S v_g)/(v_o + v_g)] \quad (6.26)$$

where ΔE_X is the potential shift resulting from the addition, to a sample of volume v_o, of a known volume v_g of a standard solution of X, with a concentration of m_S. On rearranging, equation 6.26 gives a Gran function:

$$F_1 = (v_o + v_g)\, 10 \exp \Delta E_X/kT = m_o v_o + m_S \Sigma v_g \quad (6.27)$$

A plot of F_1 versus Σv_g should give a straight line with an intercept on the volume axis (v_e) such that

$$m_o = -m_S v_e/v_o \quad (6.28)$$

(Liberti & Mascini, 1969). Antilogarithmic paper (Gran's plot paper) is commercially available (Orion Inc.) that can be used to assess the intercept volumes from a direct plot of ΔE_X against v_g. The scale graduations are corrected for dilution effects over a limited range. Alternatively, if the volume of the increment is negligible in comparison with the sample volume, equation 6.27 can be rearranged to give

$$1 - 10 \exp \Delta E_X/kT = -m_S v_g/m_o v_o \quad (6.29)$$

A least squares fit (Brand & Rechnitz, 1970) or a graphical plot (Smith & Manahan, 1973) of the left-hand side versus v_g enables m_o to be calculated.

Electrode construction

The reference electrode

The reference electrode should make a stable, reproducible electrical contact with the sample. Ideally the contribution of this electrode to the cell potential (ϵ_{RE}, Fig. 6.1) should be independent of the nature of the sample and should exhibit a rapid response to changes in temperature that is free from hysteresis if the temperature is cycled.

These demands are best met by mounting a stable reference electrode (Fig. 6.4) in a protective glass envelope containing a solution which has a constant concentration of the reference ion and is free from contaminants that might cause the reference electrode potential to drift. Contact is made between the reference half-cell and the sample via a salt bridge which prevents cross-contamination between the reference solution and the sample and minimizes the potential generated at the liquid junctions. The influence of the sample solution on the potential of the reference half-cell is thus transferred from the sensitive surface of the reference electrode itself to the less sensitive, but less well-defined, liquid junction at the interface between the salt bridge solution and the sample.

Fig. 6.4. Reference half-cells. *a*. Schematic. *b*. Practical, double junction. *c*. Practical, single junction.

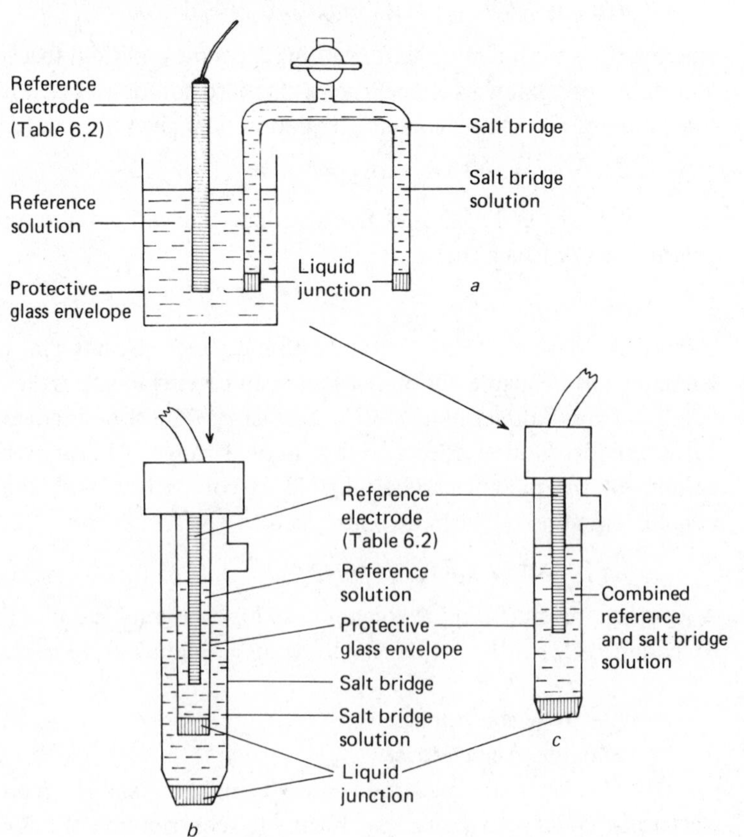

Table 6.2. *Reference electrodes*[a]

Electrode	$-\log_{10} K_S$[b]	Advantages	Disadvantages
Hg/Hg_2Cl_2 (calomel)	17.96	Strain-free liquid metal base. Gives very stable and reproducible results under controlled conditions[c]	Complex solution chemistry contributes to poor thermal characteristics. Sensitive to oxygen, bromide, iodide and impurities in the mercury
Ag/AgCl	9.72	Simple solution chemistry. Thermal hysteresis not as marked as for calomel electrode. Compact structure, easily prepared. Recommended for general use	Potential not as reproducible or as stable as calomel electrode under controlled conditions. Photosensitive[d]. Potential drifts sometimes observed because of ageing of solid metal base. Sensitive to oxygen, bromide, iodide, sulphide. Electrode fragile
Tl(Hg)/TlCl (40 % amalgam)	3.72	Strain-free liquid metal base. Simple solution chemistry. Excellent thermal characteristics. Recommended for use where large temperature variations are expected	TlCl is poisonous and has a relatively high solubility in KCl. The amalgam base is very sensitive to oxygen and the half cell must be sealed. The negative standard potential (Table 6.3) makes it inconvenient to use with most ISEs

[a] Ives & Janz (1961), Covington (1969). All reference solutions must be saturated with the relevant sparingly soluble salt.
[b] K_S = solubility product.
[c] Many commercial calomel reference electrodes exhibit appreciable drift (Covington, 1969).
[d] Electrodes prepared by dipping silver wire into molten silver chloride are not so photosensitive (Ben-Yaakov & Kaplan, 1968).

Table 6.3. *Half-cell potentials*[a] *(mV) of some practical reference electrodes*

Cell[b]	Temperature °C						
	10	15	20	25	30	35	40
Calomel (0.1 M KCl)	336.2	336.2	335.9	335.6	335.1	334.4	333.6
Calomel (3.5 M KCl)	255.6	—	252.0	250.1	248.1	—	243.9
Calomel (saturated KCl)	254.3	251.1	247.9	244.4[c]	241.1	237.6	234.0
Ag/AgCl (3.5 M KCl)	215.2	211.7	208.2	204.6	200.9	197.1	193.3
Ag/AgCl (saturated KCl)	213.8	208.9	204.0	198.9[d]	193.9	188.7	183.5
Tl(Hg)/TlCl (40 % amalgam)	−565.2	−568.7	−572.7	−576.7	−580.6	−584.6	−588.9

[a] ϵ_{RE} Fig. 6.1 including the liquid junction contribution for a sample ionic strength of < 0.1 mol l^{-1} (Bates, 1973).
[b] Combined reference solution and salt bridge solution.
[c] Mattock (1963) gives 244.6 mV at low pH, 243.4 mV at intermediate pH and 242.4 mV at high pH.
[d] Mattock (1963) gives 199 mV at low pH, 198 mV at intermediate pH and 197 mV at high pH.

The most useful reference electrodes (Ives & Janz, 1961; Covington 1969; Covington & Rebello, 1983; Tables 6.2 and 6.3) are prepared by immersing a metal electrode in a solution where the concentration of the metal ion is controlled by the dissolution of the sparingly soluble metal chloride. Of the three types of reference electrodes usually encountered (see Table 6.2), the thallium amalgam/thallium chloride electrode has the best temperature response and shows little hysteresis. However, the silver/silver chloride electrode would seem the best choice for routine work (Corwin & Conti, 1973) because it is more easily prepared and, when used with most ISEs, gives a cell potential which is compatible with conventional pH meters.

A silver/silver chloride electrode suitable for field work (reproducibility ± 0.1 mV) can be prepared by the direct chloridation of analytical grade silver wire. A fairly stout silver wire, about 1 mm in diameter and 2–3 cm long is soldered to an insulated copper wire and sealed into a 4 mm internal diameter glass tube with epoxy-resin or silicone-rubber adhesive so that about 3 cm of the silver wire protrude from the base. The copper wire is sealed into the top of the glass tube with silicone-rubber adhesive. The oxide coating is removed from the silver wire by rubbing with abrasive plastic or fine emery cloth and the electrode is then chloridized in 0.1 mol l^{-1} HCl against a platinum cathode with a current density of 10 mA cm^{-2} for 45 minutes. The electrode is next washed in distilled water for an hour and stored in the reference solution for several days before use. If several electrodes are prepared at the same time they should be short-circuited together during storage.

Working electrodes

Detailed accounts of the construction, performance and theory of the various types of ISE have been discussed in numerous reviews (see for example Koryta 1975; Midgley & Torrance 1978; Moody & Thomas 1971; Bailey 1980; Lakshminarayanaiah 1976; Covington 1979; Freiser 1978). Here we will briefly summarize those characteristics of the commercially available electrodes that are particularly relevant to estuarine studies. The structures of typical ISEs are illustrated in Fig. 6.5 and electrode characteristics are summarized in Table 6.4.

Glass electrodes

Because of the high intrinsic resistivity of glass, these electrodes are formed as fine membranes (usually 50 to 150 μm thick) in the form of a bulb, a cup, a spear or simply as a flat surface depending on the practical application (Portnoy, 1967). The electrode response depends on the

Fig. 6.5. Working electrodes. The lightly stippled zone indicates the ion-selective membrane.
a. Solution connected (all membrane types). The reference electrode is usually chloride-selective and the filling solution contains chloride as well as the selected ion.
b. Mercury connected, most commonly used with glass electrodes.
c. Silver plate contact used to give mechanical strength and electrical screening to a flow-through electrode (Portnoy, 1967).
d. Coated wire electrode, most commonly used with liquid-complexones set in a PVC matrix.
e. Fused salt connection, most commonly used with sparingly soluble salt electrodes.
f. The selectrode where contact is made via a graphite rod.
g. Gas-sensing probe in which a complete cell, consisting of an ISE and an outer reference electrode isolated from the sample by a gas permeable membrane.
Further structural details of ISEs are given in the manufacturer's literature and in Covington (1974, 1979), Durst (1969), Whitfield (1971 *a*), Koryta (1972, 1975), Freiser (1978), Bailey (1980), Lakshminarayanaiah (1976).

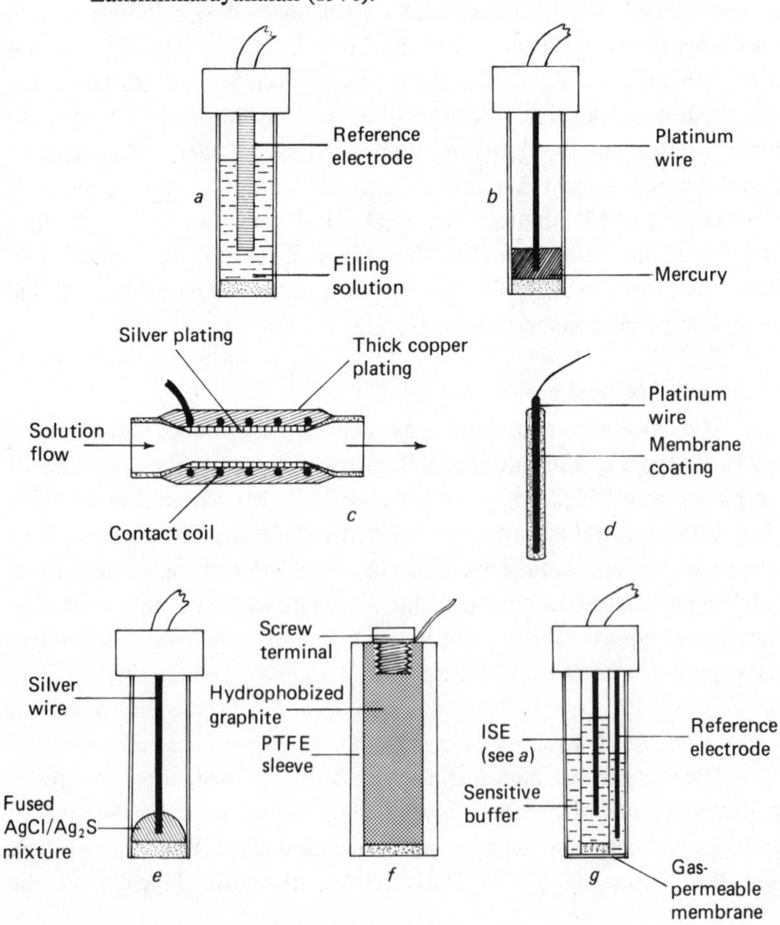

formation and stability of a delicate hydrated layer at the electrode surface. This layer should be established by soaking the electrode overnight in a conditioning solution with an ionic strength and an activity of the selected ion close to that expected in the samples. For example, a pH electrode which has been calibrated and stored in a dilute pH buffer will take five or ten minutes to settle down to a steady potential if it is transferred to sea water (Pytkowicz *et al.*, 1966; Hansson, 1973). There is evidence, however, to suggest that much of this settling time may be attributed to the liquid junction of the reference electrode. The electrode should not be rinsed with distilled water between readings but should be carefully dried with soft tissue (Almgren *et al.*, 1975). The hydrated layer is fragile and the electrode response will become erratic if the membrane is scratched.

The active surface of the glass may also be desensitized by coatings of grease or slime or by algal growths if the electrodes are immersed over long periods. These effects can be reduced without affecting the electrode response by giving the glass electrode a thin silicone coating (Bates, 1973; Eisenmann, 1967, p. 337).

Electrodes fabricated from sparingly soluble salts

Sparingly soluble salts can be used to construct ISEs provided that they exhibit:

1. appreciable ionic conductivity at ambient temperatures;
2. low solubility; and
3. rapid equilibration with the solution phase via precipitation or dissolution.

Only a few salts have been found to meet these requirements and all but one of the commercial electrodes rely on silver sulphide (Ag_2S) to provide ionic conduction and to give the membrane sufficient physical strength. The exception is the fluoride electrode which uses a lanthanum fluoride crystal doped with europium to enhance its ionic conductivity. The electrodes can be prepared as thin slices of single crystals, as sintered discs, as powder uniformally distributed through a supporting matrix of inert polymer (e.g. polyethylene or silicone rubber) or as a thin layer of powder impregnated into the end of a porous, hydrophobic graphite rod. The most commonly used commercial construction is the pressed disc, cemented into a plastic electrode body with epoxy-cement. Some manufacturers seal the membrane into a fluorocarbon body with an O-ring compression seal so that it can be easily exchanged.

Sparingly soluble-salt electrodes offer a number of advantages over glass electrodes:

Table 6.4. *Ion-selective electrodes*[a]

Ion	Electrode identification[b]	Type[c]	Material	pH range[d]	Principal interferences[e]
H^+	B/E2	G	Li_2O, BaO, SiO_2	1–14	Alkaline error +0.03 pH units 0.1 mol l^{-1} NaOH, +0.17 pH units 1 mol l^{-1} NaOH
	C/O15	G	Na_2O, CaO, SiO_2	1–9	Alkaline error +1.0 pH units 0.1 mol l^{-1} NaOH, +2.5 pH units 1 mol l^{-1} NaOH
Na^+	O/94–11	G			K^+ (1×10^{-3}), NH_4^+ (3×10^{-5}), H^+ (100)
	NAS 11–18	G	Na_2O, Al_2O_3, SiO_2		K^+ (4×10^{-4}, pH 11; 3×10^{-3}, pH 7)
K^+	NAS 27–4	G	Na_2O, Al_2O_3, SiO_2		Na^+ (5×10^{-2})
	B/39622	L	Valinomycin	2–12	H^+ (2×10^{-4}), Na^+ (5×10^{-5}), Mg^{2+} (2×10^{-5}), Ca^{2+} (2×10^{-5}), NH_4^+ (0.014)
	C/476132	L	Potassium tetra-(chlorophenyl) borate	1.3–11	Na^+ (0.012), Mg^{2+} (3×10^{-3}), Ca^{2+} (5×10^{-3}), NH_4^+ (0.023)
	P/560–K	L	Valinomycin		H^+ (5×10^{-5}), Na^+ (2.5×10^{-4}), Mg^{2+} (2×10^{-4}), Ca^{2+} (2×10^{-4}), NH_4^+ (0.012)
	O/92–19	L	Valinomycin	1–12	Na^+ (7×10^{-4}), Mg^{2+} (2×10^{-4}) Ca^{2+} (2×10^{-4}), NH_4^+ (0.05)
Ag^+	All makes	IS[f]	Ag_2S	0–14	Hg^{2+} must be absent
Ca^{2+}	B/39608	S	Didecyl or dioctyl phosphonates in collodion	5–11	H^+ (1.5×10^{-4}), Na^+ (0.029), K^+ (0.034), Mg^{2+} (0.34)
	C/476041	L		5–10	Na^+ (0.029), K^+ (0.034), Mg^{2+} (0.34)
	O/92–20	L	Didecyl phosphonate in di-*n*-octyl phosphonate	5.5–11	H^+ (10^{-7}), Na^+ (1.6×10^{-3}), K^+ (10^{-4}), Mg^{2+} (0.014)
	Griffiths[g]	S	Orion exchanger in PVC	5–9	Na^+ (2.8×10^{-3}), K^+ (1.6×10^{-4}), Mg^{2+} (2.4×10^{-2})
	Ammann[h]	S	Synthetic neutral carrier in PVC		H^+ (4.1×10^{-2}), Na^+ (5.7×10^{-3}), K^+ (7.3×10^{-2}), Mg^{2+} (3×10^{-5})
	Růžička[i]	S			H^+ (1.6×10^{-4}), Na^+ (6.3×10^{-6}), K^+ (1.99×10^{-6}), Mg^{2+} (2.51×10^{-4})

Ca^{2+}	B/39614	L		6–11	Na^{+} (0.013), K^{+} (0.13)
Mg^{2+}	O/92–32	L	Didecyl phosphonate in 1-decanol	5.5–11	Na^{+} (0.015), K^{+} (0.15)
Cd^{2+}	O/94–48 P/IS–550–Cd	IS	Ag_2S, CdS	1–14	Ag^{2+}, Hg^{2+}, $Cu^{2+} \leqslant 0.1\ \mu mol\ l^{-1}$. High levels of Pb(II), Fe(III) interfere
Cu^{2+}	O/94–29	IS	Ag_2S, CuS	0–14	Ag^{2+}, Hg^{2+}, $S^{2-} \leqslant 0.1\ \mu mol\ l^{-1}$. High levels of Cl^{-}, Br^{-}, Fe(III), Cd(II) interfere
Pb^{2+}	O/94–82 P/IS–550–Pb	IS	Ag_2S/PbS	2–14	Ag^{2+}, Hg^{2+}, $Cu^{2+} \leqslant 0.1\ \mu mol\ l^{-1}$. High levels of Cd(II), Fe(III) interfere
Br^{-}	B/39062[j]	IS	Ag_2S/AgBr	0–14	Cl^{-} (2.5×10^{-3}), I^{-} (6×10^{-3}), $S^{2-} \leqslant 0.1\ \mu mol\ l^{-1}$
Cl^{-}	C/476131	L		1–12	Br^{-} (2.5), I^{-} (15), NO_3^{-} (2.5)
	O/92–17	L	Dimethyl distearyl ammonium chloride	2–11	Br^{-} (1–6), I^{-} (17), SO_4^{2-} (0.14), NO_3^{-} (5.89), HCO_3^{-} (0.19), F^{-} (0.10)
	O/94–17[k]	IS	Ag_2S, AgCl	0–13	Br^{-} (333), I^{-} (2×10^{6}), $S^{2-} \leqslant 0.1\ \mu mol\ l^{-1}$
	Co/3–802	IS	Ag_2S, AgCl		Br^{-} (204), I^{-} (10^{6}), $S^{2-} \leqslant 1\ \mu mol\ l^{-1}$
CN^{-}	O/94–06	IS	Ag_2S, AgI	3–14	$S^{2-} \leqslant 1\ \mu mol\ l^{-1}$, I^{-} (10), Br^{-} (2×10^{-4}), Cl^{-} (10^{-6})
	P/IS–550–I	IS	Ag_2S, AgI	1–12	$S^{2-} \leqslant 1\ \mu mol\ l^{-1}$, I^{-} (3), Br^{-} (2×10^{-5}), Cl^{-} (2×10^{-4})
F^{-}	All makes	IS	LaF_3, EuF_2	4–8	OH^{-} (0.1)
I^{-}	O/94–53	IS	Ag_2S, AgI	0–14	Cl^{-} (10^{-6}), Br^{-} (2×10^{-4}), $S^{2-} \leqslant 1\ \mu mol\ l^{-1}$
	P/IS–550–I	IS	Ag_2S, AgI		Cl^{-} (6.6×10^{-6}), Br^{-} (6.5×10^{-5}), CN^{-} (0.34), CO_3^{2-} (1.2×10^{-4}), $S^{2-} \leqslant 1\ \mu mol\ l^{-1}$
NO_3^{-}	O/92–07	L	Ni(II) phenanthroline nitrate	2–12	Cl^{-} (6×10^{-3}), Br^{-} (0.1), I^{-} (20), SO_4^{2-} (6×10^{-4}), CO_3^{2-} (6×10^{-3}), HCO_3^{-} (0.02)

Table 6.4. (*Cont.*)

Ion	Electrode identification[b]	Type[c]	Material	pH range[d]	Principal interferences[e]
NO_3^-	C/476134	L	Tri-dodecyl hexadecyl ammonium nitrate	2.5–10	Cl^- (4×10^{-3}), Br^- (0.01), I^- (25), SO_4^{2-} (10^{-3}), HCO_3^- (10^{-3})
	B/39618	S		2–12	Cl^- (0.02), Br^- (0.28), I^- (5.6), SO_4^{2-} (10^{-5}), CO_3^{2-} (1.9×10^{-4})
	Davies[l]	S	Orion exchanger in PVC	2.5–8	Cl^- (4×10^{-3}), I^- (16), SO_4^{2-} (3×10^{-4})
	Davies[l]	S	Corning exchanger in PVC	2.5–8	Cl^- (5×10^{-3}), I^- (17), SO_4^{2-} (10^{-5})
SCN^-	O/94–58	IS	$Ag_2S/AgSCN$	0–14	I^- and $S^{2-} \leqslant 1\ \mu mol\ l^{-1}$, OH^- (1), Br^- (333), Cl^- (0.05), NH_3 (7.7), $S_2O_3^{2-}$ (100), CN^- (143)
CO_3^{2-}	Rechnitz	L			Cl^- (1.8×10^{-4}), SO_4^{2-} (1.49×10^{-4})
S^{2-}	All makes	IS	Ag_2S	0–14	No interferences in natural systems

[a] Data taken from Bates (1973), Covington (1974), Moody & Thomas (1971), Whitfield (1971*a*, 1975*b*), Warner (1972).
[b] Manufacturer/serial no. O, Orion; P, Philips; B, Beckman; C, Corning; Co, Coleman.
[c] G, Glass. Response time ~ 2 ms, resistance ~ 100 MΩ, pre-soaking essential. IS, sparingly soluble salt. Response time ~ 200 ms, resistance ~ 1 MΩ, pre-soaking unnecessary. L, liquid–complexone. Response time ~ 2 s, resistance ~ 30 MΩ, pre-soaking not essential. S, liquid–complexone set in PVC or other inert matrix. Response time variable from 2 s to a few minutes, resistance ~ 50 MΩ, pre-soaking not essential.
[d] Usually for 10 mmol l^{-1} of selected ion. Range becomes narrower as concentration falls. Because of H^+ and OH^- complexing, the useful range for total concentration measurements may be narrower than shown.
[e] Selectivity coefficients (K_{XJ}) values in parentheses from manufacturers' literature. Values for ISEs refer to anion ratios typically found in sea water and estuarine water (Whitfield, 1975*b*).
[f] For tabulations of solubility products see Meites (1963).
[g] Griffiths *et al.* (1972).
[h] Ammann *et al.* (1972).
[i] Růžička *et al.* (1973).
[j] Also for O/94–35, Co/3–801, C/476128.
[k] Also for B/39604, C/476126.
[l] Davies *et al.* (1972).
[m] For data on other electrode types see Covington (1979), Bailey (1980) and Midgley & Torrance (1978).

1. No hydrated surface is involved in the electrode reaction and the membrane surfaces require no pre-soaking before use.
2. Although the electrodes must be given reasonable protection against impacts that might crack the membrane, scratches do not present the difficulties experienced with glass electrodes.
3. The pitting and scarring of the electrodes which occur in use may be removed by cleaning with a fine abrasive.
4. The electrodes are less sensitive to biological growths than are glass electrodes.
5. Their low electrical resistance makes them suitable for use with a wide range of voltage detectors.

All these factors taken together make sparingly soluble-salt electrodes the most useful group of detectors for estuarine work.

Liquid–complexone electrodes

Electrodes with useful sensitivities can be prepared by dissolving a hydrophobic complexone in a water-insoluble solvent with a low dielectric constant to form an ion-exchange liquid which is held in a thin porous membrane between the internal and sample solutions. A positive head of the liquid complexone is required to provide a continually renewable and reproducible electrode surface and thus the life of the electrode is limited usually to little more than a month, by the size of the reservoir. The electrode potential depends on the uniformity of stirring of the solution and the rate of discharge of the liquid complexone such that even under the most carefully controlled conditions the reproducibility of measurements is limited to ± 0.2 mV (Briggs & Lilley, 1974). As with sparingly soluble-salt electrodes no electrode pre-soaking is required, although some conditioning is advised for the most precise work (Rechnitz, 1967). The performance of liquid–complexone electrodes can be improved by immobilizing the exchanger in a collodion, PVC or silicone-rubber matrix (see for example Baum & Lynn, 1973; Moody & Thomas, 1971, 1979; Covington, 1974). The simplest membranes to prepare and the most effective in use are those where the complexone is distributed in a PVC matrix (Fiedler & Růžička, 1973). These membranes are very robust and can be cemented on to PVC tubing to form the electrode.

Gas-sensing probes

The most widely used gas sensing probes are constructed by placing a pH-sensitive glass electrode in contact with a thin film of pH buffer solution (Fig. 6.5*g*). The buffer is trapped against the sensitive surface of the glass electrode by a porous membrane that acts as an air gap

Table 6.5. *Gas sensing probes*[a]

Species sensed	Sensor	Internal electrolyte	Approximate lower limit, mol l^{-1}	Slope	Sample preparation	Interferences	Species measured
CO_2	H^+	0.01 M $NaHCO_3$	10^{-5}	+60	< pH 4		CO_2, HCO_3^-, CO_3^{2-}, H_2CO_3
NH_3[b]	H^+	0.01 M NH_4Cl	10^{-6}	−60	> pH 11	Volatile amines	NH_3, NH_4^+
SO_2	H^+	0.01 M $NaHSO_3$	10^{-6}	+60	HSO_4 buffer	Cl_2, NO_2 must be destroyed (add N_2H_4)	SO_2, H_2SO_3, SO_3^{2-}
NO_2	H^+	0.02 M $NaNO_2$	5×10^{-7}	+60	Citrate buffer	SO_2 must be destroyed (add CrO_4^{2-}), CO_2 interferes	NO_2^-, NO_2
H_2S	S^{2-}	Citrate buffer (pH 5)	10^{-8}	−30	< pH 5	O_2 (ascorbic acid must be added to samples)	S^{2-}, HS^-, H_2S
HCN	Ag^+	$KAg(CN)_2$	10^{-7}	−120	< pH 7	H_2S (add Pb^{2+})	CN^-, HCN
Cl_2	Cl^-	HSO_4 buffer	5×10^{-3}	−60	< pH 2		Cl_2, OCl^-, Cl^-

[a] Adapted from Ross *et al.* (1973).

[b] Ammonia probes with a low electrical resistance and more useful response (100 mV per decade) can be prepared by replacing the pH electrode by a metal selective electrode and using equilibria of the form $M^{n+} + i\,NH_3 \rightleftharpoons M(NH_3)_i^{n+}$ (Anfält *et al.*, 1975).

minimizing the exchange of water molecules between the buffer and the sample solution. When the probe is placed in a solution containing a volatile acid or base (e.g. H_2S, CO_2, NH_3) the volatile solute will pass through the membrane and equilibrate with the buffer solution causing a pH change which is proportional to the concentration of the volatile component. Suitable selectivity can be conferred on the probe by carefully tailoring the properties of the buffer solution and the hydrophobic membrane (Ross *et al.*, 1973; see also Table 6.5). The internal buffer solution should have an ionic strength comparable to that of the sample to prevent undue transport of water molecules across the membrane. Where the ionic strength of the samples varies considerably, a concentrated electrolyte can be added to bring them all to a similar ionic strength to the buffer.

Losses of the volatile component must be minimized by the use of a closed measuring cell or by covering an open one with a piece of paraffin film (Gilbert & Clay, 1973). This is particularly important for carbon dioxide measurements. Electrodes using a hydrophobic membrane tend to be rather sluggish in their response, because of the relatively slow diffusion of gas through the membrane. The performance of the membrane itself is adversely affected by the presence of wetting agents in the sample (not an uncommon occurrence in polluted waters or sewage outfalls) and by the deposition of sludge or inorganic precipitates on the membrane surface.

One possible solution to these problems, the air gap electrode, has been suggested by Růžička & Hansen (1974) where the porous membrane is dispensed with and the electrode, wetted by a smear of the reference buffer solution, is mounted about 1 cm above the surface of the sample. To maintain performance, the thin layer of buffer is regularly renewed by moistening the electrode with a sponge soaked in the buffer.

General factors affecting the performance of ion-selective electrodes

Problems associated with the liquid junction

The liquid junction potential

When a junction is established between two dissimilar solutions, anions and cations with different mobilities will migrate across the junction at different rates until the mutual attraction between the positive and negative charges balances the tendency for the ions to move. The charge separation that is established when a steady state is achieved gives rise to the liquid junction potential (ϵ_J, Fig. 6.1) the size of which depends on the ionic composition of the bridge and sample solutions and the physical structure of the junction (Covington & Rebello, 1983). The equation relating ϵ_J to the mobilities and charges of the individual ions contains two terms ϵ_{JM} which is related to variations in solute concentrations across the liquid junction, and $\epsilon_{J\gamma}$ which describes the consequences of variations in the single ion activity coefficients of the solutes. A working estimate of ϵ_{JM} can be made by treating the junction as a continuous mixture between the bridge solution and the sample (the Henderson equation, MacInnes, 1939; Bates, 1973). For sea-water and estuarine samples with a salinity S this equation reduces to

$$\epsilon_{JM} = -kT[16.63\,S - am)/(77.84\,S - bm)]\log_{10}(77.84\,S/bm) \quad (6.30)$$

where a and b are functions of the charge and mobility of the component ions of the bridge solution (Table 6.6) and m is its concentration (Whitfield & Turner, 1981). This equation provides some guidance to the order of magnitude of ϵ_{JM} and its probable variation with solution composition (Fig. 6.6). The magnitude of $\epsilon_{J\gamma}$ cannot be calculated exactly, even for a simple junction, since the necessary data do not at present exist but calculations using conventional single ion activity coefficients suggest that it might be comparable to ϵ_{JM}.

ϵ_J can be minimized by using a concentrated bridge solution containing anions and cations with similar mobilities (i.e. equi-transferent solutions). ϵ_{JM} is unaffected by changes in the concentration of the bridge solution in the range 3.0 mol l^{-1} to saturation for temperatures from 10 ° to 40 °C if KCl is used (Bates, 1973, p. 312). Most reference electrodes use saturated KCl solutions (4.16 mol l^{-1} at 25 °C) although fewer problems are en-

Table 6.6. *Parameters for the Henderson equation*[a] *(equation 6.30)*

Electrolyte	a	b
KCl	−2.83	149.87
NaCl	−26.24	126.46
$NaNO_3$	−21.31	121.53
KNO_3	2.10	144.94
NH_4Cl	−2.85	149.85
NH_4NO_3	2.08	144.92
RbCl	1.45	154.15
CsCl	0.95	153.65
KNO_3/KCl (equimolar)	−0.37	147.41
HCl	273.47	426.17
Sea water[b]	−16.63	77.84

[a] $a = \Sigma\lambda_+ - \Sigma\lambda_-$; $b = \Sigma\lambda_+ z_+ + \Sigma\lambda_- z_-$ where λ is the limiting ionic conductance of the appropriate anion or cation.
[b] Use salinity rather than molarity in equation 6.30.

countered in the field if more dilute solutions are employed. A 3.5 mol l^{-1} KCl solution is widely used in Europe (see e.g. Manheim, 1961). A solution of 1.8 mol l^{-1} KCl and 1.8 mol l^{-1} KNO_3 has been used by Grove-Rasmussen (1951) for fresh waters since it gives a liquid junction potential which actually decreases as the concentration of the external solution falls. This feature should make it well suited to estuarine work and its performance using estuarine samples is worth further investigation.

Fig. 6.6. Residual liquid junction potentials (ϵ_J) in estuarine waters relative to a salinity of 17.5 ‰ (equation 6.30). Salt bridge solutions:
Curve A: 17.5 ‰ sea water.
Curve B: NaCl, 1 mol l^{-1} or 3.5 mol l^{-1}.
Curve C: Saturated KCl or 3.5 mol l^{-1} NH_4NO_3.
○: 1 mol l^{-1} KCl.
●: 1.8 mol l^{-1} KNO_3, 1.8 mol l^{-1} KCl.
All curves except those for KCl and NH_4NO_3 give very large ϵ_J values at salinities below 1 ‰.

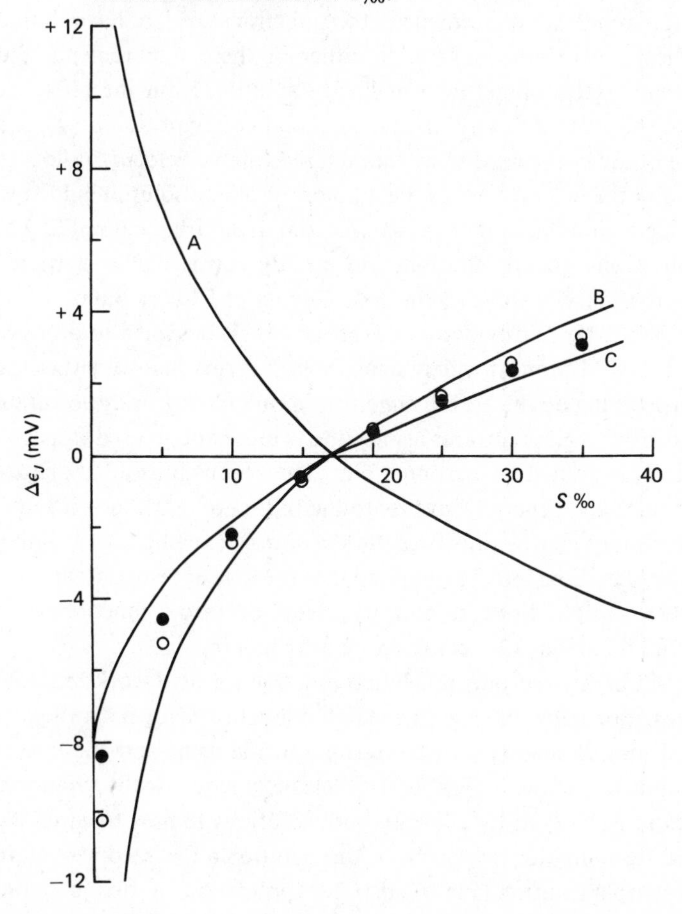

The presence of charged particles, such as colloidal or finely divided clays, are likely to affect ϵ_J by influencing charge separation at the sample/bridge solution interface (Bates, 1973; Brezinski, 1983). The magnitude is greater in low ionic strength solutions but the effect apparently decreases with the ionic strength of the bridge solution (Bower, 1961) suggesting further advantages of 3.5 mol l^{-1} KCl or 1.8 mol l^{-1} KCl and KNO_3 bridge solutions for measurements at the lower salinity end of estuarine systems.

Structure of the liquid junction

The size of ϵ_J is also dependent on the physical structure of the junction (Covington & Rebello, 1983). To achieve a junction potential which does not vary with time, the junction should:

1. be formed with the denser solution (usually the bridge solution) below; and
2. have cylindrical symmetry.

However, for practical convenience, the junctions which have found widest application do not conform to either of these requirements and rely on a steady, streamlined flow of bridge solution from the reference half-cell into the sample to stabilize ϵ_J. The final configuration involves a compromise between the need to maintain a reasonable velocity of flow to stabilize ϵ_J and the need to reduce the amount of bridge solution added to prevent undue contamination of the sample and avoid frequent refilling of the reservoir. Consequently the junction usually consists of a restricted exit port in the glass envelope (Fig. 6.4; Covington, 1974; Bates, 1973; Mattock & Band, 1967; Covington & Rebello, 1983) designed to produce a controlled flow. The most widely encountered forms involve either the use of a porous plug or the establishment of a thin film of bridge solution between two ground glass surfaces using either a closed ungreased stopcock or a glass sleeve mounted on a ground joint. The porous plug junctions are the most popular and generally give reproducibilities of ± 0.2 to ± 0.8 mV. The reproducibility can be improved by the simple expedient of bending the end of the glass tube into a J-shape so that the denser bridge solution is overlain by the sample. For routine work, sleeve or porous junctions with cross-sectional diameters of 1 cm or more are recommended.

A simple but highly reproducible liquid junction for field work has been described by Culberson (1981). Here the liquid junction is formed in a glass T-piece (< 1 mm diameter) with the sample in the arms and the denser bridge solution in the stem (Fig. 6.7). A clean junction with cylindrical symmetry can be achieved by allowing both solutions to flow towards the junction and stopping the flow of the bridge solution a few seconds before that of the sample solution (Fig. 6.7*a*.) The application of this junction,

Fig. 6.7. Liquid junction configuration for high precision work in static *a* and flowing *b* systems (Culberson, 1981; Whitfield and Knox, unpublished work). The system containing the bridge solution must be airtight and free of bubbles.

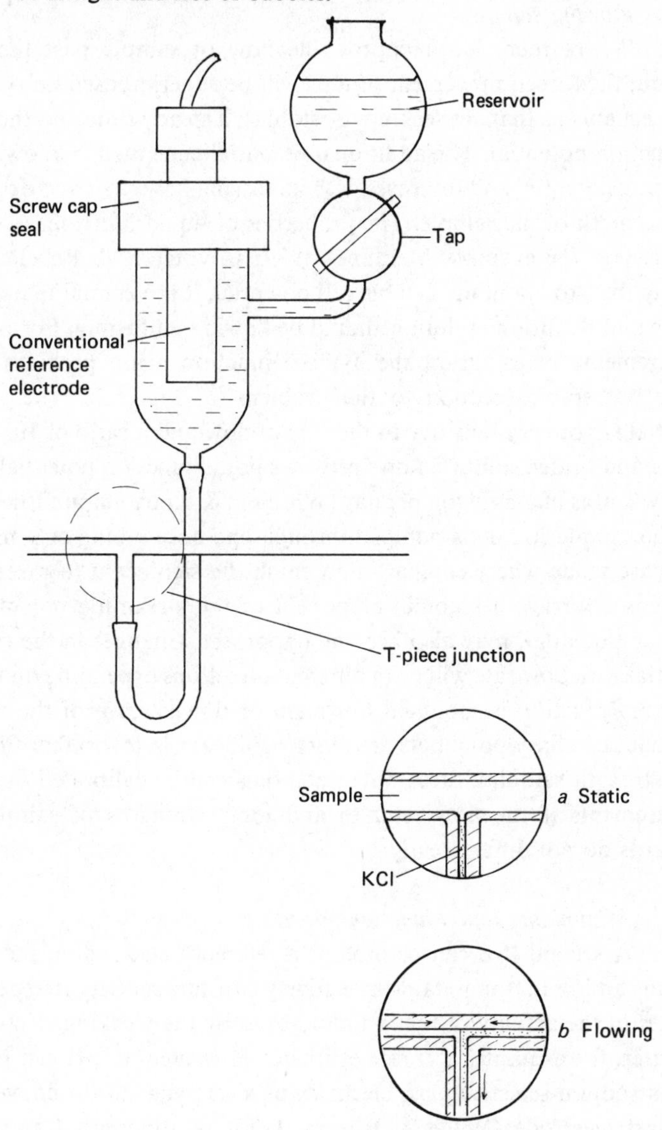

which has a long pedigree (Covington & Rebello, 1983), will be considered when the measurement of pH is discussed.

Flowing samples

Where there is an appreciable flow of sample past the liquid junction, the forced movement of ions will be superimposed on the transport mechanisms that are seeking to establish a steady state and thus affect the junction potential. The additional potential generated is known as the streaming potential and increases with increasing flow rates and decreasing ionic strength of the sample. The properties of liquid junctions in flowing systems (see for example MacInnes 1939; Covington & Rebello, 1983) indicate that, to maintain a stable cell potential, it is essential that *both* the sample and the bridge solution should be flowing uniformly. For practical measurements in estuaries, the T-piece junction again provides a very simple but stable solution to the problem (*b*, Fig. 6.7). The junction potential is not very sensitive to the rate of flow and a ratio of 10:1 in the sample and bridge solution flows provides liquid junction potentials stable to a few tenths of a millivolt per day (Whitfield & Knox, unpublished data).

If the sample stream is pumped through fine-bore tubing or if measurements are made where capillary flow might be significant (e.g. sediments and dense slurries) a second component of the streaming potential, the capillary potential, may also become important. Changes in the capillary potential with flow rate will be in different directions depending on whether the liquid junction is mounted upstream or downstream of the working electrode; capillary potentials are more significant in low rather than high ionic strength samples. Streaming potentials can be calibrated out of the measurements if the flow regimes and ionic strengths of samples and standards do not differ greatly.

Minimizing liquid junction potentials

A second ISE can be used as a reference electrode if the sample contains an ion that maintains a relatively constant activity irrespective of changes in the activity of the ion monitored by the working electrode. In sea water, for example, accurate estimates of *changes* in pH can be made using a sodium-selective glass electrode as a reference to the conventional pH glass electrode (Wilde & Rogers, 1970). A differential amplifier is required to measure the potential developed across a cell composed of two high resistance electrodes and care must be taken to ensure that *both* electrodes are adequately screened (Ben Yaakov & Ruth, 1974; Ben-Yaakov, 1981; Wilde & Rogers, 1970). An essential prerequisite for this procedure is that the activity of the reference ion should be the same in the

Table 6.7. *Multi-functional buffers*

Ion determined	Buffer name	Components: 1. Swamping electrolyte	2. pH buffer	3. Interference[a] eliminator	Reference
Cu^{2+}	Complexing anti-oxidant buffer (CAB)	Component 2	50 mmol l^{-1} HAc 50 mmol l^{-1} NaAc (pH 4.8)	20 mmol l^{-1} NaF (Fe(III) 2 mmol l^{-1} formaldehyde (antioxidant)	Smith & Manahan (1973)
F^-	Total ionic strength adjustment buffer (TISAB)	1 mol l^{-1} NaCl	0.25 mol l^{-1} HAc 0.25 mol l^{-1} NaAc (pH 5)	1 mmol l^{-1} Na citrate (Al(III), Fe(III))	Frant & Ross (1968)
F^-	TISAB IV (estuarine)	1 mol l^{-1} NaCl	0.25 mol l^{-1} HAc 0.25 mol l^{-1} NaAc (pH 5)	5 g l^{-1} DCTA (Al(III), Fe(III))	Warner (1971*a*)
F^-	Optimum fluoride buffer	1 mol l^{-1} NaCl	57 ml l^{-1} anhydrous acetic acid. Adjust to pH 5.4 with 10 M NaOH	4 g l^{-1} DCTA 243 g l^{-1} triammonium citrate (Al(III), Fe(III))	Nicholson & Duff (1981)
NO_3^-	Soil buffer	Components 2 and 3	0.2 mol l^{-1} H_2SO_4 ($pH < 3$) 0.2 mol l^{-1} boric acid	0.1 mol l^{-1} Ag_2SO_4 (Cl^-) 10 mmol l^{-1} $Al_2(SO_4)_3$ (organic acid anions) 20 mmol l^{-1} sulphuric acid (NO_2^-)	Milham *et al.* (1970)
S^{2-}[b]	Standard anti-oxidant buffer (SAOB)	Components 2 and 3	Sodium salicylate	Ascorbic acid (O_2)	Orion Research

[a] The active component is followed, in parenthesis, by the component that it effectively removes.
[b] A modified buffer (SAOB II) is described by Wilson *et al.* (1981). This consists of 35 g ascorbic acid and 67 g of Na_2EDTA in 600 ml of solution. This solution is mixed with 200 ml of 10 M NaOH and made up to 1 litre.

sample and in the standardizing solutions. The two solutions must therefore have comparable ionic strengths and ionic compositions.

If the solution is of variable composition and no convenient reference ion is available it is possible to 'dope' the solution with a swamping electrolyte containing a suitable component (Sekerka & Lechner 1979). For example, Manahan (1970) added fluoride ions to samples being analysed for nitrate and used the lanthanum fluoride electrode as a reference for the nitrate working electrode. A sodium-selective electrode can be used as a reference for the determination of fluoride in fresh waters since a large excess of sodium is added with the total ionic strength adjustment buffer (see Table 6.7). If a pH buffer is added, a pH-selective glass electrode can be used as a reference electrode. This technique is really limited to measurements on samples taken from regions with a naturally low, or effectively constant, background ionic composition. Again it is essential that standardizing solutions should be prepared with the buffer and swamping electrolyte that are used to condition the sample.

In waters with an effectively constant salinity, it is feasible to fill the reference electrode compartment with a background electrolyte which matches closely the composition of the sample solution. Hansson (1973) for example recommends the use of a silver/silver chloride reference electrode mounted in a glass tube containing sea water with a salinity within 5‰ of the sample. In estuarine work this technique is only likely to be useful where salinity changes are small (less than ± 5‰, see Fig. 6.6).

Interference effects

Glass– and liquid–complexone electrodes

A measure of the possible interfering effects encountered when using glass– or liquid–complexone ISEs is given by the selectivity coefficient K_{XJ} (equation 6.6). These coefficients vary with the solution composition and are best determined experimentally by varying the concentration of the selected ion (m_X) in the presence of a constant background level of one interfering ion (m_J). As m_X falls, the cell potential will approach a Nerstian slope so long as $m_X \gg K_{XJ}m_J$ (Fig. 6.8). As the interference becomes significant the curve will level off and eventually become horizontal when $m_X \ll K_{XJ}m_J$. When both linear sections are present they can be extrapolated to meet at the point where $K_{XJ}m_J = m_X$ and K_{XJ} can be determined from the solution composition. If the horizontal portion of the curve is not accessible experimentally the expected intersection point can be calculated since the difference between the extrapolated Nernstian section and the observed response curve is given by

$$\Delta E = kT \log_{10} [(m_X + K_{XJ}m_J)/m_X] \tag{6.31}$$

so that when $\Delta E = kT \log_{10} 2$ we know that $K_{XJ} m_J = m_X$. Where graphical determination of K_{XJ} is impracticable, mathematical techniques such as those of Srinivarsan & Rechnitz (1969) and Baum & Lynn (1973) may be used.

A realistic estimate of the lower useful limit for an electrode is given by

$$m_X = 10.(K_{XJ} m_J) \qquad (6.32)$$

that is one decade (tenfold increase in concentration) higher than the point at which the ions make equal contributions to the cell response. This results in a maximum error of 10 per cent in the measured concentration and prevents the electrode from being used in the region where excessive incursions of the interfering ion into the electrode might cause drift in its response.

When pH interference is being estimated, m_X is held constant and the pH of the solution is adjusted by the addition of strong acid or strong base. Since the solutions are poorly buffered and the influence of pH on ϵ_J is

Fig. 6.8. The response of a nitrate selective electrode to changes in nitrate concentration (m_{NO_3}) in the presence of 5 mmol l^{-1} of chloride (Curve A) and 0.5 mol l^{-1} of chloride (Curve B). The selectivity coefficients ($K_{NO_3,Cl}$) in these solutions are 1.3×10^{-2} and 4×10^{-3} respectively (Moody & Thomas, 1973). The vertical dashed lines refer to Curve A.

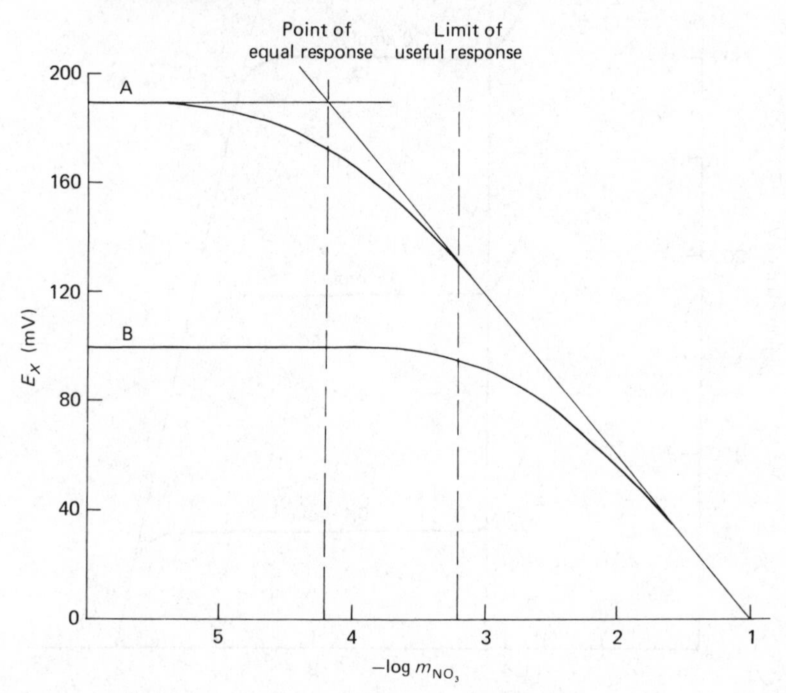

considerable, pH interference effects are seldom straightforward and even less confidence than usual can be placed in interpretations based on equation 6.6. It is preferable simply to perform a pH titration in a medium that approximates as closely as possible to the sample and to select only conditions represented by the horizontal portion of the curve (see Fig. 6.9). The pH ranges given by the electrode manufacturers (Table 6.4) should define the limits of this horizontal section but this is not always clearly stated.

Sparingly soluble-salt electrodes

Interference effects in these electrodes occur via the formation of a layer of less soluble salt on the electrode surface, e.g. for thiocyanate, interference with the bromide electrode:

$$AgBr(s) + SCN^- \rightleftharpoons Br^- + AgSCN(s) \qquad (6.33)$$

Fig. 6.9. pH interference curves for the solid state fluoride electrode at three different fluoride concentrations. The pH was varied by the addition of sodium hydroxide or perchloric acid. The upper pH limit is set by hydroxyl interferences. The lower pH limit is set by the complexation of free fluoride ions to form HF and HF_2^-.

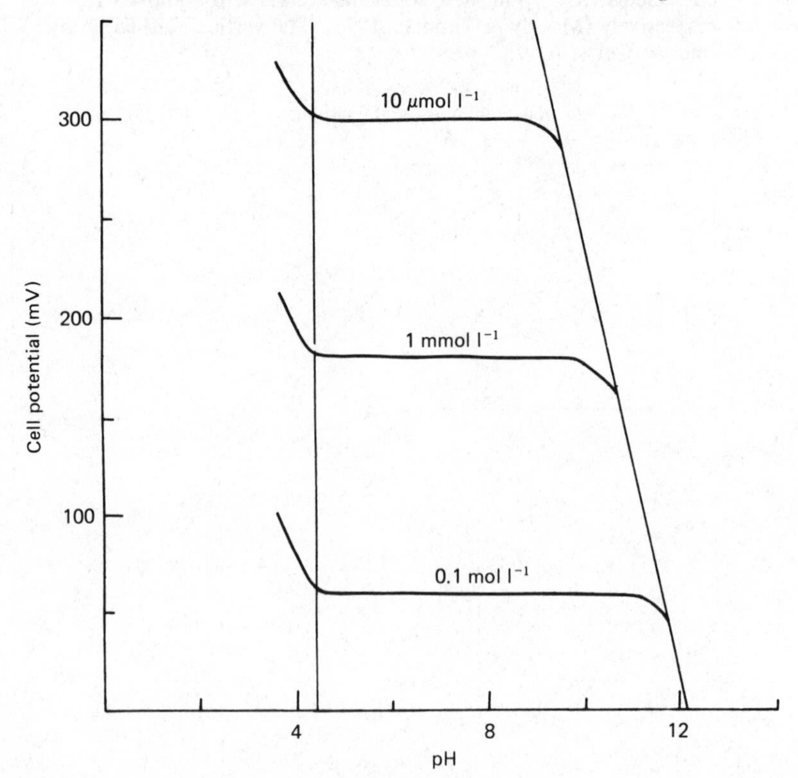

So long as $a_{Br^-}/a_{SCN^-} > K_{s\,(AgBr)}/K_{s\,(AgSCN)}$ the electrode will respond to Br^-. The ratio of the solubility products ($K_{s\,(AgBr)}/K_{s\,(AgSCN)}$) is therefore equivalent to K_{XJ} in equation 6.6 (cf., Morf *et al.*, 1973). The onset of interference is sharply defined (Fig. 6.10) and the electrode cannot be used when the solubility product ratio has been exceeded. The effect is, however, reversible and the interference can be removed by soaking the electrode in a concentrated solution of the selected ion for an hour or so.

On occasion, kinetic problems might slow down the formation of the interfering precipitate so that the electrode functions ideally at concentrations well below the predicted limit (interference of silver with divalent metal ion electrodes). On the other hand, formation of solid solutions between the salts of the selected and interfering ions can cause the interference to be noted much sooner than expected (interference of bromide ions with the silver chloride/silver sulphide electrode). In electrodes selective to divalent metal ions, a second type of interference can arise when conditions favour the formation of an insoluble silver halide at the electrode surface (Ross, 1969). For example, if sufficient chloride is present in the solution then AgCl may be precipitated on the surface of a CdS/Ag_2S membrane via the reaction

$$Ag_2S(s) + Cd^{2+} + 2Cl^- \rightleftharpoons 2AgCl(s) + CdS(s) \qquad (6.34)$$

Fig. 6.10. Interference from thiocyanate with a solid-state bromide electrode. The thiocyanate was added to a background concentration of 1 mol l^{-1} bromide. Adapted from Ross (1969).

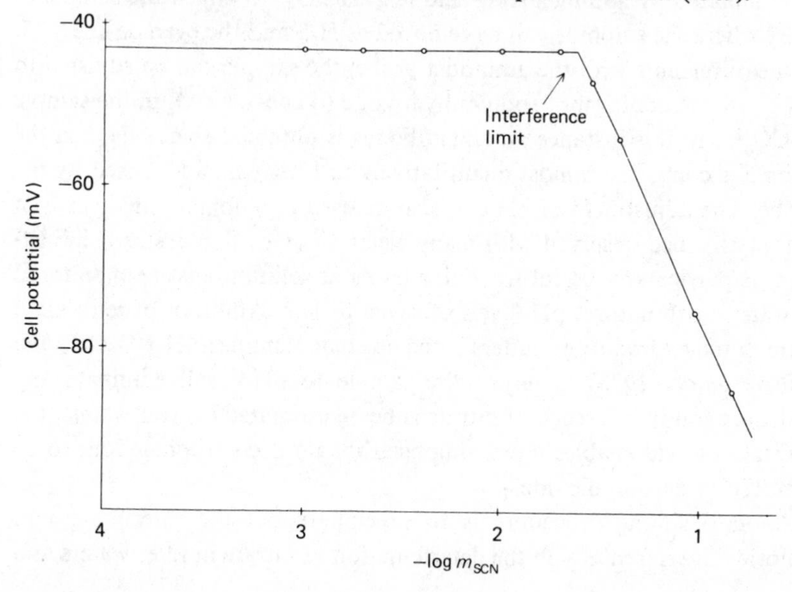

The electrode will continue to function as a cadmium sensor provided that $1/a_{Cd}a_{Cl}^2 > K_{s(Ag_2S)}/K^2_{s(AgCl)}K_{s(CdS)}$. Otherwise the electrode will respond selectively to chloride ions. The principal interferences with the fluoride electrode occur via the formation of lanthanum hydroxide at high pH.

Less well-defined interferences occur when components are present in the solution that form strong complexes with the dominant cation (Ag^+ or La^{3+}). In effect a second mechanism is taking control of the cation activity and bypassing the chain of solubility equilibria that the electrode was designed to accommodate. Interferences of this kind are noted with Ag_2S based electrodes in the presence of thiosulphate or ammonium and with the LaF_3 electrode in the presence of acetate and citrate ions. Such an interference has been put to good use to prepare a cyanide-selective electrode from a Ag_2S/AgI membrane using the reaction,

$$AgI(s)+2CN^- \rightleftharpoons Ag(CN)_2^- + I^-. \quad (6.35)$$

Eliminating electrode interferences

If the preliminary look at the electrode specifications suggests that interference effects might prevent the use of a direct potentiometric method, it is advisable to consider ways of eliminating the interference. This is most simply done by modifying the chemistry of the sample to complex or precipitate the interfering ion.

The simplest chemical adjustment that can be made is to add a strong acid or a strong base to alter the solution pH. For example, the interference of CN^- and S^{2-} ions with the performance of the $Ag_2S/AgCl$ electrode can be eliminated by adding strong acid (e.g. $HClO_4$) to adjust the sample to pH 2 where the sulphide can be removed as H_2S and the cyanide as HCN. In measurements with the ammonia probe, the sample can be adjusted to pH 11 by the addition of sodium hydroxide to convert CO_2 in the sample to CO_3^{2-}. In this instance an added bonus is obtained since NH_4^+ in the sample is converted almost quantitatively to NH_3 which is sensed by the probe. The adjustment of pH can also be used to eliminate the hydrogen ion interference observed with many electrodes (see Fig. 6.9). pH adjustment is particularly useful when low levels of sodium must be monitored in waters with natural pH levels of seven or less. Addition of tetra alkyl ammonium hydroxide or buffers based on ethanolamine ($NH_2CH_2CH_2OH$, Wilson *et al.*, 1975) to adjust the sample to pH 9 will eliminate any hydrogen ion interference. If nitrate is being estimated in river waters, the addition of acid enables interfering bicarbonate and carbonate ions to be removed as carbon dioxide.

Another simple procedure is to precipitate out the interfering ion. Chloride interference with the determination of nitrate in river waters and brackish waters can be removed by precipitation with silver sulphate. Similarly sulphide interferences with the determination of cyanide can be removed by precipitation with lead perchlorate. A modification of this procedure (Reynolds 1971) is to add an *excess* of the precipitating agent to remove the ion of interest completely from the solution. An ISE can then be used to measure the excess of the precipitating agent and hence determine the original concentration of the precipitated ion by difference.

If a suitable strong complexone is available, the interfering ion may be complexed so that its free concentration is reduced to negligible level. In the determination of total fluoride in natural waters, aluminium and ferric ions (which interfere by forming strong fluoride complexes) can be removed by complexing with citrate or with CDTA (1, 2-diaminocyclohexane-N, N, N′, N′ tetra acetic acid). Conversely, ferric ions, which interfere with the determination of copper by the CuS/Ag_2S electrode, can be complexed with fluoride ions to reduce their free concentration (Smith & Manahan, 1973). More drastic techniques can be used to convert the interfering ion to an innocuous form. For example cyanide ions, which may interfere with the analysis of cadmium using the CdS/Ag_2S electrode may be destroyed by the addition of hypochlorite; and oxygen, which interferes with sulphide ions, can itself be removed by reaction with salicylic acid. A number of multi-functional buffers are available (Table 6.7) that combine several of these techniques and also adjust the samples to a common ionic strength.

Temperature effects

Changes in the ambient temperature affect the cell by altering the intercept potential and the slope of the cell response as well as the activity of the selected ion. Differentiating equation 6.12 with respect to temperature we find

$$(\mathrm{d}E_X/\mathrm{d}T) = \underbrace{(\mathrm{d}E''_X/\mathrm{d}T)}_{\text{intercept term}} + \underbrace{k \log_{10} m_X}_{\text{slope term}} + \underbrace{kT\,\mathrm{d}\log_{10}\gamma_X/\mathrm{d}T}_{\text{solution term}} \qquad (6.36)$$

where $E''_X = E'_X - kT \log \gamma_X$ and E'_X includes the liquid junction term. Provided that the electrodes are equilibrated and both the standard and sample solutions are at the same temperature no problems arise if the slope term (equation 6.36) is correctly estimated.

If measurements are being made in the field, it is not always feasible to re-calibrate every time a temperature change is experienced and the errors introduced by the intercept term (equation 6.36) when the sample and standards are at different temperatures must be estimated. For most ISEs the intercept potential is a linear function of temperature over a 20 °C range about room temperature so that the cell equation can be rewritten as

$$E_X = a + bT - kTpX. \qquad (6.37)$$

where $pX = -\log_{10} m_X$. When the cell potential is independent of temperature

$$(\mathrm{d}E_X/\mathrm{d}T) = 0 = b - kpX \tag{6.38}$$

so that an *isopotential pX value* can be defined where $pX_i = b/k$.

Equation 6.37 can now be rewritten to give

$$E_X = a + kT(pX_i - pX) \tag{6.39}$$

Fig. 6.11. Isopotential curves for a solid state fluoride electrode (Table 6.9) used in the isothermal configuration (Fig. 6.12). When the electrode is used in the thermal configuration the curves cross in an isopotential region rather than at a single isopotential point.

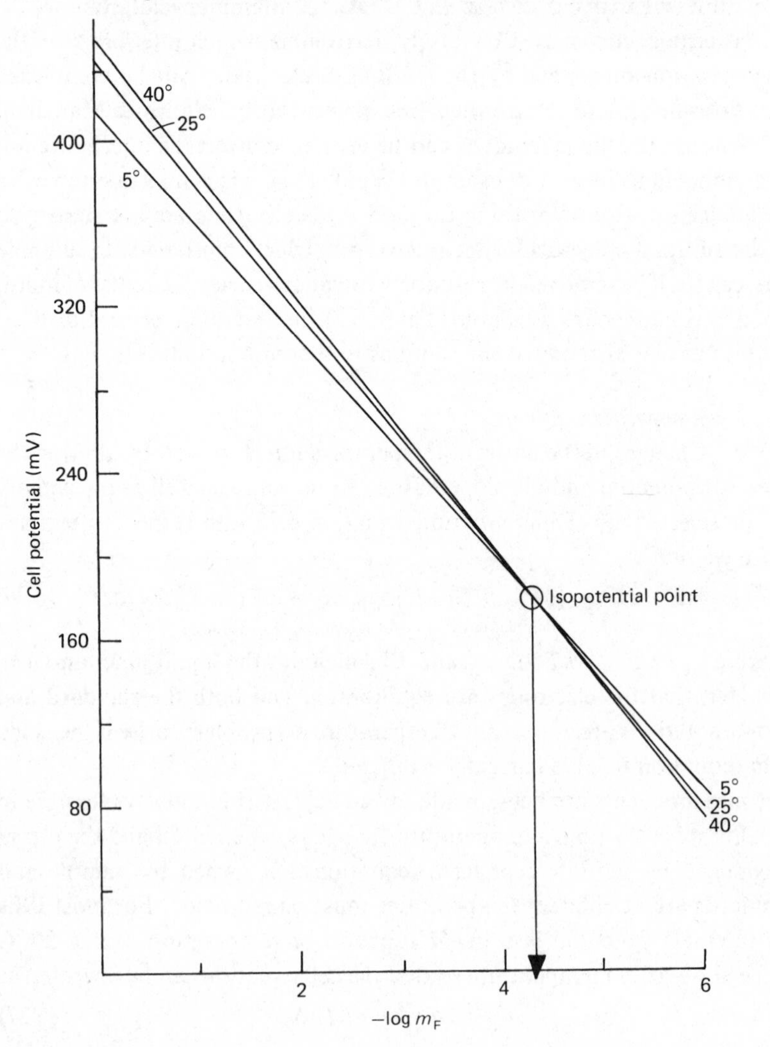

If a series of response curves are measured for a given ISE over a range of temperatures they will all intersect at, or near, pX_i (Fig. 6.11; Table 6.8). By altering the composition of the internal electrolyte in the working electrode, the cell potential can be set at zero at the isopotential point and pX_i can be fixed at some convenient level. For example, pH electrodes are usually designed to give a zero potential at $pH_i = 7.0$ and fluoride electrodes can be prepared to give a zero potential at $pF_i = 1$ ppm (0.53 mmol l^{-1}) for water supply analysis (Negus & Light, 1972; Light, 1969) with appropriate reference electrodes.

Two distinct cell configurations can be used. The thermal cell (A, Fig. 6.12) is most convenient when a series of discrete samples are being measured on deck or when the measurements are made with electrodes in a continuous flow system. The isothermal cell (B, Fig. 6.12) must be employed in submersible instrument packages. The isopotential value of the thermal cell will be independent of the structure of the reference half cell but a new pX_i value must be estimated whenever the reference electrode is changed in the isothermal cell.

If the potential of the cell is measured in a standard solution at a temperature T_1 (E_S) and in the sample at a temperature T_2 (E_X) then

$$pX = (E_S - E_X)/kT_2 + pX_S T_1/T_2 + pX_i(T_2 - T_1)/T_2 \tag{6.40}$$

The error introduced by neglecting the contributions of the intercept term (equation 6.36) to the temperature coefficient is given by the final term.

Table 6.8. *Isopotential pX values for ion selective electrodes with the reference half cell 1 mol l^{-1} KCl|AgCl|Ag*[a]

Electrode	Temperature coefficients (mV/°C)		pX_i (equation 6.39)	
	b	*c*	*b*	*c*
Fluoride	+0.85	+0.61	4.29	3.03
Iodide	−0.35	−0.59	−1.76	−2.98
Sulphide	+0.46	+0.22	4.64	2.22
Cyanide	−0.25	−0.49	−1.26	2.47
Hydrogen	−1.25	−1.48	6.31	7.47
Sodium	−0.63	−0.87	3.18	4.39
Copper(II)	−0.85	−1.09	8.58	11.01
Water hardness	−0.28	−0.52	2.82	5.25

[a] Values calculated by Covington (1974) from data given by Negus & Light (1972). Actual values depend on the internal structure of the working electrode and must be checked experimentally.
[b] Isothermal cell (Fig. 6.12). Subtract 0.03 to give values for a saturated KCl/calomel reference half-cell.
[c] Thermal cell (Fig. 6.12), values independent of reference half-cell.

This correction factor can be applied instrumentally in a similar manner to corrections for the slope factor (Taylor, 1961; Burton, 1979).

Measurement of cell potential

Most membrane electrodes have a high resistance (Table 6.4) so that the cell potential must be measured using a voltmeter with a correspondingly high input impedance (at least 10^3 times higher than the electrode resistance) and a very small input bias current (10^{-12} to 10^{-14} A) (Burton, 1979). Portable, battery-operated instruments with expanded

Fig. 6.12. Cell configurations used in the calibration of ion-selective electrodes over a range of temperature. *a*. Thermal cell. *b*. Isothermal cell. Reproduced with permission from Whitfield (1975*b*).

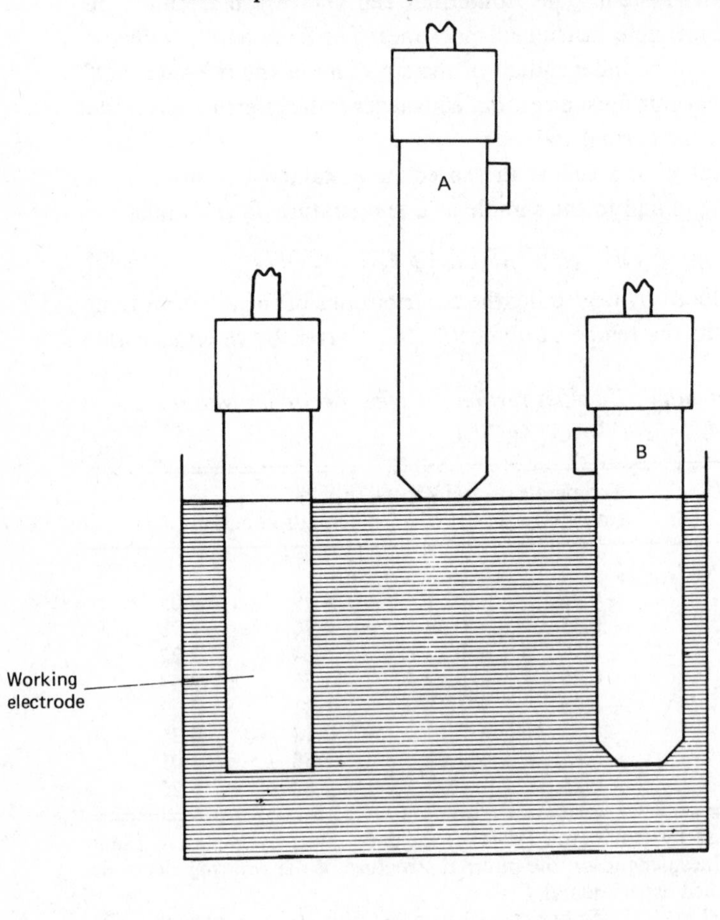

analogue pH scales capable of a resolution of ±1 mV are available but are not well suited for field work. Digital pH meters with liquid crystal displays having a resolution of ±0.1 mV can now be obtained and most manufacturers supply small portable instruments with a resolution of ±1 mV.

Because of the high electrode resistance, the cable connecting the working electrode to the voltmeter must be effectively screened to minimize the pick-up of electrical noise (Wolbarsht, 1964). The cable screening is earthed to the voltmeter casing. For precise measurement the cable should be as short as possible and loops should be avoided. The cable can be secured to the bench to minimize noise generated when it is flexed. Anti-microphonic cables, in which a conductive coating is interposed between the screening and the insulation sheath, are essential for field work. The cable insulation and the insulation of all connectors should have a resistance at least 10^3 times greater than that of the working electrode. Careful attention should be paid to the way in which earth connections are made when the cell is set up (Whitfield, 1971*a*, Wolbarsht, 1964). Earth loops, which cause noisy and drifting readings, are easily set up unless earth connections are all taken, like the branches of a Christmas tree, to a single earth wire. The individual wires should be insulated to prevent stray earthings. It is also important to avoid earthing *both* the reference electrode and the solution since a closed circuit will be established which will result in drifting and unstable readings (Burton, 1979). For *in situ* probes it is preferable to have both the reference and working electrodes isolated from earth, and to earth the electronics to the water via the instrument casing (Ben-Yaakov & Ruth, 1974; Ben-Yaakov, 1981). When measurements are made on board ship it is simpler to allow the reference electrode to be earthed at the voltmeter (this is the usual arrangement in commercial instruments) and to ensure that the solution is *not* earthed by placing the sample beaker on a thin sheet of PTFE or polystyrene, preferably underlain by an earthed metal sheet (Whitfield, 1971*a*).

A summary of common problems encountered in measurements with ISEs and their correction is given in Table 6.9. Useful practical guidelines for measurements with ISEs are also provided by Midgley & Torrance (1978), Bailey (1979), Koryta (1975) and Burton (1979).

In situ analysis

In a system as mobile as an estuary, there are distinct advantages in having an analytical technique that can give a continuous record of variations in solution composition as they occur (Table 6.10). The most direct way of obtaining such a continuous record is to mount the reference

Table 6.9. *Trouble shooting in potentiometric measurements*[a]

Symptom	Possible cause	Check
1. Unstable emf readings	(a) Open circuit or very high resistance in *WE* or *RE*	(a) Are both electrodes in contact with the solution and connected to the meter? Substitute new *RE* or *WE*. An intermittent contact in the cable will give erratic readings when the cable is moved with both electrodes connected to the meter (i) If *WE* faulty replace or reassemble ensuring that all air bubbles are expelled (ii) If *RE* faulty check for an air lock, clogged liquid junction or dried out reference element
	(b) Polarizing current passing through *RE* (usually gives drifting readings)	(b) Spray all connections with water repellent. If *in situ* check that *RE* not earthed at voltmeter. If on deck check that either *RE* or solution is not earthed. Reading will continue to drift after fault located unless *RE* replaced
	(c) Variable corrosion or other potential present	(c) Test as under (b)
	(d) Hand or body movements acting on unscreened electrode (especially in dry weather, relative humidity < 20 %)	(d) Worsened by introduction of statically charged materials (e.g. nylon) to *WE*. Screen entire *WE*, including stem
2. No response to pX change	(a) Solution leak across *WE*	(a) Test as in 1 (a). Membranes may be cracked or inadequately sealed
	(b) Short circuit between *WE* lead and cable screening, earth or *RE*	(b) Resistance between terminal pin and cable screening should be $\nless$ 1000 × *WE* resistance

3. Inadequate response to pX change (i.e. $kT \ll$ theoretical value)	(a) Inadequate insulation between *WE* lead and cable screening, earth or *RE*	(a) As 2(b). If relative humidity high (> 60 %) spray leads and connectors with water repellent
	(b) Deteriorated or coated *WE*	(b) Clean or re-make *WE*
4. Sluggish response	(a) As 3(b)	(a) As 3(b)
	(b) Insufficient pre-soaking before use (glass electrodes only)	(b) Stand in activating solution recommended (usually 0.1 M HCl) for 1–3 days then re-calibrate
	(c) Electrode working near concentration limit	(c) Obtain estimate of ion concentration by standard addition potentiometry
	(d) Electrode working in an excess of interfering ions (especially liquid complexone electrodes)	(d) Check selectivity coefficients supplied by manufacturer against expected composition of sample
5. Impossible to standardize	(a) As 2(a) or 2(b)	(a) As 2(a) or 2(b)
	(b) Shift in inner reference of *WE* or *RE* beyond backing off capabilities of meter	(b) Replace *RE* and *WE* in turn to locate source. Certain electrode combinations (especially with Tl(Hg)/TlCl *RE*) give emf values outside the compensation range of most pH meters. Insert auxiliary back-off potentiometer in *RE* lead if necessary (Whitfield, 1971 *a*)
6. Sudden kicks or violent swings in emf	(a) As 1(a) with intermittent connection	(a) As 1(a)
	(b) External interference (e.g. from outboard motor)	(b) Screen cell with earthed aluminium foil
	(c) Sudden load changes in supply voltage (mains instruments)	(c) Check stability of shipboard power supply

[a] Taken in part from Mattock (1963) and Whitfield (1971 *a*). See also Burton (1979). *RE* = reference electrode, *WE* = working electrode

and working electrodes in a submersible probe (Whitfield, 1975*b*; Ben-Yaakov, 1981; Culberson, 1981). Problems associated with the design of waterproof electrical connections in *in situ* probes have been successfully overcome in oceanographic work where the additional problem of pressure equilibration has also been solved. In the most useful design for vertical profiling (Ben-Yaakov & Kaplan, 1971; Ben-Yaakov & Ruth, 1974; Ben-Yaakov, 1981), conventional pipe fittings are used to mount the electrodes in the base of a watertight case which houses the impedance matching amplifier and the data recording system (a digital tape recorder in this instance). The amplifier employed is provided with a truly differential input so that the reference electrode itself is not earthed. Noise rejection is significantly improved by taking the electronics earth to the solution via the instrument casing (Ben-Yaakov, 1981). The working electrodes are pressure compensated simply by replacing a section of the electrode body by flexible PVC tubing. Using such a system with pH sensors Ben-Yaakov & Kaplan (1968) reported a reproducibility of $\pm$1–2 mV and Warner (private communication) has achieved an order of magnitude improvement

Table 6.10. *The relevance of* in situ *analysis to estuarine work*

Advantages	Disadvantages
1. Avoids sampling errors arising from (a) contamination (b) gas exchange and oxidation (c) patchy distribution	1. Electrode drift, contamination and possible damage necessitate frequent calibration
2. Gives finer resolution of concentration variations in time and space	2. Changes in salinity, composition and temperature complicate the calibration procedure by altering activity coefficients and liquid junction potentials
3. Gives instantaneous picture of concentration variations so that (a) unusually high concentrations can be traced to their source (e.g. pollutant spills) (b) rapid mixing and/or chemical reaction can be followed on the spot	3. Variability in salinity and composition will also affect electrode interferences by altering K_{XJ} and a_X/a_J
4. Can alter data interval to suit the particular problem rather than to match a predetermined sampling pattern	4. Relatively few probes (F^-, Cl^-, Na^+, H^+, S^{2-}, NH_3) are sufficiently specific for routine work
5. Can give remote recording, or be used for vertical or horizontal profiles on board ship	

on this when using a fluoride electrode of similar configuration for oceanic measurements. A towed vehicle incorporating ISEs for pH and pS was used by Conti *et al.* (1971) for work in shallow waters (down to 30 m). In this system pressure equilibration was achieved by means of a small neoprene disc in the electrode. Probes without *in situ* amplification have been described for measuring pH (Manheim, 1961) and pH and pS (Whitfield, 1971*b*). These designs are compact and rugged but have the disadvantage of requiring specially blown pH electrodes. A somewhat simpler device for pH measurement based on commercial electrodes without pressure compensation has been described by Lidén & Ingri (1980). It is well suited for routine estuarine measurements and has a reproducibility of ± 1–3 mV. Conti & Wilde (1972) described a hand-held probe in which the electrodes were constructed by drilling holes into the flat face of a perspex (plexiglass) wedge. Appropriate ion-selective membranes were cemented across each hole and a filling solution added to the resulting chamber. An inner silver/silver chloride reference electrode was used. The wedge, with several sensing 'windows' prepared in this way, can be used by a diver to make *in situ* measurements using a hand-held instrument package consisting of a differential amplifier, a potentiometer and a null indicator.

The wider use of ion-selective electrodes for *in situ* monitoring is therefore no longer dependent on the solution of the technical problems of probe design but on the capabilities of the working electrodes themselves (Table 6.11). The scope of submersible instruments is restricted by the need to calibrate all ion-selective electrodes at least twice a day. In addition, relatively few electrodes (H^+, F^-, Na^+, S^{2-}) have sufficient selectivity to be used for direct measurements in untreated samples under estuarine conditions and the variability of the sample can cause changes in the liquid junction potential and in the interference characteristics of the electrodes. The most likely applications of submersible probes based on direct potentiometry are for profiling (vertical or horizontal) where dramatic changes are likely to take place (e.g. at discharges of sewage, acid wastes or nitrate run off) and where registration of concentration *changes* rather than measurement of accurate absolute values is required. Such measurements can be made using probes towed behind a boat underway.

The possibility of running *in situ* titrations has been considered by Hillbom *et al.* (1983) who produced a submersible titration system for alkalinity measurements based on the 'inverse burette'. In this system (Granéli & Anfält, 1977) the barrel of the burette acts as the reaction vessel and the electrodes are mounted in the burette piston. A known volume of sample is sucked into the burette followed by known increments of acid. The potential is recorded following a suitable equilibration time after each

Table 6.11. *Characteristics of analytical procedures employing ion-selective electrodes*[a]

Procedure	Standard solutions required	Outline	Slope determination	Concentration error[b]	Notes
1. Direct potentiometry	At least two standards (S_1, S_2) with a composition and ionic strength similar to that of the sample	(a) Measure E_X in two standards. Construct a calibration graph relating E_X to m_X or a_X	Determine from cell response in two standards	±0·8 %	Only procedure for measuring a_X and for using direct reading *in situ* probes. Frequent re-calibrations required
		(b) Measure E_X in standard and in sample and use equation 6.10 to calculate m_X or a_X. Occasionally measure E_X in a second standard to check the electrode slope			
2. Standard addition (SA) potentiometry	Single concentrated standard (S_1)	Measure potential change observed when an aliquot of standard is added to the sample (equation 6.23). Check agreement between	Determine as in 1(a) or by dilution of spiked sample (equation 6.25)	±1.8 % (15 mV change)	Very simple procedure for determination of m_X. No complicated standards required. Recommended procedure where sample is

		addition of five or more aliquots. The initial concentration is then determined by Gran's plot (equation 6.16)			experimentally cumbersome. Autotitrator required for convenient analysis
4. Potentiometric titration	Titrant (T_1) to complex or precipitate ion being estimated	Use electrode selective to T_1 or to ion being removed to follow the course of the reaction. Note the volume of T_1 added when the reaction is complete	As in 3	$\pm 0.1\,\%$	As in 3. Widens scope of potentiometric method since ions for which a suitable electrode is not available can be determined indirectly

[a] Interference effects influence *all* procedures adversely and must be eliminated by chemical manipulation.
[b] Error resulting in the determination of monovalent ions if there is an uncertainty of ± 0.2 mV in E_X. The error is twice as large for divalent ions (methods 1 and 2). In field measurements errors are more typically ± 1 mV and the associated concentration errors are correspondingly five times larger.

acid addition. When the titration is completed, the sample is discharged to waste, the burette rinsed and the cycle repeated. The whole sequence of events is controlled from a microcomputer at the surface which also analyses the data using the Gran extrapolation method. The total time required for each analysis is less than five minutes. Such an approach is particularly useful where the sampling procedure itself might introduce a significant error into the analyses.

Continuous flow systems

Many of the problems associated with the use of submersible probes for *in situ* measurements can be overcome by the use of continuous flow systems where water is pumped from a known depth and delivered to an electrode assembly at the surface (Fig. 6.13). Such a system undoubtedly causes more disturbance to the sample than the use of an *in situ* probe but ensures that an equitable and controlled environment can be maintained at the electrode assembly. If the volume of the flow system is sufficiently small, the solution can be chemically conditioned and thermostated before it reaches the electrode. This greatly increases the range of

Fig. 6.13. Schematic picture of a flow-through system. The functions of the various components are described in the text. Flow rates are usually around 10 ml min^{-1}. The treated solution may be segmented with air bubbles before entering the mixing coil. Metering pumps are marketed by, *inter alia*, Technicon Corporation, Basingstoke, Hants and EIL, Chertsey, Surrey (Bailey, 1979, 1980; Hulanicki & Trojanowicz, 1979).

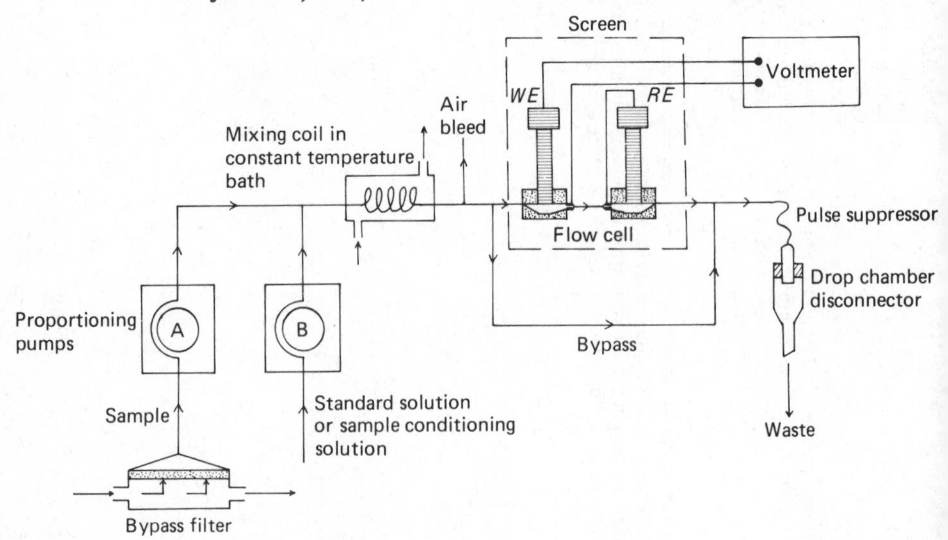

analyses that can be carried out (Růžička & Hansen, 1981; Pecsok *et al.*, 1976; Bauer *et al.*, 1978).

Proportioning pumps can be used to ensure accurately reproducible mixing ratios and to introduce standard solutions. All these operations are readily automated and, by using microprocessors, relatively complex analyses may be performed. Since all samples and standards are treated identically and conditions are strictly controlled, a high precision can be achieved. If the electrodes are thermostated, very stable baselines with noise levels of less than ± 0.1 mV can be attained. Since the flow system is closed it is possible to carry out analyses of samples that are sensitive to gas exchange (e.g. total CO_2 measurements) or to oxidation (e.g. sulphide measurements in anoxic waters). Contamination can be greatly reduced since the sample is taken directly to the sensors and need not be stored for any length of time or transferred from one container to another. A one-point calibration every hour and a two-point calibration twice a day is usually sufficient to attain reproducibilities of a few per cent.

A variety of approaches have been taken to the design of sensors for flow systems. For working electrodes (Fig. 6.5*c* and 6.14, see also Bailey, 1979; Hulanicki & Trojanowicz, 1979; Nagy *et al.*, 1977) the main emphasis has been on minimizing the sample volume required and ensuring that the solution in the electrode chamber is well mixed and that no dead spaces remain. The liquid junction to the reference electrode must be designed so as to minimize streaming-potential artefacts (Fig. 6.7). A continuous flow junction based on a capillary T-piece is the simplest and most reproducible although a number of designs have been produced based on porous plug junctions (see for example Whitfield, 1971*a*; Blaedel & Laessig, 1964). The contribution to the cell potential associated with the movement of the solution through the tubing can be minimized by using a background electrolyte with a relatively high ionic strength ($> 0.2\,mol\,l^{-1}$) and by judicious connection of the reference electrode to the voltmeter (Van den Winkel *et al.*, 1974). The sample must be filtered before it enters the line otherwise the small bore tubes used (2 mm internal diameter maximum) will rapidly clog and the liquid junction potential will become erratic. This is best done with a bypass filter where the sample stream is directed across the surface of the filter rather than passing directly through it (Morris *et al.*, 1978; Fig. 6.13). The resulting washing action prevents the filter from clogging and only a proportion of the sample stream actually passes into the line. The build-up of fouling organisms can be prevented by the occasional injection of biocides into the flow stream. If fouling does become a serious problem, the simplest solution is to remove the electrodes

Fig. 6.14. Working electrode configurations for flow-through systems (see also Figs. 6.5 and 6.7).

a. Conventional electrodes fitted into a perspex or fluorocarbon block via O-ring seals (Sawyer & Foreman, 1969; Webber & Wilson, 1969; Zipper *et al*., 1974).

b. Screw cap fitted on to conventional electrodes to give a small volume solution chamber in contact with the membrane (Van den Winkel *et al*. 1972; Fleet & von Storp, 1971; Thompson & Rechnitz, 1972*b*; Hussein & Guilbault, 1975).

c. Radially drilled crystal membrane with silver wire contact cemented on to outside by fused $AgCl/Ag_2S$ (Thompson & Rechnitz, 1972*a*).

d. Axial version of *c* with holder attached. If the crystal is split and holes are drilled in each half, a 'dual beam' electrode is obtained that is ideally suited for null-point potentiometry. Dahms (1967) has described the preparation of a tubular metal-based silver–silver chloride electrode.

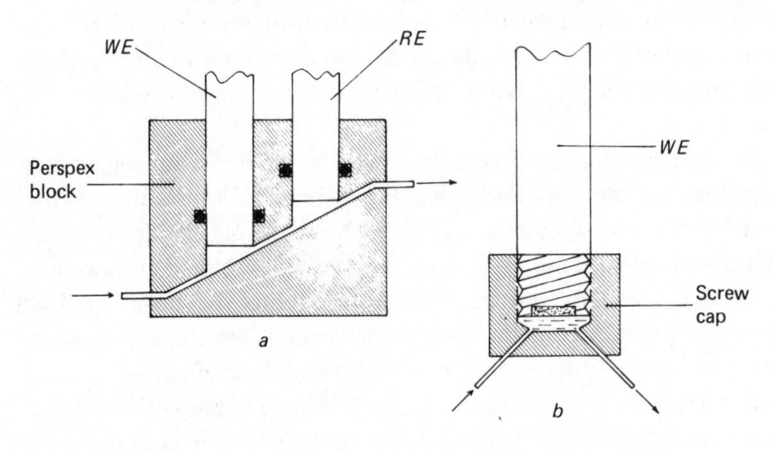

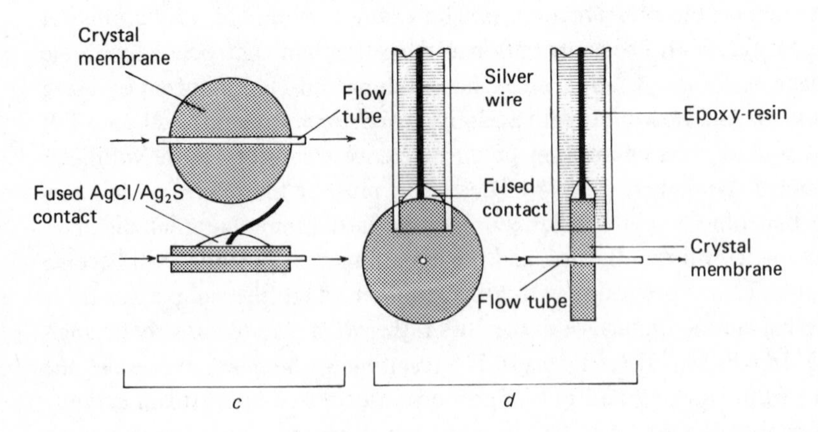

from the assembly and give them a careful wipe with a soft tissue after a brief soaking in strong (1 M) acid.

To minimize electrical interference, silicone rubber tubing should be used throughout. The electrodes can be screened by wrapping them in an earthed layer of aluminium foil. The sample solution should itself be earthed and noise pick-up from the circulation pump can be minimized if the earth point is placed in the tubing between the pump and the electrodes. In some systems (Bauer *et al.*, 1978; Pecsok *et al.*, 1976) electrostatic noise is reduced by introducing bubble segmentation into the flow stream between the pump and the electrodes. The bubbles must be bled off before the sample stream reaches the electrodes otherwise uncontrollably erratic readings will be obtained. The 'AutoAnalyzer' system (Technicon, Basingstoke, Hants) is generally used for constructing the flow line in such systems since it enables samples to be successively aspirated from individual containers into the flow system where they become part of a continuously moving bubble segmented stream. Samples are typically processed at the rate of 20 per hour. Even in the absence of bubble segmentation the flow system should be arranged so that a general 'uphill' slope is maintained between the inlet and the outlet so that bubbles appearing in the system can quickly find their own way out. Problems caused by pulsing from the pump can be reduced by introducing a pulse suppressor (a 25 cm length of 1 mm internal diameter tubing) into the line at the outlet from the electrode chamber and by allowing solution from the suppressor to fall freely into the waste line. The proportioning pumps used to circulate samples in commercial continuous ion monitors (e.g. Bailey, 1979; Hulanicki & Trojanowicz, 1979) are essentially pulse free. Zipper *et al.* (1974) have made a detailed study of noise rejection problems in flow systems and of the difficulties encountered when several ISEs are incorporated into a flow system and their emfs must be scanned in rapid succession.

Standard addition potentiometry can be used in flow systems if a series of standard solutions are fed sequentially into the buffered sample stream at a uniform rate. Fleet & Ho (1973*a*, *b*) describe such an arrangement for the determination of cyanide where the sample is pretreated with sodium hydroxide prior to passing through the mixing coil (Fig. 6.13). Standard additions are metered into the sample solution from a series of standards held in an 'AutoAnalyzer' turntable. The relative flow rates of the sample and the standard determine the increase in the concentration of the selected ion.

Techniques for the potentiometric determination of trace components in flow systems have been reviewed briefly by Braat (1980) and an automated standard addition procedure for the determination of fluoride has been described by Landry *et al.* (1980). The application of computer control to standard addition potentiometry in flow systems has been studied in some detail (Mascini, 1980) and a system capable of monitoring five different sensors simultaneously has been developed (Zipper *et al.*, 1974).

Flow injection analysis (Růžička & Hansen, 1975, 1980, 1981) enables high sampling rates to be achieved. In this technique, a discrete sample is injected into a carrier stream in which the degree of dispersion can be closely controlled by paying careful attention to the hydrodynamics of the flowing solution (Fehér *et al.*, 1978). Under conditions of turbulent flow, dispersion is minized and the samples remain segregated without the need for bubble segmentation. The technique lends itself readily to automation (Slanina *et al.*, 1980). A range of applications employing potentiometric sensors has been described (see for example Růžička & Hansen, 1981). A useful recent example illustrates the potentiometric determination of chloride in waters of low salinity (Trojanowicz & Matuszewski, 1983) at rates in excess of 100 samples per hour.

Titrations for continuous-flow analysis

Titrimetric procedures are certainly precise but they are also tedious because a large number of data points must be collected and analysed for each determination. Computer-controlled titrators (Squirrel, 1964; Jagner, 1970, 1974; Anfält & Jagner, 1971*c*; Frazer *et al.*, 1975; Jagner, 1981) can ease this problem somewhat but full automation is difficult because the cell and the electrodes must be carefully rinsed after each titration and flushed with the new sample to minimize the carry over of titrant. Rather elaborate solution-handling facilities are therefore required to provide continual analysis by titrimetric procedures in conventional cells (Light, 1969). To overcome these difficulties, an ingenious procedure has been developed using a flow cell in an autoanalyser system (Fleet & Ho, 1973*a*, 1974). The change in potential of the cell is followed as a stream of sample is mixed with a stream of titrant whose concentration is increasing linearly with time. If the flow rates are stable and carefully calibrated, the time that elapses between the initiation of the concentration gradient and the measurement of the endpoint break by the electrode is directly proportional to the concentration of the analyte in the sample solution. Using a conventional autoanalyser system, the feasibility of the method was demonstrated by titrating sodium sulphide solution in the range 10–100 μmol l^{-1} with acidified mercuric nitrate solution (0.1 to 0.5 mmol l^{-1} $Hg(NO_3)_2$ in 0.2 mol l^{-1} HNO_3). A coefficient of variation of ± 1.16 per cent was achieved using a drilled silver sulphide membrane as the sensor (Fig. 6.14*c*). Samples were thermostated to 24 °C immediately prior to entering the cell. The sample and gradient flows were presented for four minutes followed by a two-minute distilled water wash. With further refinement, this procedure should enable the simplicity of direct measurement to be combined with the precision of titrimetry. It would be particularly interesting to use this technique for precise alkalinity and chlorinity titrations under estuarine conditions.

Applications to estuarine waters

Developments in the construction and application of ISEs are reviewed biennially (see e.g. Buck, 1978; Fricke, 1980; Meyerhoff & Fraticelli, 1982). A number of more specialized reviews are also available that provide a useful introduction to the application of ISE techniques to water analysis. Detailed reviews of analytical techniques for use in sea water have been provided by Warner (1972), Whitfield (1975*b*), Culberson (1981) and Jagner (1981). Simeonov (1980) provides some useful indications of the problems encountered when commercial ISEs are used in sea water. The analysis of fresh waters using potentiometric techniques has been reviewed by Hulanicki & Trojanowicz (1979) and step-by-step instructions for the determination of a wide range of components are provided in an informative book by Midgley & Torrance (1978). Applications more specifically concerned with waste waters and industrial effluents are considered by Bailey (1979) and in the Proceedings of the Aachen Workshop (1979). The characteristics of analytical procedures employing ISEs are summarized in Table 6.11 and a small selection of applications that have a particular relevance to estuarine studies are summarized in Table 6.12 (in alphabetical order of determinand). In the following discussion a number of applications will be considered briefly to illustrate the range of techniques available. The original references should be consulted for the practical details.

Measurement of pH

pH is a master variable controlling chemical speciation in natural waters and it also provides useful indications of the progress of photosynthesis and respiration in biologically productive areas. To study these effects on a routine basis, pH measurements with a tolerance of ± 0.02 pH units are usually quite adequate. For quantitative studies involving, for example, an assessment of the status of the carbon dioxide system, a tolerance of ± 0.002 pH units or better is required. The variable ionic strength of estuarine waters and the high ionic strength of the sea-water end-member place severe demands on the performance of the glass

Table 6.12. *Representative applications of ion-selective electrodes with relevance to estuarine chemistry*[a]

Component determined	Electrode	Sample	Procedure[b]	Notes	References
Alkalinity	H^+ glass	Sea water	PT	Can be fully automated	Edmond (1970) Almgren *et al.* (1977)
		Sea water and estuarine water	Single addition	Rapid and simple procedure	Culberson *et al.* (1970)
C – total inorganic – total organic	CO_2 probe	Waste waters	DP	0.1 mol l^{-1}	Fiedler *et al.* (1975)
Ca^{2+}, Mg^{2+}	Ag/Hg	Sea water	PT	Indicator titration	Lebel & Poisson (1976)
	Ca^{2+} PVC membrane	Fresh water	Flow injection	200 samples $hour^{-1}$	Hansen *et al.* (1978)
Cd^{2+}	Cd^{2+} insoluble salt	Sea water	DP	In experimental systems enriched with Cd^{2+}	Sunda *et al.* (1978)
Cl^-	AgCl insoluble salt	River water	Flow injection	100 samples $hour^{-1}$	Trojanowicz & Matuszewski (1983)
		Sea water and estuarine water	PT	Fully automated	Jagner & Åren (1970)
		Estuarine water	DP	Continuous analysis on pumped samples	Morris *et al.* (1978)
		Fresh water	Gradient titration	Up to 300 samples $hour^{-1}$	Bound & Fleet (1980)
Cu^{2+}	Cu^{2+} insoluble salt	River water	MSA	Close to electrode limit	Smith & Manahan (1973)
		Sea water	SA		Jasinski *et al.* (1974), Oglesby *et al.* (1977), Zirino *et al.* (1983)

CN^-	CN^- insoluble salt	Waste waters	MSA flow	Autoanalsyer system	Fleet & Ho (1973*a*, *b*)
		Waste waters	DP	Continuous measurement	Galster (1979)
F^-	F^- insoluble salt	Sea water	DP, SA	Shipboard	Warner (1969), Warner *et al.* (1975)
		Sea water	MSA	Fully automated	Anfält & Jagner (1971*a*)
		Estuarine water	SA	20 h conditioning time	Warner (1971*a*)
		Drinking water	Flow	On-site monitor	Webber & Wilson (1969), Collis & Diggens (1969)
		Fresh water	MSA	Fully automated	Phillips & Rix (1981)
H^+	H^+ glass	Estuarine water	DP	*in situ*	Manheim (1961), Whitfield (1971*b*)
		Sea water, lake water	DP	*in situ*	Ben-Yaakov & Kaplan (1968, 1971)
		Sea water	DP	Ionic medium scale NBS scale	Almgren *et al.* (1975), Culberson (1981)
K^+	K^+ liquid complexone	Sea water	MSA	Fully automated	Anfält & Jagner (1973)
Na^+	Na^+ glass	Closed basin waters	DP	Measurements to high salinities	Truesdell *et al.* (1965)
		Fresh water	DP flow	Suited for low Na^+ values	Midgley & Torrance (1978)
NH_4^+	NH_3 gas electrode	Waste waters	DP, SA		Thomas & Booth (1973), Růžička *et al.* (1974)
		Sea water	SA		Gilbert & Clay (1973)
		Saline waters	SA	Down to 0.2 μmol l^{-1}	Garside *et al.* (1978)

Table 6.12 (cont.)

Component determined	Electrode	Sample	Procedure[b]	Notes	References
NO_3^-	NO_3^- liquid complexone	Brackish waters	DP	Ingenious sample pretreatment	Paul & Carlson (1968)
	NO_3^- PVC matrix	Drinking water	DP flow	80 samples hour^{-1}	Tietz *et al.* (1980)
HS^-/S^{2-}	Ag_2S electrode	Natural waters	DP	Using modified SAOB buffer	Wilson *et al.* (1981)
		Saline waters	DP	3 μmol l^{-1} detection limit	Frevert (1980)
		Interstitial water in estuarine sediments	DP	Modified for *in situ* measurement	Berner (1963), Whitfield (1971*b*), Frevert & Galster (1978)
		Interstitial water in estuarine sediments	PT	0.01 μmol l^{-1} sensitivity	Green & Schnitker (1974)
Reduced sulphur species	H^+	Fresh water	PT	Determines polysulphides, thiols, thiolsulphate, sulphite	Boulègue & Popoff (1979)
SO_4^{2-}	Pb^{2+} insoluble salt	Sea water	PT	Tedious sample pretreatment	Mascini (1973)
		Saline water	Flow	Differential flow technique ($\pm 5\%$)	Trojanowicz (1980)
	Ag/Hg	Sea water	PT	$\pm 0.5\%$	Lebel & Belzile (1980)
	Fe^{3+} semi-conducting glass	Sea water	PT	Indirect titration	Jasinski & Trachtenberg (1973)

[a] See also summaries by Riseman (1969), Warner (1972), Koryta (1972), Whitfield (1971*a*, 1975*b*), Hulanicki & Trojanowicz (1979)

[b] DP – direct potentiometry; SA – standard addition; MSA – multiple standard addition; PT – potentiometric titration; Flow – using a flow system rather than a static sample.

electrode/reference electrode couple employed, particularly if the conventional dilute National Bureau of Standards (NBS) buffers are used in the calibration (Bates, 1973). To overcome such problems, and to achieve the necessary accuracy and precision, marine chemists have had to look carefully at the problems of defining pH scales and at the design and function of the electrodes used (Culberson, 1981). In all, three pH scales have been proposed for use in sea water but the one most widely used is the NBS scale. In the application of this scale, the cell potential is measured in a dilute standard pH buffer (subscript *S*) and in the sample (subscript *X*) at the same temperature. The values are combined according to the equation (cf. equation 6.8).

$$\mathrm{pH}_X = -\log_{10} a_{H,X} = \mathrm{pH}_S - \Delta E/kT \qquad (6.41)$$

If a pH meter is used, two buffers might be employed to check the electrode slope (kT). This simple relationship is only correct if the buffer and the sample are matched sufficiently closely in ionic strength and composition for the residual liquid junction potential ($\Delta\epsilon_J$, equation 6.8) to be neglected. Since the ionic strength of the NBS standard buffers is less than 0.1, this assumption is only likely to be correct at salinities of 5‰ or less. Measurements made in sea water indicate that variations in liquid junction potentials associated with different designs of commercial reference electrode can give rise to systematic errors in pH of up to 0.05 pH units if dilute NBS buffers are used (Culberson, 1981). This implies that the pH values reported by two workers, each achieving an internal consistency of ±0.01 pH units, may differ by 0.05 pH units because of differences in the liquid junctions used in their pH cells. This problem can be minimized by using the free diffusion liquid junction shown in Figure 6.7. When such a system is used for pH measurement the potential drift is usually equivalent to less than 0.003 pH units per day at constant temperature (Culberson, 1981). Furthermore the systematic error associated with measurements in different assemblies is less than 0.005 pH units at a salinity of 35‰.

The measurement system is shown schematically in Fig. 6.15. The glass electrode can be a conventional electrode mounted in a flow cell (Fig. 6.14, see also Bailey, 1979), an electrode mounted in a ground glass joint (Culberson, 1981) or a capillary glass electrode (Fig. 6.5*c*, Whitfield & Knox, unpublished work). The free diffusion junction (Fig. 6.7) is generated in a glass T-piece (1 mm internal diameter) and the flow line (which may be constructed of glass or of flexible plastic tubing 1–2 mm internal diameter) is closed at either end with a tap. A syringe containing the buffer (potassium hydrogen phthalate or 1:1 phosphate; Bates, 1973) is connected to the 3-way tap (A, Fig. 6.15) and any air bubbles entrained are vented to waste. The taps A and B are then opened to allow solution to flow through the cell from the syringe. While the solution is flowing, a small amount of salt bridge solution is injected into the T-junction. The flow of buffer is stopped, taps A and B are closed and the potential is read. The whole operation takes only a minute or two and need only involve the use of a few tenths of a millilitre of buffer. The sample (~ 10 ml) can also be conveniently held in a syringe which, if fitted with a cap, can be used to store the sample for up to 6 hours. To measure the sample pH, the buffer syringe is exchanged for a sample syringe and the process repeated. It is only necessary to renew the liquid junction every fifth or sixth measurement. The procedure is readily automated. For best results the whole cell should be thermostated. The advantages of this technique are summarized in Table 6.13 (see also Culberson, 1981).

If, for reasons of custom or expense, the measurements are to be made with a conventional pH cell it is best to use a beaker with a close-fitting lid. The electrodes are mounted in the lid which also has a small vent hole. The

Fig. 6.15. Cell configuration for precise pH measurements (after Culberson, 1981). The sample and buffer solutions are held in syringes which can be connected to the cell via a three-way tap (A) which enables air bubbles to be expelled. A free-diffusion liquid junction is employed (see Fig. 6.7). Tap B closes off the system during measurement.

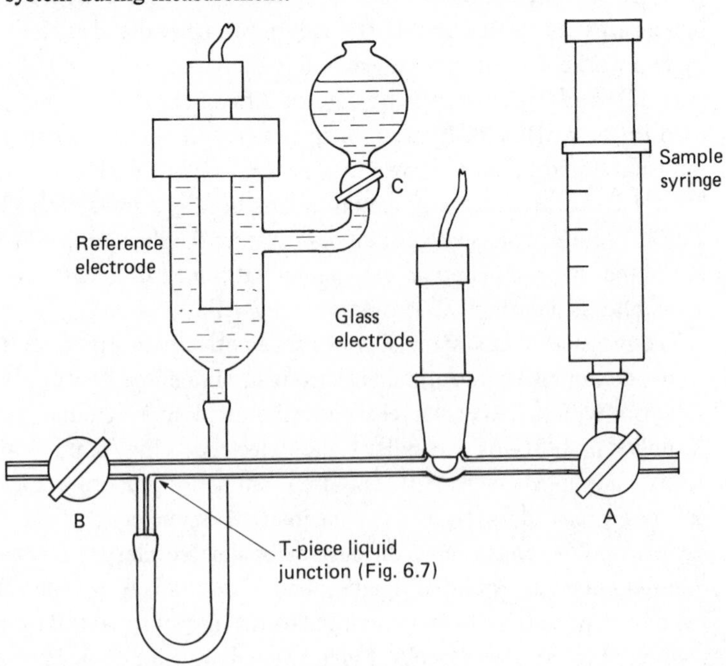

beaker is filled to overflowing with sample and the lid is pushed gently but firmly on, expelling air and excess solution out of the vent hole. The solution should be gently stirred while the measurement is made. Combination electrodes with very small porous liquid junctions should be avoided. Problems due to systematic shifts in the liquid junction potential can be assessed by measuring the response of the cell in an acidified sample. By definition we can write

$$\text{pH (NBS)} = -\log_{10} m_{\text{H},t} - \log_{10} f_{\text{H}} \tag{6.42}$$

where $$\log_{10} f_{\text{H}} = \log_{10} \gamma_{\text{H},t} + \Delta\epsilon_J/kT \tag{6.43}$$

The subscript t indicates *total* hydrogen ion concentrations and activity coefficients.

An estuarine sample is acidified to pH 3–4 and the excess CO_2 is removed by bubbling. Several aliquots of acid are then added and the calculated hydrogen ion concentration (from the acid strength and solution volumes) is plotted against $10^{-\text{pH(NBS)}}$ measured at each point. A straight line will be obtained whose slope is f_{H}. Since $\gamma_{\text{H},t}$ will be a constant at a given temperature and salinity, the differences between $\log f_{\text{H}}$ values measured for different electrode pairs will provide a direct estimate of the systematic pH errors related to differences in the liquid junction potential (Culberson, 1981). If we are to improve the quality of pH measurements made in estuaries, the determination of f_{H} must become a routine part of the commissioning of each new reference electrode. Since the same procedure can also be used for assessing the sample alkalinity (see the following section), the additional effort will be amply repaid.

Table 6 13. *Advantages of the flow pH cell with a free diffusion junction*

1. The renewable and highly reproducible liquid junction makes inter-laboratory comparisons feasible.
2. The characteristics of the cell do not change when the reference electrode is changed.
3. The glass electrode is continuously immersed and therefore exhibits more stable behaviour.
4. If the cell is thermostated it need only be calibrated once a day.
5. The measurement is made in a closed system so that pH drifts due to gas exchange are avoided.
6. The cell response is more rapid because there is no need to wait for complex flow patterns around the reference electrode to stabilize.
7. The flow system is readily automated.

(Culberson, 1981).

Alkalinity and the carbon dioxide system

Titrimetric procedures can be used to estimate the total alkalinity (m_{At}) and total carbon dioxide (m_{Ct}) content of estuarine samples (Almgren & Fonselius, 1976; Culberson, 1981; Jagner, 1981). By subtracting the contribution of boric acid from m_{At} the carbonate alkalinity (m_{AC}) can be estimated. m_{Ct} and m_{AC} may be defined as:

$$m_{Ct} = m_{H_2O.CO_2} + m_{HCO_3^-} + m_{CO_3^{2-}} \tag{6.44}$$

and $$m_{AC} = m_{HCO_3^-} + 2m_{CO_3^{2-}} \tag{6.45}$$

m_{Ct} and m_{AC} together can be used to calculate the total concentration of the various components of the carbon dioxide system ($H_2O.CO_2$, HCO_3^- and CO_3^{2-}) and the pH of the sample. If the pH is measured separately, the three parameters m_{AC}, m_{Ct} and pH can be combined in five different ways (Park, 1969) to calculate the concentrations of these components, thus providing a rigorous test of the internal consistency of the measurements.

The most effective way of determining the endpoint of these titrations is by the use of Gran functions. Details of automatic titration and calculation procedures are given by Anfält & Jagner (1971*b*), Almgren *et al.* (1977) and Jagner (1981). If the aim is simply to measure m_{At}, the Gran technique can be speeded up by initially adding sufficient acid to exceed the endpoint. Several successive additions of acid can then be made to provide sufficient data for a Gran plot (Hillbom *et al.*, 1983; Lidén, 1983). The pH electrode will respond very rapidly in the well-buffered acid solutions and an analysis can be completed in five minutes or so. The rapid acidification also reduces the risk of iron precipitation in anoxic samples. The possible contribution of iron and other components to the sample alkalinity is considered by Midgley & Torrance (1978) and by Lidén (1983).

A simpler but less precise procedure for estimating m_{At} has been developed by Culberson *et al.* (1970). In this procedure a single addition (v_a) of 0.01 M HCl is made to a volume of sea water (v_{sw}) so that the sample is taken beyond the bicarbonate endpoint to a pH of 3 to 4. The carbon dioxide generated is released by purging the sample with air and the pH (NBS) is measured. The total alkalinity is then given by:

$$m_{At} = [H^+] \text{ added} - [H^+] \text{ remaining} \tag{6.46}$$

i.e.

$$m_{At} = \frac{1000 v_a . 0.01}{v_{sw}} - \frac{1000(v_{sw}+v_a)\, a_H'}{v_{sw} f_{H,t}} \tag{6.47}$$

where $a_H' = 10^{-pH\,(NBS)}$. The analysis is rapid and requires only simple equipment.

The carbon dioxide air gap probe can be used for estimating total inorganic and total organic carbon in natural waters (Fiedler *et al.*, 1975).

In this procedure it is necessary to hermetically seal the probe into the sample vessel to prevent loss of CO_2. The useful linear range of the probe is restricted to total carbon concentrations equivalent to 0.1 mmol l^{-1} or more. Total organic carbon is determined by measuring the CO_2 content of samples before and after boiling with persulphate and sodium hydroxide under pressure.

Major constituents

Sodium

Sodium concentrations in marine and estuarine waters may be determined directly by means of a sodium-selective glass electrode. (Kester, 1970; Platford, 1965). The electrode is not subject to interference from K^+ or H^+ at the concentrations they occur in natural sea water and gives a very stable potential if handled with reasonable care. Continuous flow systems have been described by Van den Winkel *et al.* (1972) and Sekerka & Lechner (1974). Details of a direct potentiometric procedure designed specifically for fresh waters are given by Midgley & Torrance (1978) and Bailey (1979) reviews industrial applications of continuous flow systems.

Chloride and total halides

The total halide concentration of estuarine waters may be determined most precisely by the precipitation of the silver halides by titration with silver nitrate solution. The titration is monitored by means of a silver billet electrode and a saturated calomel reference electrode with a double junction to prevent contamination of the sample by the KCl filling solution (Jagner & Årén, 1970). The solubility of the silver halides is reduced by adding ethanol and the pH is maintained in range where only silver–halide interactions are significant by the addition of acidic KNO_3 solution. The endpoint is located using a simple Gran function (F_1 equation 6.27).

More rapid but less precise analytical procedures are based on direct potentiometry using insoluble salt electrodes (Pungor *et al.*, 1979). Flow injection techniques can be used with sampling rates of up to 100 samples hour^{-1} and a precision of a few per cent (Trojanowicz & Matuszewski, 1983). Continuous systems in which water is pumped through a flow cell on board ship have proved useful for the determination of estuarine mixing parameters in low salinity regions where conductimetrically derived salinity measurements are unreliable (Morris *et al.*, 1982). The liquid complexone chloride electrode (Table 6.5) shows a lower selectivity for S^{2-}, Br^- and I^- than the corresponding insoluble salt electrode and should be better suited for work in interstitial waters particularly if the liquid complexone is incorporated into a PVC matrix (Moody & Thomas, 1979).

Sulphate

Sulphate may be determined in sea water by an indirect titration procedure using an Fe^{3+} selective electrode (Jasinski & Trachtenburg, 1973). The sample is acidified and ferric chloride is added to complex the sulphate ions and provide a slight excess of Fe^{2+} as sensed by the electrode. The solution is then titrated with barium chloride which breaks up the complex by precipitating $BaSO_4$ and hence releasing Fe^{3+} ions. The endpoint is indicated by the levelling off in the response of the Fe^{3+} electrode. A higher precision (± 0.5 per cent) has been reported using an alternative indicator titration technique with silver amalgam electrodes (Lebel & Belzile, 1980).

An ingenious differential technique has been described which provides a much greater sampling rate (or even continuous monitoring) but at a somewhat lower precision (± 5 per cent, Trojanowicz, 1980). A stream of standard lead perchlorate and a stream of sample are allowed to converge in a flow system. One lead-sensitive electrode is mounted in the lead perchlorate stream prior to mixing and the second is mounted in the mixed stream after the confluence point. The potential difference between the two electrodes is directly related to the quantity of lead precipitated by sulphate in the mixture. The procedure was tested over the range 30 to 400 mg l^{-1} (0.3–4 mmol l^{-1}).

Fluoride

After the pH electrode, the fluoride electrode is the ISE most widely used in sea-water analysis. Although its selectivity allows it to be used directly, greater precision can be achieved with single and multiple standard addition procedures.

Standard addition potentiometry has been used to determine fluoride in estuarine waters (Warner, 1971*a*). 5 ml of the sample are added to 1 ml of TISAB IV buffer (Table 6.11) and left for 20 hours to allow for slow Al(III) complexation (see also Nicholson & Duff, 1981). Sufficient standard was added to give a potential shift of 18 mV. Unchanging potentials were generally achieved within 15 minutes of the addition of the spike. In river waters with very low fluoride levels ($< 5\ \mu$mol l^{-1}), smaller spikes were added (to give $\Delta E = 4$ to 8 mV) and readings were taken arbitrarily after 15 minutes. The method was tested over the salinity range 7‰ to 35‰ and, at the 50 μmol l^{-1} level, the results were accurate to within 2 per cent. At the 5 μmol l^{-1} level the results were in general 5 to 30 per cent high.

A similar procedure used in sea water (Warner, 1971*b*) employed spikes of 0.17 ml of 10^{-2} M sodium fluoride in a sample containing 25 ml of sea water and 5 ml of TISAB buffer (Table 6.11). The precision of this method at sea is ± 1.2 μmol l^{-1} at the 70 μmol l^{-1} level. This procedure compares favourably with alternative spectrophotometric techniques (Warner *et al.*, 1975).

Total fluoride concentrations may also be measured by a multiple standard addition procedure as described by Anfält & Jagner (1971*a*). For this determination, advantage is taken of the buffering capacity of the natural carbonate system at pH 6 and a Gran linearization procedure is used to relate the volume of titrant added to the original fluoride concentration of the sample. With standard sea water a precision of ± 0.6 per cent was achieved but this rose to ± 7 per cent for dilute estuarine waters with salinities around 2‰. The Gran function used had to be modified to take due account of the complexation of fluoride ions by calcium and magnesium in sea water (see equation 6.19).

Continuous flow standard addition procedures have also been described for the analysis of fluoride (Erdmann, 1975; Phillips & Rix, 1981). In estuarine samples some preheating is required (to 55 °C) to ensure adequate reaction between the TISAB buffer and Al(III) prior to analysis by the electrode. A conventional flow system has been described by Sawyer & Foreman (1969). In all instances the precision of fluoride measurements decreases with a fall in salinity because of lower fluoride levels and the greater probability of significant complexing of fluoride by Al(III).

Nutrients and trace constituents

Nitrate

The use of nitrate selective electrodes for direct determinations in estuarine waters of appreciable salinity is precluded by the interfering effects of Cl^-, SO_4^{2-} and HCO_3^-. However it is possible to remove these interferences by precipitation of SO_4^{2-} as $BaSO_4$ and adsorption of Cl^- and HCO_3^- on to suitable ion exchange resins. There are kits commercially available which can deal with up to a thousandfold excess of interfering ions (Orion model 92-07-51) and it would be interesting if results obtained using these could be compared with spectrophotometric determinations. Continuous monitors for nitrate in drinking waters have been described by Tietz *et al.* (1980) and by Trojanowicz & Lewandowski (1981) and the use of nitrate ISEs in fresh-water analysis has been thoroughly reviewed (Hulanicki & Trojanowicz, 1979; Bailey, 1979). A potentially more suitable approach for estuarine waters involves the preliminary reduction of the nitrate to ammonium with Devarda's alloy and the subsequent determination of the ammonia released by an ammonia gas electrode Hansen *et al.*, 1975; Mertens *et al.*, 1975).

Ammonium

Ammonium can be determined in saline samples at concentrations down to 0.2 μmol l^{-1} by using an ammonia gas-sensing electrode and a standard addition procedure (Garside *et al.*, 1978). By raising the pH of the sample to between 10 and 11, all the ammonium is converted to NH_3 and its concentration determined by means of a modified NH_3 gas-sensing probe in which the normal filling solution is replaced by a solution containing 5 mmol l^{-1} NH_4Cl and 0.5 mol l^{-1} $NaNO_3$. The electrode requires re-calibrating no more frequently than once per day but does require cleaning every few hours to remove a hydroxide precipitate which forms on the membrane. The method exhibits no salt effect and there is no interference from other ions at the concentrations at which they normally occur in natural waters. The potentiometric procedure involves far less sample manipulation than do the conventional spectrophotometric methods and should give a better estimate of the true ammonium concentration than the chemical methods which all suffer some interference from other nitrogen forms.

Details of the use of ammonia sensing electrodes at higher concentrations are given by HMSO (1982) and Růžička *et al.*, (1974). The latter workers used an air gap probe to measure ammonia concentrations over the range 60 to 900 μmol l^{-1} and also, after modification, urea in the range 0.1 to 10 mmol l^{-1}. A flow injection procedure using the air gap probe is described by Růžička & Hansen (1975). Details of commercial on-line ammonium monitors are provided by Bailey (1979) and Hulanicki & Trojanowicz (1979).

Cyanide

A flow technique involving additions metered into the sample stream from a series of standards (Fleet & Ho 1973*a*, & *b*) has been used for determining cyanide in saline waters. By plotting the Gran function (F_1, equation 6.27) against the concentration of the standard (m_{Xt}) a straight line is obtained with an intercept on the volume axis such that the concentration in the sample $m_X = -m_{Xt}/2$. A complete four-standard cycle takes six minutes with each standard being pumped for 60 s followed by a 30 s wash. The flow system was constructed from standard Technicon AutoAnalyzer components.

Trace metals

By definition, the concentrations of trace metals in sea water are less than the lower limit traditionally associated with ISEs (1 μmol l^{-1}). However, the factors controlling the detection limits of ISEs are becoming more clearly defined (Midgley, 1981) and procedures for extending the range of useful measurements are being developed (Moody & Thomas, 1981; Midgley & Torrance, 1978). Moreover the concentrations of trace metals are frequently elevated in estuaries and the response of ISEs to the free metal ion concentrations makes them especially suitable for measuring the fraction of the total metal most readily available for biological uptake (Turner & Whitfield, 1980). The cadmium electrode behaves well in sea water (Sunda *et al.*, 1978) and could be used for monitoring purposes. However, detailed studies have revealed a number of problems in the response of the insoluble-salt copper electrodes in chloride media. The electrode response is sensitive to light and to stirring (Jasinski *et al.*, 1974; Rice & Jasinski, 1976) and the electrode slope is variable but is closer to that expected for a monovalent ion than a divalent ion (see also Oglesby *et al.*, 1977). Despite these difficulties, a somewhat laborious method, requiring frequent electrode calibration, for copper analysis in sea water has been described (Jasinski *et al.*, 1974). The procedure for analysis in fresh water is more straightforward (Hulanicki & Trojanowicz, 1979; Midgley & Torrance, 1978). For example, Smith & Manahan (1973) describe a standard addition method for determining copper in fresh waters. The procedure involves treating samples with a complexing antioxidant buffer (CAB, Table 6.11) containing sodium acetate, acetic acid, sodium fluoride and formaldehyde. The acetate buffer adjusts the sample to an optimum pH of 4.8 and the acetate ions complex Cu ions and prevent removal of copper by precipitation or sorption on to the container walls. The fluoride complexes Fe^{3+} which would otherwise interfere and the formaldehyde prevents interference from oxidizing agents. A flow system for copper determination in fresh water is described by Blaedel & Dinwiddie (1975).

Conclusions

It is worth while taking a fresh look back at the attractive properties of ISEs that introduced this discussion (page 201). Taking them in turn:

1. The direct readout is affected by drifts in the cell potential, by variations in the liquid-junction potential and by interference from other ions in solution. The first two effects can be minimized if the cell is calibrated at frequent intervals in solutions that have an ionic strength close to that of the sample. Interference effects are

not so easily avoided and they must be assessed experimentally and removed by chemical manipulation where they are serious. Useful direct reading monitors are feasible for Cl^-, Na^+, H^+, F^-, CN^- and S^{2-}.

2. Charged particles in low salinity waters can make a significant contribution to the cell potential.

3.–6. These claims are well substantiated although many of the sensors available do not have sufficient selectivity for direct measurement.

7. A number of on-site analysers are now available (EIL, Richmond, Surrey, England; Foxboro-Yoxall, Redhill, Surrey, England; Orion Research Inc., Boston, Massachusetts, USA) and pH, fluoride and ammonium concentrations are monitored on a routine basis (Hulanicki & Trojanowicz, 1979; Bailey, 1979, 1980). On-line titrators are also available commercially (Ionics Inc., 65 Grove Street, Watertown, Massachusetts, USA) and developments are proceeding rapidly in the application of autoanalyser techniques and microprocessor systems to on-line analysis (Slanina *et al.*, 1979; Rigdon *et al.*, 1979). If due regard is taken of the problems associated with potentiometric measurements that have been discussed in detail here and if care is taken to select the appropriate method for the problem in hand then ISEs can provide a simple and convenient solution to many analytical problems.

References

Aachen Workshop (1979). Ion-selective electrodes for measurements in natural and waste waters. *Gewässerschutz, Wasser, Abwasser*, **39**, 1–295.

Almgren, T. & Fonselius, S. H. (1976). Determination of alkalinity and total carbonate. In *Methods of Seawater Analysis*, ed. K. Grasshoff, pp. 97–115. Weinheim: Verlag Chemie.

Almgren, T., Dyrssen, D. & Strandberg, M. (1975). Determination of pH on the moles per kg seawater scale (M_w). *Deep-Sea Research*, **22**, 635–46.

Almgren, T., Dyrssen, D. & Strandberg, M. (1977). Computerized high precision titrations of some major constituents of sea water on board R.V. *Dmitry Mendeleev. Deep-Sea Research*, **24**, 345–64.

Ammann, D., Pretsch, E. & Simon, W. (1972). A calcium ion-selective electrode based on a neutral carrier. *Analytical Letters*, **5**, 843–50.

Anfält, T. & Jagner, D. (1969). Interpretation of titration curves by means of the computer program Haltafall. *Analytica Chimica Acta*, **47**, 57–63.

Anfält, T. & Jagner, D. (1971*a*). A standard addition titration method for the potentiometric determination of fluoride in sea water. *Analytica Chimica Acta*, **53**, 13–22.

Anfält, T. & Jagner, D. (1971*b*). The precision and accuracy of some current methods for potentiometric end point determination with reference to a computer calculated titration curve. *Analytica Chimica Acta*, **57**, 177–83.

Anfält, T. & Jagner, D. (1971*c*). A computer processed semi-automatic titrator for high precision analysis. *Analytica Chimica Acta*, **66**, 152–3.

Anfält, T. & Jagner, D. (1973). Computation of intrinsic end-point errors in titrators with ion-selective electrodes. *Analytical Chemistry*, **45**, 2412–14.

Anfält, T., Graneli, A. & Jagner, D. (1975). Potentiometric gas sensors for ammonia based on ion-selective electrodes for silver(I), copper (II) and mercury(II). *Analytica Chimica Acta*, **76**, 253–9.

Bailey, P. L. (1979). Industrial applications for ion-selective electrodes. *Ion-Selective Reviews*, **1**, 81–137.

Bailey, P. L. (1980). *Analysis with Ion-Selective Electrodes*. London: Heyden.

Bates, R. G. (1973). *Determination of pH. Theory and Practice*. London: John Wiley & Sons.

Baum, G. & Lynn, M. (1973). Polymer membrane-electrodes. Part II. A potassium ion-selective membrane electrode. *Analytica Chimica Acta*, **65**, 393–403.

Bauer, H. H., Christian, R. D. & O'Reilly, I. E. (1978). *Instrumental Analysis*. Boston: Allyn & Bacon.

Ben-Yaakov, S. (1981). Electrochemical instrumentation. In *Marine Electrochemistry*, ed. M. Whitfield & D. Jagner, pp. 99–122. Chichester: J. Wiley & Sons.

Ben-Yaakov, S. & Kaplan, I. R. (1968). High pressure pH sensor for oceanographic applications. *Review of Scientific Instruments*, **39**, 1133–8.

Ben-Yaakov, S. & Kaplan, I. R. (1971). An oceanographic instrumentation system for *in situ* application. *Marine Technology Society Journal*, **5**, 41–6.

Ben-Yaakov, S. & Ruth, E. (1974). An improved *in situ* pH sensor for oceanographic and limnological applications. *Limnology and Oceanography*, **19**, 144–51.

Berner, R. A. (1963). Electrode studies of hydrogen sulphide in sediments. *Geochimica et Cosmochimica Acta*, **27**, 563–75.

Blaedel, W. J. & Laessig, R. H. (1964). Continuous automated buretless titrator with direct readout. *Analytical Chemistry*, **36**, 1617–23.

Blaedel, W. J. & Dinwiddie, W. E. (1975). Behaviour of a micro flow-through copper ion-selective electrode system in the millimolar to submicromolar concentration range. *Analytical Chemistry*, **47**, 1070–3.

Boulègue, J. & Popoff, G. (1979). Methods of determining the principal ionic species of sulphur in natural water. *Journal Français d'Hydrologie*, **10**, 83–90.

Bound, G. P. & Fleet, B. (1980). A semi-automated single-cell gradient tritation system using ion-selective electrodes as end-point sensors. *Talanta*, **27**, 257–61.

Bower, C. A. (1961). Studies on the suspension effect with a sodium electrode. *Soil Science Society Proceedings*, **25**, 18–25.

Braat, A. (1980). Continuous trace level analysis using ion-selective electrodes. In *Analytical Techniques in Environmental Chemistry*, ed. J. Albaiges, pp. 563–9. Oxford: Pergamon Press.

Brand, M. J. D. & Rechnitz, G. A. (1970). Computer approach to ion-selective-electrode potentiometry by standard-addition methods. *Analytical Chemistry*, **42**, 1172–7.

Brezinski, D. P. (1983). Influence of colloidal charge on response of pH and reference electrodes: the suspension effect. *Talanta*, **30**, 347–54.

Briggs, C. C. & Lilley, T. H. (1974). A rigorous test of a calcium ion-exchange membrane electrode. *Journal of Chemical Thermodynamics*, **6**, 599–607.

Buck, R. P. (1978). Ion-selective electrodes. *Analytical Chemistry*, **50**, 17 R–29 R.

Buffle, J. (1972). Errors in the Gran addition method. Part II. Theoretical calculation of systematic errors. *Analytica Chimica Acta*, **59**, 439–45.

Buffle, J., Parthasarathy, N. & Monnier, D. (1972). Errors in the Gran addition method. Part I. Theoretical calculation of the statistical errors. *Analytica Chimica Acta*, **59**, 427–38.

Burton, P. R. (1979). Instrumentation for ion-selective electrodes. In *Ion-selective Electrode Methodology*, ed. A. K. Covington, vol. 1, pp. 21–41. Boca Raton: CRC Press.

Carr, P. W. (1971). Intrinsic end-point errors in precipitation titrations with ion-selective electrodes. *Analytical Chemistry*, **43**, 425–30.

Collis, D. E. & Diggens, A. A. (1969). The use of the fluoride responsive electrode for 'on-line' analysis of fluoridated water supplies. *Water Treatment and Examination*, **18**, 192–202.

Conti, U. & Wilde, P. (1972). Diver operated *in situ* electrochemical measurements. *Marine Technology Society Journal*, **6**, 17–23.

Conti, U., Wilde, P. & Richards, T. L. (1971). Towed vehicle for constant depth and bottom contouring operations. Paper No. OTC 1456, *Offshore Technology Conference*, Houston, Texas.

Corwin, R. F. & Conti, U. (1973). A rugged silver/silver chloride electrode for field use. *Review of Scientific Instruments*, **44**, 708–11.

Covington, A. K. (1969). Reference electrodes. In *Ion-Selective Electrodes*, ed. R. A. Durst, pp. 107–41. National Bureau of Standards Special Publication No. 314. Washington, D.C.: United States Government Printing Office.

Covington, A. K. (1974). Ion-selective electrodes. *CRC Critical Reviews in Analytical Chemistry*, **4**, 355–406.

Covington, A. K. (1979). *Ion-Selective Electrode Methodology*, vol. 1. Boca Raton: CRC Press.

Covington, A. K. & Rebello, M. J. F. (1983). Reference electrodes and liquid-junction effects in ion-selective potentiometry. *Ion-Selective Electrode Reviews*, **5**, 93–128.

Culberson, C. H. (1981). Direct potentiometry. In *Marine Electrochemistry*, ed. M. Whitfield & D. Jagner, pp. 187–261. Chichester: J. Wiley & Sons.

Culberson, C. H., Pytkowicz, R. M. & Hawley, J. E. (1970). Seawater alkalinity determination by the pH method. *Journal of Marine Research*, **28**, 15–21.

Dahms, H. (1967). Automated potentiometric determination of inorganic blood constituents (Na^+, K^+, H^+, Cl^-). *Clinical Chemistry*, **13**, 437–50.

Davies, J. E. W., Moody, G. J. & Thomas, J. D. R. (1972). Nitrate ion-selective electrodes based on poly(vinyl chloride) matrix membranes. *Analyst, London*, **97**, 87–94.

Durst, R. A. (ed.) (1969). *Ion-Selective Electrodes*. National Bureau of Standards Special Publication No. 314. Washington, D.C.: United States Government Printing Office.

Edmond, J. M. (1970). High precision determination of titration alkalinity and total carbon dioxide determination in sea water by potentiometric titration. *Deep-Sea Research*, **17**, 737–50.

Eisenmann, G. (ed.) (1967). *Glass Electrodes for Hydrogen and Other Cations*. New York: Marcel Dekker.

Erdmann, D. E. (1975). Automated ion-selective electrode method for determining fluoride in natural waters. *Environmental Science and Technology*, **9**, 252–3.

Fehér, Zs., Nagy, G., Tóth, K. & Pungor, E. (1978). A detailed study of sample injection into flowing streams with potentiometric detection. *Analytica Chimica Acta*, **98**, 198–203.

Fiedler, U. & Růžička, J. (1973). Selectrode – the universal ion-selective electrode. Part VII. A valinomycin-based potassium electrode with nonporous polymer membrane and solid state inner reference system. *Analytica Chimica Acta*, **67**, 179–93.

Fiedler, U., Hansen, E. H. & Růžička, J. (1975). Measurements of carbon dioxide with the air-gap electrode. Determination of the total inorganic and total organic carbon content in waters. *Analytica Chimica Acta*, **74**, 423–35.

Fleet, B. & Ho, A. Y. W. (1973*a*). Applications of ion-selective electrodes in continuous analysis. In *Ion-Selective Electrodes*, ed. E. Pungor, pp. 17–35. Budapest: Akdemiai Kiadó.

Fleet, B. & Ho, A. Y. W. (1973*b*). An ion-selective electrode system for continuously monitoring cyanide ion, based on a computerized Gran plot technique. *Talanta*, **20**, 793–8.

Fleet, B. & Ho, A. Y. W. (1974). Gradient titration – a novel approach to continuous monitoring using ion-selective electrodes. *Analytical Chemistry*, **46**, 9–11.

Fleet, B. & von Storp, H. (1971). Analytical evaluation of a cyanide-ion-selective membrane-electrode under flow-stream conditions. *Analytical Chemistry*. **43**, 1575–81.

Frant, M. S. & Ross, J. W. (1968). Use of a total ionic strength adjustment buffer for electrode determination of fluoride in water supplies. *Analytical Chemistry*, **40**, 1169–71.

Frazer, J. W., Kray, A. M., Selig, W. & Lim, R. (1975). Interactive experimentation employing ion-selective electrodes. *Analytical Chemistry*, **47**, 869–75.

Freiser, H. (ed.) (1978). *Ion-Selective Electrodes in Analytical Chemistry*, vol. 7. New York: Plenum Press.

Frevert, T. (1980). Determination of hydrogen sulphide in saline solutions. *Schweize Zeitschrift für Hydrologie*, **42**, 255–308.

Frevert, T. & Galster, H. (1978). Rapid and simple *in situ* determination of hydrogen sulphide in water and sediment. *Schweize Zeitschrift für Hydrologie*, **40**, 199–208.

Fricke, G. H. (1980). Ion-selective electrodes. *Analytical Chemistry*, **52**, 259R–75R.

Galster, H. (1979). Ion-selective electrodes for continuous measurement of chloride and cyanide ions. *Gewässerschutz, Wasser, Abwasser*, **39**, 143–8.

Garside, C., Hull, G. & Murray, S. (1978). Determination of submicromolar concentrations of ammonia in natural waters. *Limnology and Oceanography*, **23**, 1072–6.

Gilbert, T. R. & Clay, A. M. (1973). Determination of ammonia in aquaria and in sea water using the ammonia electrode. *Analytical Chemistry*, **45**, 1759–9.

Gran, G. (1952). Determination of the equivalence point in potentiometric titrations. *Analyst, London*, **77**, 661–71.

Graneli, A. & Anfält, T. (1977). A simple automatic phototitrator for the determination of total carbonate and total alkalinity of sea water. *Analytica Chimica Acta*, **91**, 175–80.

Green, E. J. & Schnitker, D. (1974). The direct titration of water soluble sulphide in estuarine muds of Montsweag Bay, Maine. *Marine Chemistry*, **2**, 111–24.

Griffiths, G. H., Moody, G. J. & Thomas, J. D. R. (1972). An investigation of the optimum composition of poly(vinyl chloride) matrix membranes used for selective calcium-sensitive-electrodes. *Analyst, London*, **97**, 420–7.

Grove-Rasmussen, K. V. (1951). Equitransferent salt bridge. *Acta Chimica Scandinavica*, **5**, 422–30.

Hansen, E. H., Růžička, J. & Larsen, N. R. (1975). Determination of total inorganic nitrogen by means of air-gap electrode. *Analytica Chimica Acta*, **79**, 1–7.

Hansen, E. H., Růžička, J. & Ghose, A. K. (1978). Flow-injection analysis of calcium in serum, water and waste waters by spectrophotometry and ion-selective electrode. *Analytica Chimica Acta*, **100**, 151–65.

Hansson, I. (1973). A new pH scale and set of standard buffers for sea water. *Deep-Sea Research*, **20**, 479–91.

Hansson, I. & Jagner, D. (1973). Evaluation of the accuracy of Gran plots by means of computer calculations. Applications to the potentiometric titration of total alkalinity and carbonate content of sea water. *Analytica Chimica Acta*, **65**, 363–73.

Hillbom, E., Liden, J. & Pettersson, S. (1983). Probe for *in situ* measurement of alkalinity and pH in natural waters. *Analytical Chemistry*, **55**, 1180–2.

HMSO (1982). *Ammonia in waters* 1981. Methods for the examination of waters and associated materials. London: Her Majesty's Stationery Office.

Hulanicki, A. & Trojanowicz, M. (1979). Application of ion-selective electrodes in water analysis. *Ion-Selective Electrode Reviews*, **1**, 207–50.

Hussein, W. R. & Guilbault, G. G. (1975). Nitrate and ammonium ion-selective electrodes as sensors. Part II. Assay of nitrate ion and nitrate and nitrite reductases in stationary solutions and under flow stream conditions. *Analytica Chimica Acta*, **76**, 183–92.

Isbell, A. F., Pecsok, R. L., Davies, R. H. & Purnell, J. H. (1973). Computer analysis of data from potentiometric titrations using ion-selective electrodes. *Analytical Chemistry*, **45**, 2363–9.

Ives, D. J. G. & Janz, G. J. (1961). *Reference Electrodes*. London: Academic Press.

Jagner, D. (1970). A semi-automatic titrator for precision analysis. *Analytica Chimica Acta*, **50**, 15–22.

Jagner, D. (1974). Computers in titrimetry. *Microchemical Journal*, **19**, 406–15.

Jagner, D. (1981). Potentiometric titrations. In *Marine Electrochemistry*, ed. M. Whitfield & D. Jagner, pp. 263–300. Chichester: J. Wiley & Sons.

Jagner, D. & Årén, K. (1970). A rapid semi-automatic method for the determination of the total halide concentration in sea water by means of a potentiometric titration. *Analytica Chimica Acta*, **52**, 491–502.

Jasinski, R. & Trachtenberg, I. (1973). Application of a sulphate sensitive electrode to natural waters. *Analytical Chemistry*, **45**, 1277–9.

Jasinski, R., Trachtenberg, I. & Andrychuk, D. (1974). Potentiometric measurements of copper in sea water with ion-selective electrodes. *Analytical Chemistry*, **46**, 364–9.

Johansson, A. (1974). Titrations using an apparatus for recording the antilogarithm of pH or pM. *Talanta*, **21**, 1269–80.

Karlberg, B. (1971). Nomograph for known addition methods in analysis with selective-ion electrodes. *Analytical Chemistry*, **43**, 1911–3.

Kester, D. R. (1970). *Ion association of sodium, magnesium and calcium with sulphate in aqueous solutions*. Ph.D. Thesis, Oregon State University, Corvallis, Oregon.

Koryta, J. (1972). Theory and applications of ion-selective electrodes. *Analytica Chimica Acta*, **61**, 329–411.

Koryta, J. (1975). *Ion-selective Electrodes*. Cambridge: Cambridge University Press.

Lakshminarayanaiah, M. (1976). *Membrane Electrodes*, New York: Academic Press.

Landry, J.-Cl., Michal, C. & Cupelin, F. (1980). Fluoride determination by continuous flow analysis. In *Analytical Techniques in Environmental Chemistry*, ed. J. Albaiges, pp. 571–9. Oxford: Pergamon Press.

Lebel, J. & Poisson, A. (1976). Potentiometric determination of calcium and magnesium in sea water. *Marine Chemistry*, **4**, 321–32.

Lebel, J. & Belzile, N. (1980). A simplified, automated chelatometric method for the determination of sulphate in interstitial water and in sea water. *Marine Chemistry*, **9**, 237–41.

Liberti, A. & Mascini, M. (1969). Anion determination with ion-selective electrodes using Gran's plots. Application to fluoride. *Analytical Chemistry*, **41**, 676–9.

Lidén, J. (1983). Equilibrium approaches to natural water systems. Ph.D. Thesis, University of Umeå.

Lidén, J. & Ingri, N. (1980). A probe for the measurement of pH and pE *in situ* in natural water systems. *Chemica Scripta*, **15**, 203–5.

Light, T. S. (1969). Industrial analysis and control with ion-selective electrodes. In *Ion-Selective Electrodes*, ed. R. A. Durst, pp. 349-74. National Bureau of Standards Special Publication No. 314. Washington, D.C.: United States Government Printing Office.

McCallum, C. & Midgley, D. (1973). Improved linear titration plots for potentiometric precipitation and strong acid strong base titrations. *Analytica Chimica Acta*, **65**, 155–62.

MacInnes, D. A. (1939). *The Principles of Electrochemistry*. New York: Reinhold Publishing Corporation.

Manahan, S. E. (1970). Fluoride electrode as a reference in the determination of nitrate ion. *Analytical Chemistry*, **42**, 128–9.

Manheim, F. T. (1961). *In situ* measurements of pH and Eh in natural waters and sediments. *Stockholm Contributions in Geology*, **8**, 27–36.

Mascini, M. (1973). Titration of sulphate in mineral waters and sea water by using the solid-state lead electrode. *Analyst, London*, **98**, 325–8.

Mascini, M. (1980). Uses of known addition, Gran's plots and related methods with ion-selective electrodes. *Ion-Selective Electrode Reviews*, **2**, 17–71.

Mattock, G. (1963). Laboratory pH measurements. In *Advances in Analytical Chemistry and Instrumentation*, ed. C. N. Reilly, vol. 2, pp. 35–121. New York: Interscience.

Mattock, G. & Band, D. M. (1967). Interpretation of pH and cation measurements. In *Glass Electrodes for Hydrogen and other Cations*, ed. G. Eisenmann, pp. 9–49. New York: Marcel Dekker.

Meites, L. (ed.) (1963). *Handbook of Analytical Chemistry*. New York: McGraw-Hill.

Mertens, J., van de Winkel, R. & Massart, R. L. (1975). Determination of nitrate in water with an ammonia probe. *Analytical Chemistry*, **47**, 522–6.

Meyerhoff, M. E. & Fraticelli, Y. M. (1982). Ion-selective electrodes. *Analytical Chemistry*, **54**, 27R–44R.

Midgley, D. (1981). Detection limits of ion-selective electrodes. *Ion-Selective Electrode Reviews*, **3**, 43–104.

Midgley, D. & Torrance, K. (1978). *Potentiometric Water Analysis*. Chichester: J. Wiley & Sons.

Milham, P. J., Awad, A. S., Paull, R. E. & Bull, J. N. (1970). Analysis of plants, soils and waters for nitrate using an ion-selective electrode. *Analyst, London*, **95**, 751–7.

Moody, G. J. & Thomas, J. D. R. (1971). *Selective Ion-Sensitive Electrodes*. Watford: Merrow Publishing Company.

Moody, G. J. & Thomas, J. D. R. (1973). Selectivity and sensitivity of ion-selective electrodes. In *Ion-Selective Electrodes*, ed. E. Pungor, pp. 77–113. Budapest: Akademiai Kiadó.

Moody, G. J. & Thomas, J. D. R. (1979). Poly(vinyl chloride) matrix membrane ion-selective electrodes. In *Ion-Selective Electrode Methodology*, ed. A. K. Covington, vol. 1, pp. 111–49. Boca Raton: CRC Press.

Moody, G. J. & Thomas, J. D. R. (1981). Ion-selective electrodes of extended linear range. *Ion-Selective Electrode Reviews*, **3**, 189–206.

Morf, W. E., Amman, D., Pretsch, E. & Simon, W. (1973). Carrier antibiotics and model compounds as components of selective ion-sensitive electrodes. *Pure and Applied Chemistry*, **36**, 421–39.

Morris, A. W., Howland, R. J. M. & Bale, A. J. (1978). A filtration unit for use with continuous autoanalytical systems applied to highly turbid waters. *Estuarine and Coastal Marine Science*, **6**, 105–9.

Morris, A. W., Bale, A. J. & Howland, R. J. M. (1982). Chemical variability in the Tamar estuary, south-west England. *Estuarine, Coastal and Shelf Science*, **14**, 649–61.

Nagy, G., Fehér, Zs., Tóth, K. & Pungor, E. (1977). Critical survey of flow-through analytical systems employing ion-selective electrodes as detectors. *Hungarian Scientific Instruments*, **41**, 27–39.

Negus, I. E. & Light, T. S. (1972). Temperature coefficients and their compensation in ion-selective systems. *Instrument Technology*, **19**, 23–9.

Nicholson, K. & Duff, E. J. (1981). Fuoride determination in water: an optimum buffer system for use with the fluoride-selective electrode. *Analytical Letters, Part A*, **14**, 887–912.

Oglesby, G. B., Duer, W. C. & Millero, F. J. (1977). Effect of chloride ion and ionic strength on the response of a copper(II) ion-selective electrode. *Analytical Chemistry*, **49**, 887–9.

Park, P. K. (1969). Oceanic CO_2 systems. An evaluation of ten methods of investigation. *Limnology and Oceanography*, **14**, 179–86.

Paul, J. L. & Carlson, R. M. (1968). Nitrate determination in plant extracts by the nitrate electrode. *Journal of Agricultural and Food Chemistry*, **16**, 766–8.

Pecsok, R. L., Shields, L. D., Cairns, T. & McWilliam, I. G. (1976). *Modern Methods of Chemical Analysis*. New York: J. Wiley & Sons.

Phillips, K. A. & Rix, C. J. (1981). Micropressor-controlled determination of fluoride in environmental and biological samples by a method of standard additions with a fluoride-selective electrode. *Analytical Chemistry*, **53**, 2141–3.

Platford, R. F. (1965). Activity coefficient of sodium ion in sea water. *Journal of the Fisheries Research Board of Canada*, **22**, 885–9.

Portnoy, H. D. (1967). The construction of glass electrodes. In *Glass Electrodes for Hydrogen and Other Cations*, ed. G. Eisenmann, pp. 241–68. New York: Marcel Dekker.

Pungor, E., Fehér, Zs., Linder, E., Nagy, G. & Tóth, K. (1979). Use of ion-selective electrodes in determination of halides. *Zeitschrift für Chemie*, **19**, 367–71.

Pytkowicz, R. M., Kester, D. R. & Burgener, B. C. (1966). Reproducibility of pH measurements in sea water. *Limnology and Oceanography*, **11**, 417–19.

Ratzlaff, K. (1979). Optimizing precision in standard-addition measurement. *Analytical Chemistry*, **51**, 232–5.

Rechnitz, G. A. (1967). Ion-selective electrodes. *Chemical and Engineering News*, **45**, 146–58.

Reynolds, R. C. (1971). Analysis of Alpine waters by ion-electrode methods. *Water Resources Research*, **7**, 1333–6.

Rice, G. K. & Jasinski, R. J. (1976). Monitoring of dissolved copper in sea water by means of ion-selective electrodes. In *Accuracy in Trace Analysis: Sampling, Sample Handling, Analysis*, ed. P. D. LaFleur, vol. 2, pp. 899–915. National Bureau of Standards Special Publication No. 422. Washington, D.C.: United States Government Printing Office.

Rigdon, L. P., Pomernacki, C., Balavan, D. J. & Frazer, J. W. (1979). Automated potentiometric analysis with ion-selective electrodes. *Analytica Chimica Acta*, **112**, 397–405.

Risemann, J. M. (1969). Measurement of inorganic water pollutants by specific-ion electrodes. *American Laboratory*, **7**, 32–9.

Ross, J. W. (1969). Solid-state and liquid-membrane ion-selective electrodes. In *Ion-Selective Electrodes*, ed. R. A. Durst, pp. 57–88. National Bureau of Standards Special Publication No. 314. Washington, D.C.: United States Government Printing Office.

Ross, J. W., Riseman, J. H. & Krueger, J. A. (1973). Potentiometric gas-sensing electrodes. *Pure and Applied Chemistry*, **36**, 473–87.

Růžička, J. & Hansen, E. H. (1974). A new potentiometric gas sensor – the air gap electrode. *Analytica Chimica Acta*, **69**, 129–41.

Růžička, J. & Hansen, E. H. (1975). Flow-injection analysis. Part I. A new concept in fast continuous flow analysis. *Analytica Chimica Acta*, **78**, 145–57.

Růžička, J. & Hansen, E. H. (1980). Flow-injection analysis, principles, applications and trends. *Analytica Chimica Acta*, **114**, 19–44.

Růžička, J. & Hansen, E. H. (1981). *Flow-Injection Analysis*. Chichester: J. Wiley & Sons.

Růžička, J., Hansen, E. H., Bisgaard, P. & Reyman, E. (1974). Determination of ammonia content in waste waters by means of the air-gap electrode. *Analytica Chimica Acta*, **72**, 215–19.

Růžička, J., Hansen, E. H. & Tjell, J. C. (1973). Selectrode – the universal ion-selective electrode. Part VI. The calcium(II) selectrode employing a new ion exchanger in a non-porous membrane and a solid-state reference system. *Analytica Chimica Acta*, **67**, 155–78.

Sawyer, R. & Foreman, J. K. (1969). The development of electrometric methods for analysis. *Laboratory Practice*, **18**, 35–43.

Sekerka, I. & Lechner, J. F. (1974). Simultaneous determination of sodium, potassium and ammonium ions by automated direct potentiometry. *Analytical Letters, Part A*, **7**, 463–72.

Sekerka, I. & Lechner, J. F. (1979). Applications of ion-selective electrodes as reference electrodes. *Analytical Letters, Part A*, **12**, 1239–48.

Simeonov, V. (1980). Critical considerations on the practical application of Orion ion-selective electrodes to sea water and other natural water samples. *Fresenius Zeitschrift für Analytische Chemie*, **301**, 290–3.

Slanina, J., Bakker, F., Möls, J. J., Ordelman, J. E. & Bruyn-Hes, A. G. M. (1979). Computer automation of potentiometric analysis with selective-ion electrodes. *Analytica Chimica Acta*, **112**, 45–54.

Slanina, J., Lingerak, W. A. & Bakker, F. (1980). The use of ion-selective electrodes in manual and computer-controlled flow-injection systems. *Analytica Chimica Acta*, **117**, 91–8.

Smith, M. J. & Manahan, S. E. (1973). Copper determination in water by standard-addition potentiometry. *Analytical Chemistry*, **45**, 836–9.

Squirrell, D. C. M. (1964). *Automatic Methods in Volumetric Analysis*. London: Hilger & Watts.

Srinivarsan, K. & Rechnitz, G. A. (1969). Selectivity studies on liquid-membrane ion-selective electrodes. *Analytical Chemistry*, **41**, 1203–8.

Sunda, W. G., Engel, D. W. & Thuotte, R. M. (1978). Effects of chemical specification on toxicity of cadmium to grass shrimp, *Palaeomonetes pugio:* importance of free cadmium ions. *Environmental Science and Technology*, **12**, 409–12.

Taylor, G. R. (1961). pH measuring instruments. In *pH Measurements and Titration*, ed. G. Mattock, Chapter 10. London: Heywood.

Thomas, R. F. & Booth, R. L. (1973). Selective-electrode measurement of ammonia in water and wastes. *Environmental Science and Technology*, **7**, 523–6.

Thompson, H. & Rechnitz, G. A. (1972*a*). Ion-selective flow-through electrodes: construction and properties of heavy metal sensors. *Chemical Instrumentation*, **4**, 239–53.

Thompson, H. & Rechnitz, G. A. (1972*b*). Fast reaction flow system using crystal-membrane ion-selective electrodes. *Analytical Chemistry*, **44**, 300–5.

Tietz, U., Gruerke, U. & Krause, I. (1980). Determination of nitrate in drinking water by the flow-stream principle by means of an ion-sensitive measurement cell. *Acta Hydrochimica Hydrobiologica*, **8**, 291–8.

Trojanowicz, M. (1980). Continuous potentiometric determination of sulphate in a differential flow system. *Analytica Chimica Acta*, **114**, 293–301.

Trojanowicz, M. & Lewandowski, R. (1981). Multiple potentiometric system for continuous determination of chloride, fluoride, nitrate and ammonia in natural waters. *Fresenius Zeitschrift für Analytische Chemie*, **38**, 7–10.

Trojanowicz, M. & Matuszewski, W. (1983). Potentiometric flow-injection determination of chloride. *Analytica Chimica Acta*, **151**, 77–84.

Truesdell, A. J., Jones, B. F. & Van den Burgh, A. S. (1965). Glass electrode determination of sodium in closed basin waters. *Geochimica et Cosmochimica Acta*, **29**, 725–36.

Turner, D. R. & Whitfield, M. (1980). Chemical definition of the biologically available fraction of trace metals in natural waters. *Thalassia Jugoslavica*, **16**, 231–41.

Van den Winkel, P., Mertens, J., De Baenst, G. & Massart, D. L. (1972). Automatic potentiometric analysis of sodium in river and mineral waters. *Analytical Letters*, **5**, 567–77.

Van den Winkel, P., Mertens, J. & Massart, D. L. (1974). Streaming potentials in automatic potentiometric systems. *Analytical Chemistry*, **46**, 1765–8.

Van der Meer, J. M., Den Boef, G. & van der Linden, W. E. (1975). Solid-state ion-selective electrodes as end-point detectors in compleximetric titrations. Part I. The titration of mixtures of two metals. *Analytica Chimica Acta*, **76**, 261–8.

Wagner, W. & Hull, C. J. (1971). *Inorganic Titrimetric Analysis* (*Contemporary Methods*). New York: Marcel Dekker.

Warner, T. B. (1969). Fluoride in sea water: measurement with a lanthanum fluoride electrode. *Science, New York*, **165**, 178–80.

Warner, T. B. (1971*a*). Electrode determination of fluoride in ill-characterized natural waters. *Water Research*, **51**, 459–65.

Warner, T. B. (1971*b*). Normal fluoride content of sea water. *Deep-Sea Research*, **18**, 1255–63.

Warner, T. B. (1972). Ion-selective electrodes – properties and uses in sea water. *Marine Technology Society Journal*, **6**, 24–33.

Warner, T. B., Jones, M. M., Miller, G. R. & Kester, D. R. (1975). Fluoride in sea water: intercalibration study based on electrometric and spectrophotometric methods. *Analytica Chimica Acta*, **77**, 223–8.

Webber, H. M. & Wilson, A. L. (1969). The determination of sodium in high purity water with sodium responsive glass electrodes. *Analyst, London*, **94**, 209–20.

Whitfield, M. (1971*a*). *Ion-selective Electrodes for the Analysis of Natural Waters*. AMSA Handbook No. 2, Australian Marine Sciences Association. Sydney: Australian Museum.

Whitfield, M. (1971*b*). A compact potentiometric sensor of novel design. *In situ* determination of pH, pS^{2-} and Eh. *Limnology and Oceanography*, **16**, 829–37.

Whitfield, M. (1975*a*). Sea water an electrolyte solution. In *Chemical Oceanography*, ed. J. P. Riley & G. Skirrow, 2nd edn., vol. 1, pp. 43–171. London: Academic Press.

Whitfield, M. (1975*b*). The electroanalytical chemistry of sea water. In *Chemical Oceanography*, ed. J. P. Riley & G. Skirrow, 2nd edn., vol. 4, pp. 1–154. London: Academic Press.

Whitfield, M. (1979). Activity coefficients in natural waters. In *Activity Coefficients in electrolyte solutions*, ed. R. M. Pytkowicz, pp. 153–299. Boca Raton: CRC Press.

Whitfield, M. & Leyendekkers, J. V. (1969). Liquid ion-exchange electrodes as end-point detectors in compleximetric titrations. Determination of calcium and magnesium in the presence of sodium. Part I. Theoretical considerations. *Analytica Chimica Acta*, **45**, 383–98.

Whitfield, M., Leyendekkers, J. V. & Kerr, J. D. (1969). Liquid ion-exchange electrodes as end-point detectors in compleximetric titrations. Part II. Determination of calcium and magnesium in the presence of sodium. *Analytica Chimica Acta*, **45**, 379–410.

Whitfield, M. & Turner, D. R. (1981). Sea water as an electrochemical medium. In *Marine Electrochemistry*, ed. M. Whitfield & D. Jagner, pp. 3–66. Chichester: J. Wiley & Sons.

Wilde, P. & Rodgers, P. W. (1970). Electrochemical meter for activity measurements in natural environments. *Review of Scientific Instruments*, **41**, 356–9.

Wilson, B. L., Schwarzer, R. R., Chukwuenye, C. D. & Cyrous, J. (1981). Determination of sulphide in the aqueous environment. *Microchemical Journal*, **26**, 402–10.

Wilson, M. F., Haikala, E. & Kivalo, P. (1975). An evaluation of some sodium ion selective glass electrodes in aqueous solution. Part I. Electrode calibration characteristics and selectivity with respect to hydrogen ions. *Analytical Chimica Acta*, **74**, 395–410.

Wolbarsht, M. (1964). Interference and its elimination. In *Physical Techniques in Biological Research*, ed. W. L. Nastuk, pp. 335–72. New York: Academic Press.

Zipper, J. J., Fleet, B. & Perone, S. P. (1974). Computer controlled monitoring and data reduction for multiple ion-selective electrodes in a flowing system. *Analytical Chemistry*, **46**, 2111–8.

Zirino, A., Clavell, C. & Seligman, P. F. (1983). Copper and pH in the surface waters of the eastern tropical Pacific Ocean and the Peruvian upwelling system. *Marine Chemistry*, **12**, 25–42.

7

Data presentation and interpretation

P. C. HEAD

In planning any estuarine survey programme, the organizer should have a clear idea of how the data collected are to be interpreted and presented in order to yield the maximum amount of information about the processes controlling the observed distributions. Useful reviews of the major processes involved are given by Burton & Liss (1976), Olausson & Cato (1980) and Morris (1983).

In most estuaries, the dominant factor controlling the distribution of dissolved and suspended particulate material is the degree of mixing between the fresh-water inflows and the coastal sea water. The tidal rise and fall in an estuary causes the movement of water back and forth within the estuary and augments the diffusive mixing of fresh and saline water resulting from the difference in their densities. Such density differences result principally from differences in concentrations of dissolved solids in the fresh and saline water, but very high concentrations of particulate material can also have a significant effect.

Most estuarine surveys yield data taken from different stations and depths throughout the estuary and usually at different states of tide. One of the first steps in attempting to interpret such data is usually to try to allow for variations related to tidal movements. This may be achieved in one of two ways: either by calculating the ratio of the concentration of a particular constituent to that of an index of mixing; or by adjusting the position of the sample to allow for the distance the water mass is estimated to have moved relative to a particular tidal state such as high water. Such tidal adjustment is an important consideration when attempting to construct profiles, instantaneous distributions along the length of an estuary; or sections, instantaneous distributions across the width of an estuary, from non-synoptic data.

The information and examples given in this chapter are an attempt to introduce the reader to some of the more frequently used methods of presenting data obtained from estuarine surveys so that some of the more obvious pitfalls may be avoided and the reasons for the methods adopted appreciated. Information is also given on the units commonly used for the reporting of marine and estuarine data and where necessary the reasons for the preferred units and how they are related to other commonly used units.

Units and conversion factors

The Système International d'Unités (SI) and estuarine chemistry

Most of the units commonly employed in estuarine chemistry came into use before the SI system was adopted as the primary international standard for scientific units and thus often differ from those recommended under the system. With the gradual spread of SI-based units through all branches of science their use in marine work is increasing, particularly for reporting the more evenly distributed and precisely determinable physical quantities. From the point of view of the estuarine chemist, the main changes associated with the adoption of SI-based units involve expressing volumes in dm^3 rather than litres (1 dm^3 = 1.0003 litres) and amounts of substances as moles instead of weights. For chemical measurements in estuaries, sampling and analytical variations are far greater than variations in the numerical value of concentrations resulting from the difference between volumes measured in dm^3 and litres and thus the two units may be directly equated. The use of molar quantities is discussed on p. 282.

Salinity, chlorinity and chlorosity

Salinity

The measurement of salinity is one which is routinely carried out on the majority of marine and estuarine samples as it can be determined with high precision and provide much information about the origin and history of the water mass sampled. Originally it was intended that the salinity of a sample would be the concentration of dissolved inorganic matter determined gravimetrically. In practice it was found that this quantity was extremely difficult to determine accurately and could certainly not be carried out routinely with the necessary precision. After considerable international discussion at the end of the last century it was decided that the only practical way of determining salinity was by relating it to another property which could be determined precisely and routinely. It was, therefore, agreed that salinity should be calculated from the total precipitable halide concentration as measured by titration with silver nitrate. More recently it has been possible to define salinity in terms of electrical conductivity. The internationally accepted figure for the salinity of a water is very slightly less than its total salt content. Further information about the historical development of the salinity concept and the measurement of salinity is given by Wallace (1974), Cox (1965), Wilson (1975), Grasshoff *et al.* (1983) and the background papers supporting the recent adoption of the Practical Salinity Scale 1978, published by UNESCO (UNESCO, 1981).

For waters which are subject to substantial fresh-water influence, as occur in most estuaries, it can be argued that the concept of salinity is of limited value because its estimation is dependent on a constant relationship between the ions present in solution, and this relationship begins to significantly break down when the fresh-water content exceeds about 90 per cent. Thus during the very early stages of the estuarine mixing process the salinity of the water can differ substantially from its true sea-salt content. The degree of deviation between the fresh-water content of a sample and its salinity as determined by a number of indices has been discussed in Chapter 1. In practice, however, salinity can be used as a mixing index in most cases because the sea salts very soon become dominant in the mixing process and the measurements of most other constituents are subject to analytical errors which are substantially greater than the uncertainty inherent in the estimation of salinity from a measured parameter (Burton, 1976).

So as to make it independent of temperature, salinity is expressed as the mass of dissolved inorganic matter contained in a particular mass of water. Prior to 1982, salinities were expressed as parts per thousand parts of solvent (abbreviated as ‰) which is equivalent to the mass of solids contained in 1 kg of the solvent, namely sea water.

Since 1982, salinities determined using the Practical Salinity Scale 1978 have been expressed as dimensionless numbers equivalent to 1000 times the weight ratio of solids to solvent. The salinity of open ocean water is usually about 35 on the Practical Salinity Scale which is equivalent to 35‰ for pre-1982 data.

Some of the implications for biologists and chemists, of adopting the Practical Salinity Scale 1978 have been discussed by Parsons (1982), Sharp & Culberson (1982) and Gieskes (1982).

Chlorinity

For the original definition of salinity in terms of the precipitable halide concentration of the sample it was necessary to express this in terms of a new unit – chlorinity (Cl ‰) – which was defined as the mass in g of chlorine, equivalent to the mass of precipitable halogens contained in 1 kg of sea water. From measurements made by an International Commission set up in 1899 it was found that salinity was related to chlorinity by the following relationship:

$$\text{Salinity ‰} = 0.03 + 1.8050 \text{ Chlorinity ‰}.$$

As originally defined, the chlorinity of a sample is dependent on the atomic weight used in calculating the mass of chlorine equivalent to that of the other precipitable halogens. To overcome any difficulties which might occur with the revision of atomic weights, chlorinity was re-defined in 1937 as the mass in g of pure silver necessary to precipitate the halogens in 3238.5233 g of sea water.

With the introduction of very precise conductimetric methods for the determination of salinity it became necessary to redefine chlorinity and hence salinity in terms of electrical conductivity. In 1962 it was decided by an international Joint Panel on Oceanographic Tables and Standards (JPOTS) that the newly defined chlorinity should be related to salinity by the equation

$$\text{Salinity ‰} = 1.80655 \text{ Chlorinity ‰}.$$

As with salinity determined prior to 1982, chlorinity is conventionally expressed as parts per thousand (‰) which is equivalent to g kg^{-1} of solvent. The chlorinity of sea water with a salinity of 35‰ is 19.374‰ and consists of 19.354‰ Cl and 0.032‰ precipitable halogen equivalent. The discrepancy of 0.012‰ between the chlorinity and the sum of the concentrations of chloride and the precipitable halogen equivalent reflects changes in the accepted atomic weights since the first standard sea waters were set up at the turn of the century (Wilson, 1975).

In 1981 it was agreed by JPOTS that the Practical Salinity Scale 1978 should be adopted from 1982 for the determination of salinity (see UNESCO, 1981). As a result of this, salinities are now determined directly from conductivity measurements and a knowledge of the chlorinity of standards or samples is not required.

Chlorosity

Although salinity and chlorinity are defined in terms of the mass of dissolved matter per unit mass of solvent the majority of determinations are made on samples which are measured out by volume. The volumetric methods determine the mass of chlorine, and chlorine equivalent to the mass of halogens, contained in one litre of sea water. This quantity is known as the chlorosity. In line with the units used for salinity and chlorinity, chlorosity is usually expressed as g l^{-1} but for most fresh-water work and some estuarine work mg l^{-1} is used. The chlorinity of a water sample is equal to its chlorosity divided by the specific gravity of sea water with that particular chlorosity. Before the introduction of the Practical Salinity Scale 1978 chlorinities and salinities could be determined from chlorosities by means of tables such as those given by Strickland & Parsons (1972).

The chlorosity of sea water with a salinity of 35 is 19.854 g l^{-1} and is made up of 19.834 g l^{-1} Cl^- and 0.033 g l^{-1} precipitable halogen equivalent. As with chlorinity the discrepancy between the chlorosity and the sum of the chloride and precipitable halogen equivalent concentrations reflects changes in the accepted atomic weights since the original standard sea waters were set up. The chlorosity of water found in estuaries can range from less than 0.010 g l^{-1} (10 mg l^{-1}) for the fresh-water inputs to more than 19 g l^{-1} (19000 mg l^{-1}) where virtually full strength sea water occurs.

Chlorosity provides a convenient measure of the degree of mixing between fresh and saline waters for very many estuarine investigations.

Molar units

Early investigations into the composition of sea water were concerned to establish its general composition and determine whether systematic variations were apparent. To this end it was convenient to express the results as the mass or volume of the particular constituent contained in a known mass or volume of sea water. Once reliable estimates of the concentrations had emerged it was possible to start to try to draw the information together and use it in conjunction with comparable information about the composition of river waters, the atmosphere, and suspended and deposited particulate material to try to understand the processes which determine the observed distributions. As, to all intents and purposes, the system can be likened to a large chemical reactor, there was much to be gained from expressing the concentrations of constituents in molar terms which give an immediate measure of the concentration of potentially reactive entities. Thus, for example, in normal sea water although the weight concentration of sulphate is nearly double that of magnesium the fourfold greater mass of the sulphate ion means that the molar concentration of sulphate is about half that of magnesium. This means that there are many more potentially reactive magnesium ions in sea water than there are sulphate ions, a fact which is not immediately apparent from their mass concentrations. The first suggestions that molar units should be used in marine science were put forward in the early 1930s with regard to nutrient data (see Cooper, 1933) and, using the nomenclature common at the time, it was proposed that nutrients should be expressed as mg-atom m^{-3} (often abbreviated mg-at m^{-3}) where 1 mg-atom is equal to the mass of the nutrient element present in mg divided by its atomic weight.

Subsequently it was suggested that terms such as gram-molecule and gram-atom should be replaced by the mole which is defined as the amount, of a substance that contains as many entities (atoms, molecules, electrons photons etc.) as there are in 12 g of ^{12}C and corresponds to the mass, in grams, of the atoms, molecules or ions in question which is numerically equal to their atomic or molecular weight. Providing the species considered contain only one atom of the element in question quantities expressed in moles will be numerically equivalent to those expressed in gram-atoms of the individual elements making up an ion or molecule. Where association of atoms takes place to form molecules, as is the case of the dissolved gases oxygen and nitrogen, concentrations expressed in moles will be simple submultiples of those expressed in gram-atoms.

Major elements

The salinity of sea water is essentially a measure of the concentration of nine elements (Cl, Na, Mg, Ca, K, Br, Sr, B and F) and two radicals (SO_4^{2-} and HCO_3^-) which are normally present at concentrations greater than 1 mg kg^{-1} of solution and in virtually constant proportions.

Table 7.1. *Concentrations of major elements in oceanic water*

Constituent	Molinity mol kg^{-1} of water of salinity 35	g kg^{-1} of water of salinity 35	Molarity mol l^{-1} of water of salinity 35 at 20 °C	g l^{-1} of water of salinity 35 at 20 °C
Chloride	0.546 mol kg^{-1}	19.354	0.559 mol l^{-1}	19.834
Sodium	0.468 mol kg^{-1}	10.762	0.479 mol l^{-1}	11.050
Magnesium	53.2 mmol kg^{-1}	1.293	54.6 mmol l^{-1}	1.326
Sulphate	28.2 mmol kg^{-1}	2.709	28.9 mmol l^{-1}	2.780
Calcium	10.3 mmol kg^{-1}	0.411	10.5 mmol l^{-1}	0.422
Potassium	10.2 mmol kg^{-1}	0.399	10.5 mmol l^{-1}	0.416
Bicarbonate[1]	2.33 mmol kg^{-1}	0.140	2.38 mmol l^{-1}	0.142
Bromide	0.842 mmol kg^{-1}	0.0673	0.870 mmol l^{-1}	0.068
Boron	0.411 mmol kg^{-1}	0.00445	0.416 mmol l^{-1}	0.0045
Strontium	90.1 μmol kg^{-1}	0.0079	97.0 μmol l^{-1}	0.0085
Fluoride	71.5 μmol kg^{-1}	0.00136	73.5 μmol l^{-1}	0.0014
Ionic strength	0.697 mol kg^{-1}		0.713 mol l^{-1}	
Chlorinity		19.374	Chlorosity	19.854

[1] Inorganic carbon expressed as bicarbonate.

The concentration of these constituents is usually determined on samples measured volumetrically and expressed in g l^{-1} although, when comparing the ionic concentration ratios, the data are usually expressed as g kg^{-1} so as to facilitate comparison with chlorinities, determined for or derived from salinity measurements. The differences between the gravimetric and volumetric concentrations of the major elements in oceanic water with a salinity of 35‰ are given in Table 7.1. For much oceanographic work, particularly that concerned with biogeochemical cycling of elements and studies of the evolution of the oceans, molar units are frequently used. Of the conventionally accepted major elements, all except B, Sr and F are present in ocean water at concentrations greater than 0.5 mmol kg^{-1}.

In estuaries, the composition of the water is so dominated by the major elements contributed by the marine input that the virtually constant relationship between them is maintained down to quite low salinities. In some cases, however, there can be significant inputs of major elements from fresh-water streams which will alter the usual ratios and provide an indicator of the time and distance involved in the thorough mixing of this water with the receiving water.

Minor elements

All elements usually present in estuarine waters at concentrations of less than 1 mg kg^{-1} of solution are conveniently termed minor elements or trace elements. Unlike the major elements the concentration of minor elements often varies considerably both temporally and spatially in open oceanic and coastal sea water, and the concentrations found in estuaries will vary with quantities of the element contributed from the various inputs to the estuary. The most important of these are usually from the fresh-water inflow, the sea-water inflow, the sediments and the atmosphere. The concentrations of minor elements in estuarine and sea-water samples may be expressed on a volumetric or a gravimetric basis although the former is more common. Most of the minor elements so far detected in sea water are present at concentrations ranging from about 100 mg l^{-1} down to about 0.1 ng l^{-1} although some of the naturally occurring radioactive nuclides can be detected at concentrations of less than 0.1 pg l^{-1}. For oceanographic work, the use of molar units is becoming more and more usual. Whitfield (1982) has suggested that the term minor elements be reserved for those such as Li, Rb, Ba and Mo which have typical concentrations in the range 0.1 to 500 μmol l^{-1}. Elements such as Al, Fe, Ni, Cu and Zn which usually occur at between 1 and 100 nmol l^{-1} he designates trace elements and all elements such as Cd, Hg, Ag and Sn etc. with concentrations of less than 1 nmol l^{-1} ultra-trace elements.

Nutrients

The minor elements which are intimately involved in biological processes, and are usually only present in sea water in concentrations which limit plant growth, namely nitrogen, phosphorus and silicon, are usually considered separately as nutrient or micro-nutrient elements. Silicon is an important nutrient element in most marine and estuarine waters because of the dominance of planktonic or benthic diatoms in the flora. The concentration of these elements varies both spatially and temporally depending on the extent of plant growth and local inputs. Below the depth at which plant growth is restricted by insufficient light, concentrations are much higher and more uniform although there are significant variations in the different oceanic basins.

As explained above, nutrient concentrations were the first type of oceanographic data to be routinely reported in molar units. Following the early use of mg-at m^{-3} data were conventionally expressed as μg-at l^{-1}. Because of the precision of the methods involved in determining the nutrients, quantities expressed in mg-at m^{-3} and μg-at l^{-1} may be taken as being exactly equivalent. Most nutrient data collected from marine and estuarine waters are now reported as μmol l^{-1} and are numerically equivalent to concentrations expressed as μg-at l^{-1} of the individual nutrient elements. Concentrations expressed as μmol l^{-1} may be converted to μg l^{-1} of the individual elements by multiplying by atomic weights or to μg l^{-1} of ions or molecules by multiplying by the ratio of the molecular weight of the ion or molecule to the number of atoms of the element it contains. The relationships between the commonly used units for expressing nutrient concentrations in marine, estuarine and fresh waters are given in Table 7.2.

Table 7.2. *Collected factors for the interconversion of nutrient data*

Phosphorus

1 μmol P = 1 μg-at P = 30.9738 μg P = 94.9714 μg PO_4^{3-}

0.03229 μmol P = 0.03229 μg-at P = 1.0 μg P = 3.0662 μg PO_4^{3-}

0.01053 μmol P = 0.01053 μg-at P = 0.3261 μg P = 1.0 μg PO_4^{3-}

Silicon

1μmol Si = 1 μg-at Si = 28.086 μg Si = 60.085 μg SiO_2

0.03560 μmol Si = 0.03560 μg-at Si = 1.0 μg Si = 2.139 μg SiO_2

0.01664 μmol Si = 0.01664 μg-at Si = 0.4674 μg Si = 1.0 μg SiO_2

Nitrogen

1 μmol N = 1 μg-at N = 14.007 μg N

0.07139 μmol N = 0.07139 μg-at N = 1.0 μg N

Dissolved gases

A knowledge of the concentration of dissolved gases in sea water is important in understanding marine chemical processes but it is unlikely that most estuarine investigations will routinely involve measurements of the concentrations of those not directly involved in biological processes. In many estuarine areas, plentiful supplies of nutrients and organic material, from both natural and human sources, can lead to the removal of all the dissolved oxygen from water masses and the production of hydrogen sulphide, and it is with these two gases that estuarine investigators are usually involved.

Oxygen

As with all gases the concentration of oxygen which will dissolve in a given volume of water is dependent on the temperature and salinity of the water, and the pressure to which it is subjected. The concentration of oxygen found in estuarine waters may be expressed in a number of

Fig. 7.1. The solubility of oxygen in water with the range of temperature and chlorosity usually found in estuaries. Data from Riley & Chester, 1971.

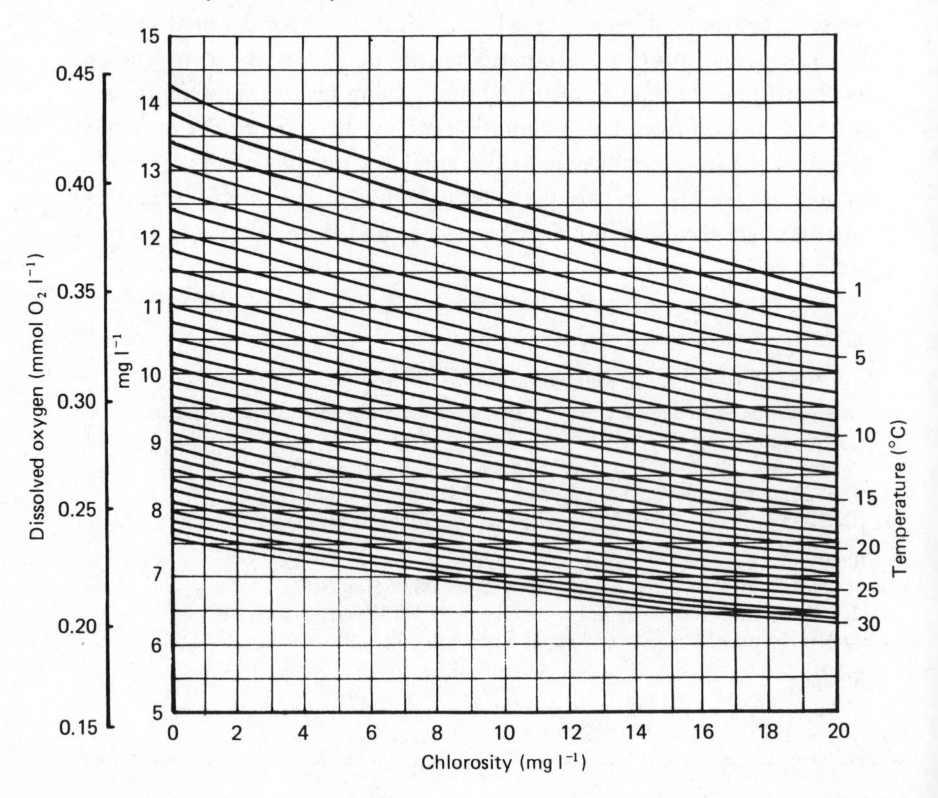

units such as ml l^{-1}, mmol l^{-1} or per cent saturation. The choice of unit used is dependent very largely on the use to which the information is to be put. For most purposes it is probably best to initially record the concentration in mg l^{-1} and then convert this value into other units if required. Kester (1975) has discussed the desirability of using molar units for dissolved gas data. Concentration in mg l^{-1} may be converted to ml O_2 (STP) by multiplying by 0.70 (reciprocal 1.429). For ease of comparing samples collected in estuaries where salinities and temperatures can vary considerably both spatially and temporally, it is often convenient to express the concentration as a percentage of its solubility in water of the appropriate *in situ* temperature, pressure and salinity, but an indication of the concentrations involved should also be given. References to publications containing nomograms and tables for determining the solubility of oxygen in water for the usually encountered ranges of temperature and salinity are given in Chapter 3. The solubility of oxygen in water within the range of temperatures and chlorosities usually found in estuaries is illustrated in Fig. 7.1.

For studies involving the amounts of oxygen liberated during photosynthesis or taken up during the decomposition of biological material, molar units are convenient as stoichiometric relationships are usually assumed. For gases such as oxygen with diatomic molecules concentrations expressed as mmol l^{-1} will be half those expressed as mg-at O l^{-1}.

A summary of the relationship between the various units currently used to express dissolved oxygen concentrations is given in Table 7.3.

Hydrogen sulphide

As with dissolved oxygen the concentrations of hydrogen sulphide present in water samples may be expressed in a number of units depending on the purpose of the investigation. Hydrogen sulphide is usually produced in estuarine waters by bacterial reduction of sulphate and occurs in the water mainly as undissociated H_2S and HS^-. At pH values below 11 only minute amounts of S^{2-} are present and H_2S dominates at pH values less than about 7 (Almgren *et al.*, 1976). Thus the concentrations of the dissolved species are often expressed as in terms of the volume of the un-

Table 7.3. *Factors for the interconversion of dissolved oxygen data*

1 mmol O_2 = 2 mg-at O = 31.9988 mg O_2 = 22.3916 ml O_2 (STP)
0.5 mmol O_2 = 1 mg-at O = 15.9994 mg O_2 = 11.1958 ml O_2 (STP)
0.03125 mmol O_2 = 0.0625 mg-at O = 1.0 mg O_2 = 0.70 ml O_2 (STP)
0.04466 mmol O_2 = 0.08931 mg-at O = 1.429 mg O_2 = 1.0 ml O_2 (STP)

dissociated gas at STP per litre of solution, although mg S^{2-} l^{-1} is also used. As with oxygen and nutrient data, however, there is much to be gained from using molar units, in this case mmol S^{2-} l^{-1}. Table 7.4 provides a summary of the relationships between the various units to be found in the literature.

Adjustment for tidal state

As explained in detail in Chapter 1 the chemical composition of the water at any particular point in an estuary will usually change markedly during the course of a tidal cycle. This variation often makes it difficult to determine the geographical distribution of substances within an estuary if the data have been collected from different stations at different states of the tide. If a synoptic picture of the distribution is required, it is desirable to try to collect all the data at the same state of tide; but in practice this is often not possible because of insufficient manpower or equipment. If a synoptic survey is not feasible it is often possible to allow for tidal movement by adjusting the actual sampling positions to the position that particular water mass would be expected to occupy at a standard tidal state such as high water, low water or half tide. In coastal waters, adjustments for water movements during a tidal cycle can often be made by referring to tidal stream data given on navigational charts or issued separately in the form of a tidal stream atlas. For most estuaries such data are usually lacking and it is necessary to use other methods to estimate tidal movements. The methods employed are designed to determine the tidal excursion or distance travelled by a water mass between successive slack waters in the absence of fresh-water flow and the partial tidal excursions for suitable intervals during the cycle.

To determine tidal adjustments for an estuary requires a considerable amount of information about variations in the water level or some other parameter related to tidal state over a tidal cycle at a number of stations along the length of the estuary. The methods employed can only give reliable information for well-mixed water masses and in well-stratified estuaries it may be necessary to determine separate tidal adjustments for

Table 7.4. *Factors for the interconversion of dissolved hydrogen sulphide data*

1 mmol H_2S =	34.076 mg H_2S =	32.06 mg S^{2-} =	22.40 ml H_2S (STP)
0.2935 mmol H_2S =	1 mg H_2S =	0.9408 mg S^{2-} =	0.657 ml H_2S (STP)
0.3119 mmol H_2S =	1.0629 mg H_2S =	1 mg S^{2-} =	0.699 mg H_2S (STP)
0.04464 mmol H_2S =	1.5213 mg H_2S =	1.4313 mg S^{2-} =	1 ml H_2S (STP)

the water above and below the halocline. Near the upstream limit of the estuary, the direction of flow is often more influenced by the flow of fresh water in the non-tidal river than by the tidal rise and fall and the estimation of tidal excursions in this region is less meaningful although it is still sometimes a useful concept in determining the possible distributions of pollutants.

The determination of tidal adjustments is an integral part of the construction of most mathematical models used for the study of the dispersion of pollutants in estuaries as they allow predictions of possible distributions to be made for different mixing conditions within the estuary. The most usual methods of determining tidal adjustments involve calculating, at various states of the tide, the cumulative volume of the estuary at various distances along its length or determining the longitudinal distribution of an indicator of the degree of mixing between the fresh and saline water, such as salinity or chlorosity.

Where it is not possible to gather enough data to determine tidal adjustments by these methods it is possible to allow for tidal variations by calculating the ratio between the constituent in question and a conservative mixing indicator as will be described later but in this case it is not possible to translate the information to a geographical location.

Volumetric method

The use of volumetric data to calculate tidal corrections in an estuary is fully described in the report on the effects of polluting discharges on the Thames Estuary by the Water Pollution Research Laboratory (Department of Scientific and Industrial Research, 1964), and it has subsequently been applied to a number of British estuaries. In essence the method consists of calculating the volume of the estuary throughout the tidal cycle and adjusting the position of the sampling stations so that there is a constant volume upstream. The main assumptions made when applying the method are that the estuary is well mixed, both vertically and horizontally and that there is no longitudinal mixing as the water moves back and forth along the estuary.

To establish the volume of the estuary upstream and downstream, limits must be set. Where there is an artificial upper limit such as a weir, it is usually convenient to use this as the point from which the volumes are calculated. Where no such limit exists it is convenient to use the point to which mean tides flow, although this will necessarily mean a small underestimate of the volume for spring tides and a small overestimate for neap tides. The lower limit is often more difficult to define as estuaries tend to merge gradually into coastal waters. In practice it is usually found that there is a point beyond which it is difficult to calculate the volume at either high water or low water because of the rapidly increasing distance between the banks of the estuary proper or the low water channel. At about this position the concept of a purely longitudinal flow in the estuary is usually untenable as the water masses will also be subject to coastal tidal currents which tend to move the water across the mouth of the estuary. Thus a practical lower limit to the estuary can usually be established.

In addition to the upstream and downstream limits of the estuary, it is necessary to establish the water levels along the estuary at intervals through the tidal cycle so that the cross-sectional areas at a number of points can be calculated. For the bigger estuaries and smaller ones with ports near their upper limit, some of these data are usually available in the form of tidal reduction tables giving predictions of the local high and low water times and tidal heights. Where there are insufficient tidal reduction data it is necessary to estimate the water levels from the observed depths of water at the time of sampling. The distances between the cross-sections should be such that the cross-sectional areas change more or less linearly from one section to the next.

To calculate the volume of the estuary, it is divided into a number of segments from the upper limit downstream and the volume of each of

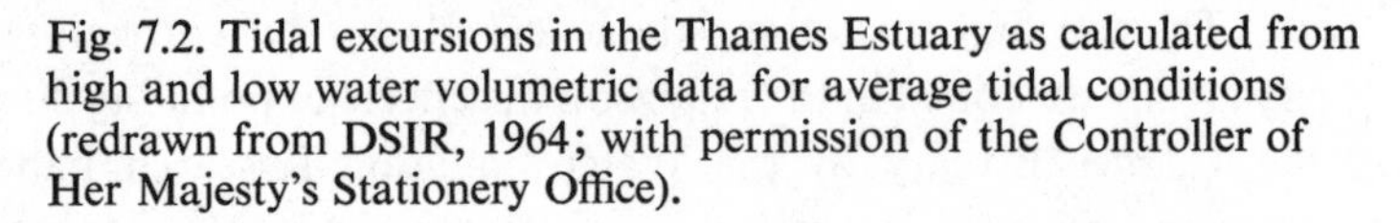
Fig. 7.2. Tidal excursions in the Thames Estuary as calculated from high and low water volumetric data for average tidal conditions (redrawn from DSIR, 1964; with permission of the Controller of Her Majesty's Stationery Office).

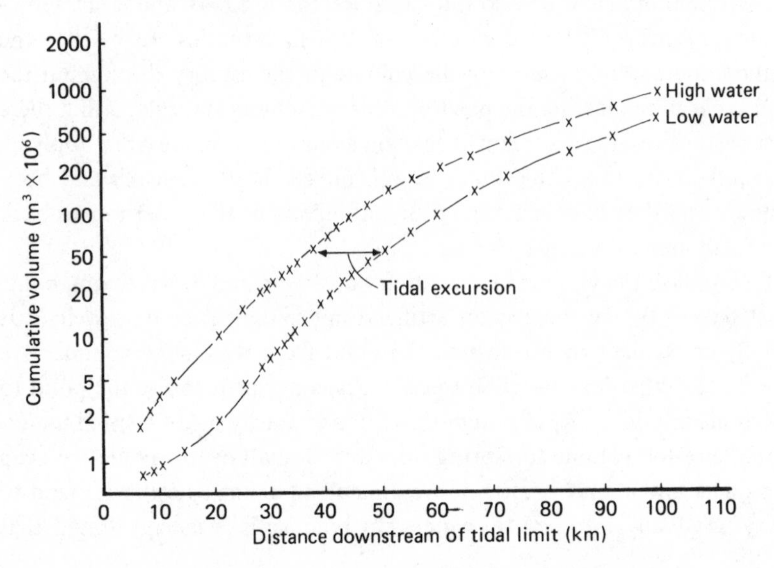

Fig. 7.3. Effect of using volumetric data to adjust the position of dissolved oxygen sampling in the Thames Estuary *a* to equivalent half-tide position *b*. The data points are average values for the period October to December 1954. ○ – high water samples, × – low water samples, ● – intermediate samples (redrawn from DSIR, 1964; with permission of the Controller of Her Majesty's Stationery Office).

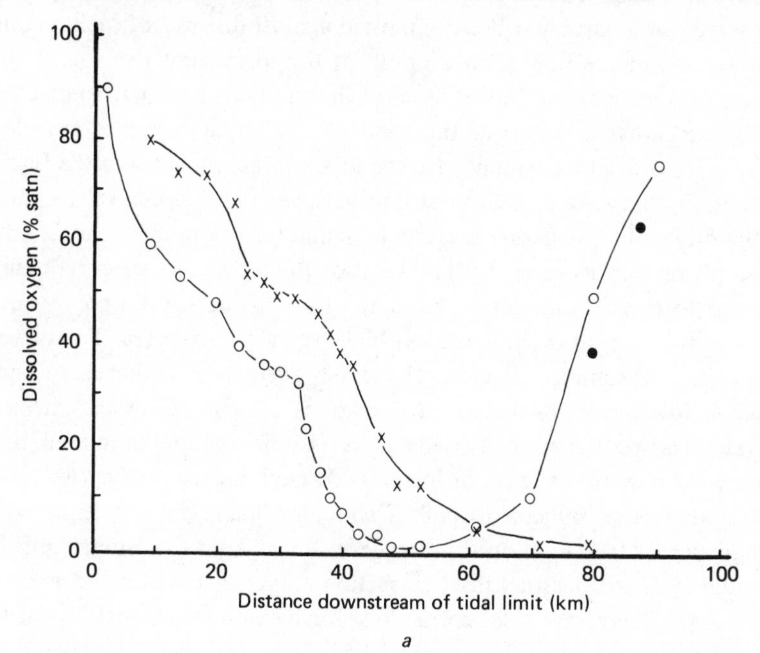

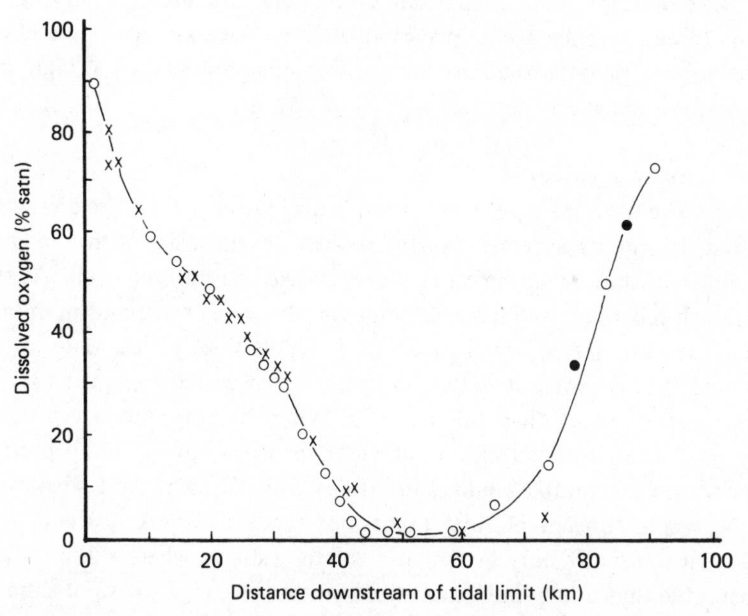

these segments is calculated at a given tidal state by multiplying the cross sectional area by the distance between the cross-section and one immediately upstream. By summing the volumes of each segment and those upstream of it, it is possible to arrive at the cumulative volume of the estuary upstream of each section. If these cumulative volumes at high and low water for a particular tide are plotted against distance of each section downstream of the tidal limit a graph of the form shown in Fig. 7.2 is obtained. A logarithmic scale is used for the cumulative volume so that the small cumulative volumes at the head of the estuary are not crowded together in order to accommodate the much larger volumes in the lower reaches. By drawing a horizontal line between the two curves the tidal excursion for that particular tide can be found for any point in the estuary except those very close to the tidal limit or the mouth. These curves may be used to correct samples taken at the time of local low water to the position they would occupy at local high water or vice versa. To correct the position of samples taken at other times during the tidal cycle to high water or low water positions, or to positions occupied at other states of the tide, it is necessary to construct curves or tables relating the cumulative estuary volumes to distances at intervals throughout the tidal cycle.

It is very often difficult to collect enough reliable data on either the bathymetry or the variation in water level throughout the estuary during the tidal cycle to construct tidal correction curves or tables and it is only possible to use average tidal corrections for mean tides or possibly mean spring and neap tides. Such tidal corrections are useful in trying to compare data averaged over a period of time. An example of the effect using volumetric data to adjust the position of samples taken at high and low water is shown in Fig. 7.3.

Salinity method

The movement of water in an estuary over a tidal cycle may be estimated from measurements of the salinity or chlorosity of the water at a number of stations in the estuary over at least half a tidal cycle. As the relationship between salinity and chlorosity becomes more and more uncertain at salinities of less than about 1 (chlorosities of less than about 500 mg l^{-1}) it is probably better, in order to cover the whole estuary, to use chlorosity rather than salinity data. Where salinity data are used, it may be difficult to determine tidal excursions near to the tidal limit as there are no measurable changes in salinity over the tidal cycle. However, as the whole concept of tidal excursions tends to break down in this region, it is usually only in that part of the estuary where salinities are measurable that tidally influenced flows always occur. The use of salinity or chlorosity measurements in estimating tidal movements is dependent on their conservative behaviour during the mixing processes (see p. 312).

To establish the tidal excursion, the chlorosity recorded at each station at the times of local high and low water is plotted against the distance downstream from the tidal limit. Where chlorosity values are used, it is convenient to use a logarithmic scale to accommodate data extending over four orders of magnitude. For salinity data, an arithmetic scale is used as the values will not extend over more than two orders of magnitude. Figure 7.4 shows examples of the chlorosity distributions at high and low water for a spring tide in the Ribble Estuary. As with the curves derived from volumetric data the tidal excursion is determined by the horizontal separation of the two curves.

Fig. 7.4. Distribution of chlorosity in the Ribble Estuary at high and low water, July 1976. Tidal excursions are determined from the horizontal separation of the two curves. ○ – high water samples, × – low water samples.

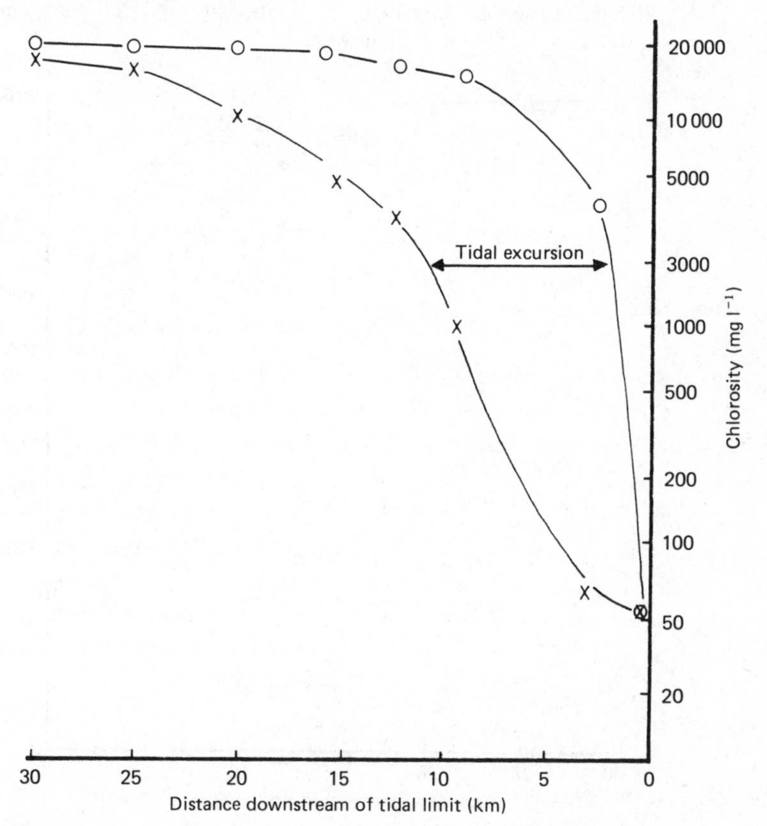

To determine tidal adjustments for samples taken at times other than local, high or low water information about the chlorosity or salinity variations at each station over the tidal cycle is required. This information can then be used for constructing diagrams of the longitudinal distribution of chlorosity or salinity at intervals during the tidal cycle, and the horizontal separation between the curves and that for a particular tidal state used to adjust the position of samples to their probable position at the tidal state required. Figure 7.5 shows the chlorosity distributions for the Ribble Estuary 5 hours and 2 hours before high water as compared with that at high water. These curves have been used to adjust the position of phosphate samples taken 5 hours and 2 hours before high water to their probable position at high water. The tidally adjusted data lie close to those recorded at high water, as shown in Fig. 7.6. Differences between the high water values and those adjusted to high water position are probably a reflection of the precision of the method used for the phosphate estimations.

Fig. 7.5. Distribution of cholorosity in the Ribble Estuary at various states of the tide, July 1976. ○ – high water, ● – 2 h before high water, × – 5 h before high water.

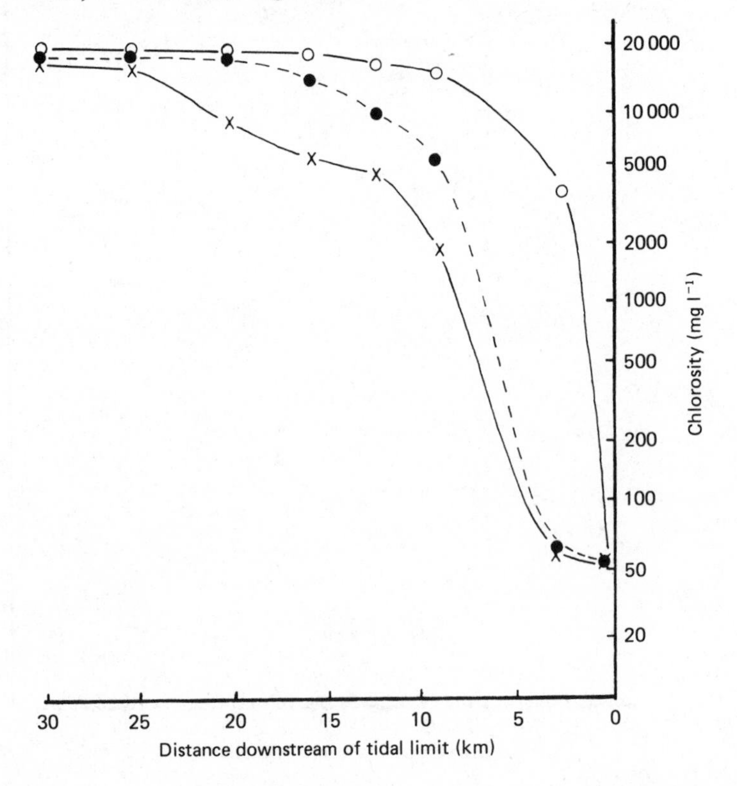

Fig. 7.6. Effect of using chlorosity data to adjust the position of phosphate samples taken from the Ribble Estuary to equivalent high water position *a* samples in original position *b* samples in adjusted position. ○ – high water, ● – 2 h before high water, × – 5 h before high water.

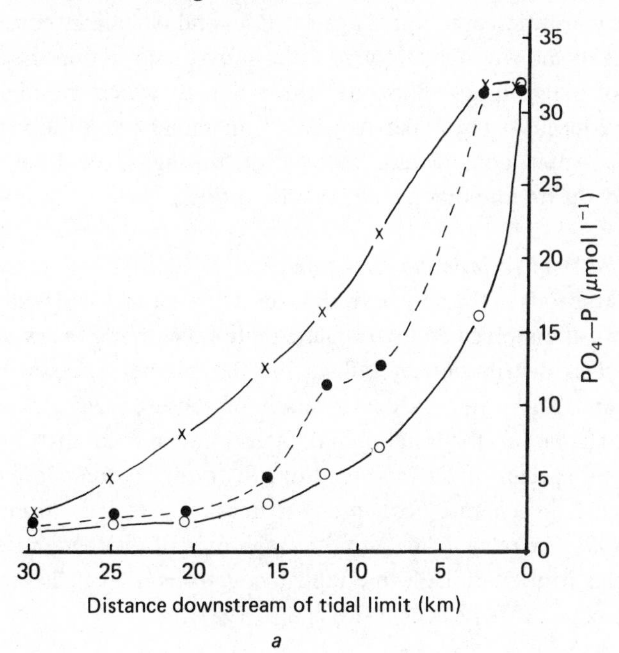

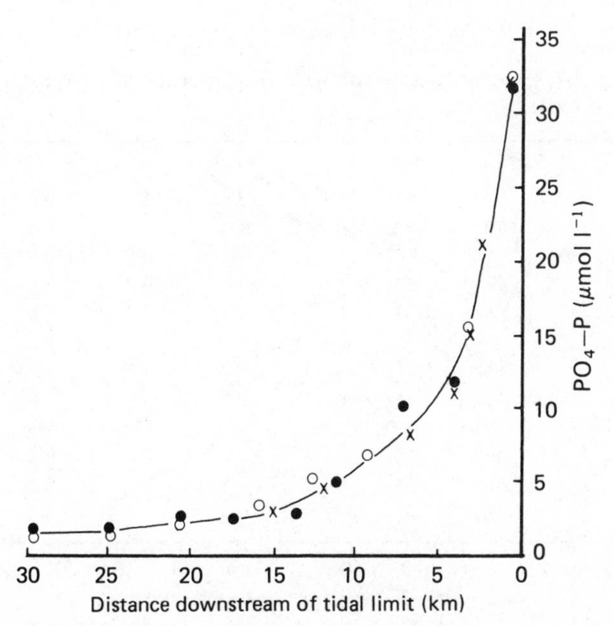

Comprehensive sets of salinity or chlorosity data collected from several stations covering the length of an estuary over a tidal cycle may also be used to construct diagrams illustrating the movement of different water masses for the period studied. As shown in Fig. 7.7 the salinity or chlorosity figures at each station are plotted against time and isohalines constructed. In addition to showing the extent of water movements within the estuary, this type of diagram also illustrates the extent to which the high water period is reduced in the upper reaches of an estuary and the very large changes in water composition that occur during short time periods particularly during the first part of the flood tide.

Effects of variation in river flow

Variations in the flow of fresh water to an estuary will tend to alter the amounts of dissolved and particulate material entering the estuary and also alter their distribution by influencing the mixing processes between the fresh and salt water. A combination of these effects can result in significant changes in the longitudinal, lateral and vertical distribution of any chemical species under investigation. In trying to compare chemical data collected on separate occasions it is important to bear in mind that differences may be related solely to the magnitude of the fresh-water input although this is often difficult to establish conclusively (see Chapter 1 and also Boyle *et al.*, 1974; Loder & Reichard, 1981).

Fig. 7.7. Chlorosity distribution in the Wyre Estuary for a spring tidal cycle.

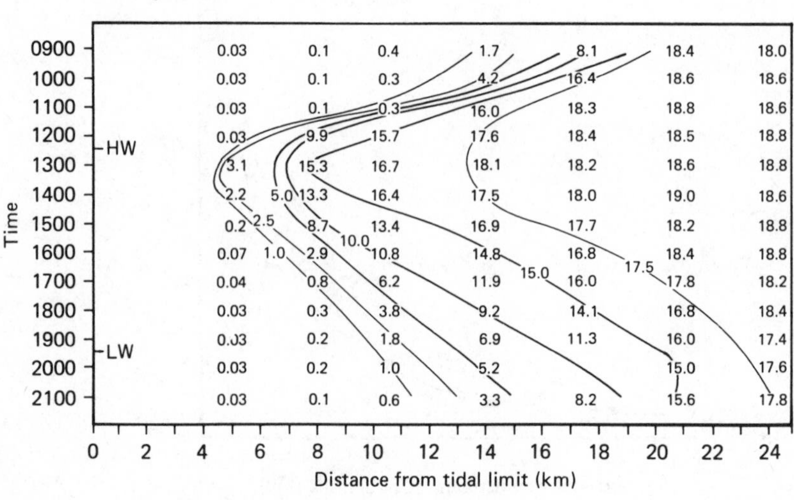

Effect of concentrations in river water

The relationship between the amount of river water discharged to an estuary and the amounts in both dissolved and particulate material carried by it, is complex and is related to the chemical behaviour of the particular species and the nature of the catchment area of the river. Figure 7.8 shows the relationship between the concentration of NO_3–N and PO_4–P and the river flow at the time of sampling for the River Mersey at its tidal limit measured over three years. From this it can be seen that the concentration of PO_4–P tends to decrease with increasing flow whereas that for NO_3–N tends to increase although there is considerable scatter for both sets of data. Thus during periods of high river flow proportionately more NO_3–N is discharged to the estuary than PO_4–P and this will tend to alter the ratio of the two elements in the waters of the inner estuary unless other mechanisms exist to remove the excess NO_3–N or supply more PO_4–P. In trying to establish the relationship between river flow and concentration of dissolved or particulate material, it is important to bear in mind that the observed concentration may better reflect the magnitude of the river flow for a period prior to the sampling time rather than that at the moment of sampling. Where continuous flow records are available, it is always worth investigating whether the instantaneous flow, the mean daily flow or the mean flow for a number of days prior to sampling are significantly correlated with the observed concentrations (see also pp. 322–324).

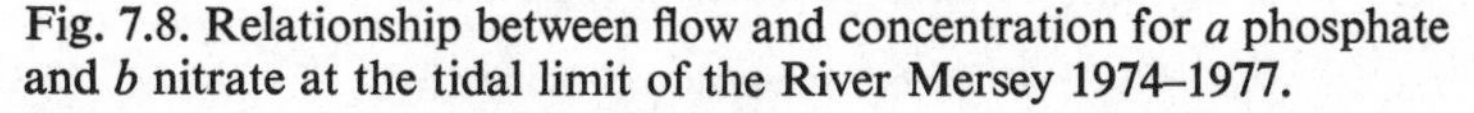

Fig. 7.8. Relationship between flow and concentration for *a* phosphate and *b* nitrate at the tidal limit of the River Mersey 1974–1977.

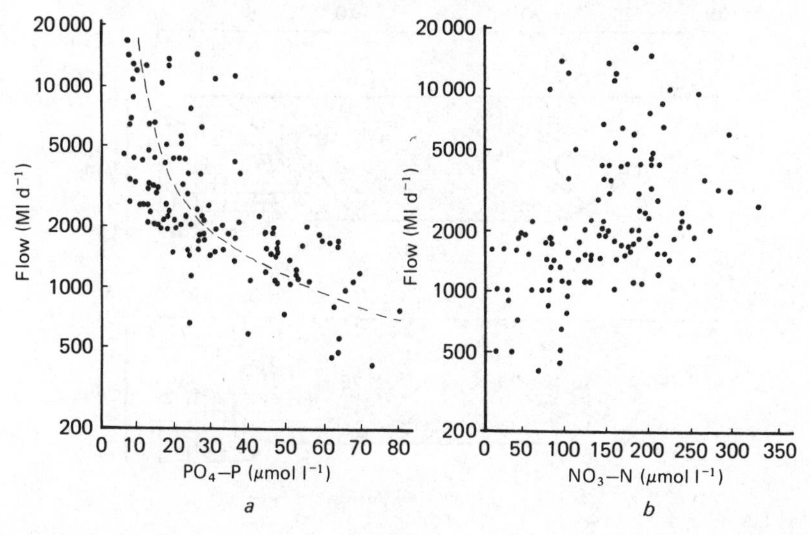

Effect of mixing processes and residence times

The distribution of fresh and salt water in an estuary is the result of the combined effects of river flow and tidal currents. In situations where river flow is dominant, vertically inhomogeneous estuaries will result with the river water forming an upper layer flowing seawards over a wedge of salt water which advances and retreats with the tidal rise and fall of the coastal water. As the strength of the tidal currents increase, relative to the river flow, more and more mixing occurs between these two layers of water until, in estuaries where tidal currents are dominant, almost complete mixing occurs and vertical and lateral variations become very small. In most estuaries, both river flow and tidal currents fluctuate considerably, the former usually irregularly the latter regularly in phase with the spring tide–neap tide cycle, and this can result in a change in the degree of stratification. In addition, the interaction between tidal currents and river flow can alter the geographical position of the salt water intrusion. Figure 7.9 illustrates the effects of different river flows and tidal ranges on the salinity structure of the Tyne Estuary.

In addition to altering the geographical location of the main mixing zone between the fresh and saline water, variations in fresh-water flow tend to alter the rate of transport of material through the estuary. A useful indication of the rate at which material is transported through an estuary

Fig. 7.9. Effect of river flow on the salinity structure of the River Tyne Estuary. *a* Winter conditions, fresh-water flow 72 $m^3 s^{-1}$. *b* Summer conditions, fresh-water flow 5 $m^3 s^{-1}$.

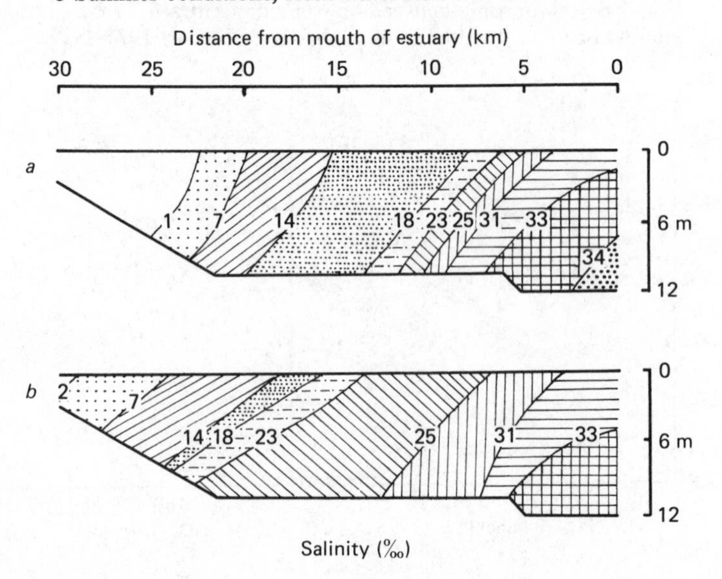

is the flushing time or time required to replace the existing fresh water at a rate equal to the river discharge. Flushing times can be calculated from a knowledge of the fresh-water flow into, and its distribution within, an estuary. Details of the methods available are given by Dyer (1973, 1981) and Bowden (1967, 1980). Flushing times decrease with increasing river flow but the relationship is not linear as the flushing time changes proportionately less with increasing river discharge at times of high flow than during periods of low flow as is shown in Fig. 7.10. In addition to calculating the flushing time for the whole estuary, it is possible to determine individual flushing times for various parts of it. In general, shorter flushing times are found for equivalent lengths of the upper reaches of an estuary than for the middle and lower reaches. Figure 7.11 shows the variation in the individual flushing times for 2 km segments of the Mersey Estuary and the cumulative flushing time for the estuary as a whole calculated from long-term mean salinity distributions and fresh-water flows. The short residence time of water in the inner estuary is evident. Variations in the flushing time of an estuary will influence the extent to which physical, chemical and biological processes can go towards equi-

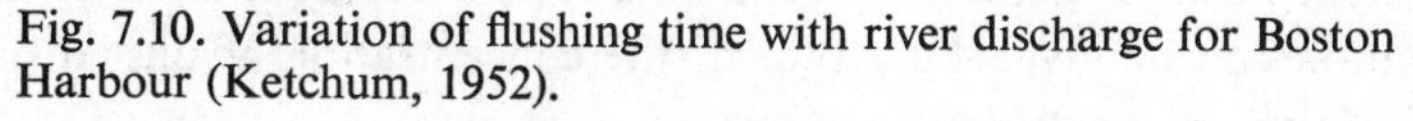

Fig. 7.10. Variation of flushing time with river discharge for Boston Harbour (Ketchum, 1952).

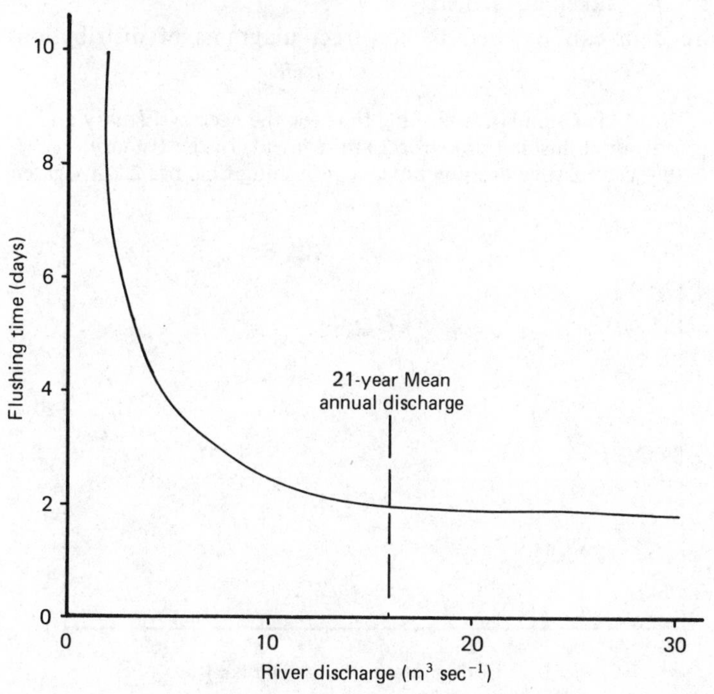

librium before the water is lost from the estuary. With the shorter flushing times for the inner reaches of an estuary, it is only the faster reactions which can alter the composition of the water to a greater extent than physical mixing processes.

Distributions of chemical constituents in estuaries

For most chemical surveys of estuaries, data are collected from a number of stations and depths within the estuary; usually at different states of the tide. As a first step in examining such data, it is useful to construct diagrams showing the distributions of the various constituents in relation to distance along or across the estuary and, where significant differences occur, in relation to depth. The latter types of diagrams are variously called longitudinal and transverse sections or longitudinal and transverse profiles. On occasions, the words 'section' and 'profile' are used alone to signify transverse and longitudinal representations respectively but this convention is by no means universally accepted.

From the values for the individual data points it is usually possible to construct isopleths indicating the distribution of water masses with a particular concentration of the constituent in question. Where vertical and lateral variations are both small, an adequate representation of the distribution can be given by a plot of the concentration against distance for single samples taken at each station.

Before data can be used to construct diagrams of distributions in

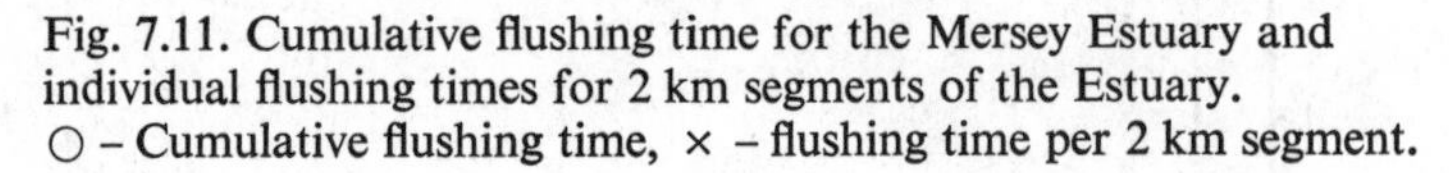
Fig. 7.11. Cumulative flushing time for the Mersey Estuary and individual flushing times for 2 km segments of the Estuary. ○ – Cumulative flushing time, × – flushing time per 2 km segment.

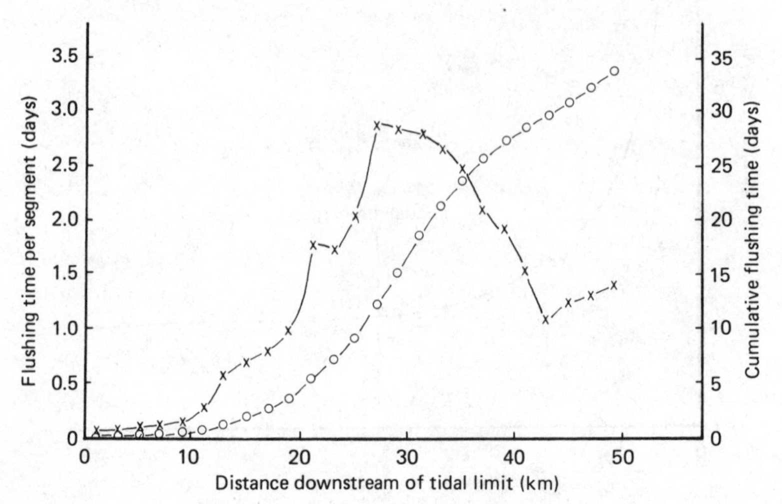

estuaries, a knowledge of the general hydrography of the estuary, or section of it, is required in order to assess the need for applying tidal corrections to data collected at different times and locations. Where the tidal excursion is large in relation to the length of the estuary it will usually be necessary to apply corrections to data not collected at very nearly the same state of tide, whereas in areas where the tidal movements are small or relatively small compared to the size of the estuary, tidal corrections may not be necessary at all.

Areal distributions

A good example of the use of areal distributions to examine chemical data from an estuary is given by Abdullah *et al.* (1973) for their investigation of the Bristol Channel. Figure 7.12 shows the distribution of temperature, salinity and various chemical constituents for 44 stations sampled in April 1971.

For the salinity data, corrections were made to salinity values at local low water. No corrections were applied to the other data. The corrected salinity data and uncorrected nutrient data were then plotted on a chart of the estuary at a position corresponding to each sampling station and isopleths at suitable intervals drawn through points with the same concentration. For this outer region of the Bristol Channel it is probable that tidal movements are small relative to the distance and time between the stations so that it is possible to compare the corrected and uncorrected data.

The influence of the fresh-water inflow to the area is shown by the closely spaced isohalines in the inner Channel which in the area upstream of about 3° 30′ W are more or less parallel and normal to the coast. Further downstream the isohalines take up a NW–SE orientation with a slight bulge in the region of the deep channel. The closer spacing of the isohalines along the southern shore of the Channel suggests an intrusion of more saline water from the Irish Sea with the main outflow of lower salinity water along the northern side of the Channel.

The distributions of the nutrient salts complement that of salinity and suggest a similar situation although the greater influence of fresh-water inputs to the most downstream section of the northern shore are suggested by the nitrate data. Water entering this section would appear to have a higher nitrate concentration and its influence can be seen as a plume of water with a nitrate concentration greater than 20 μmol l^{-1} extending southwards towards the middle of the channel. This plume of high nitrate water is also indicated by its high nitrate to phosphate ratio as is also shown in Fig. 7.12*f*. From an investigation of the nitrate to salinity

Fig. 7.12. Distributions of temperature *a*, salinity *b*, nitrate *c*, silicate *d* and total inorganic phosphorus *e* in the Bristol Channel, April 1971 along with the nitrogen to phosphorus ratio (by atoms) *f* and the relationship between nitrate and salinity *g* (Abdullah *et al.*, 1973).

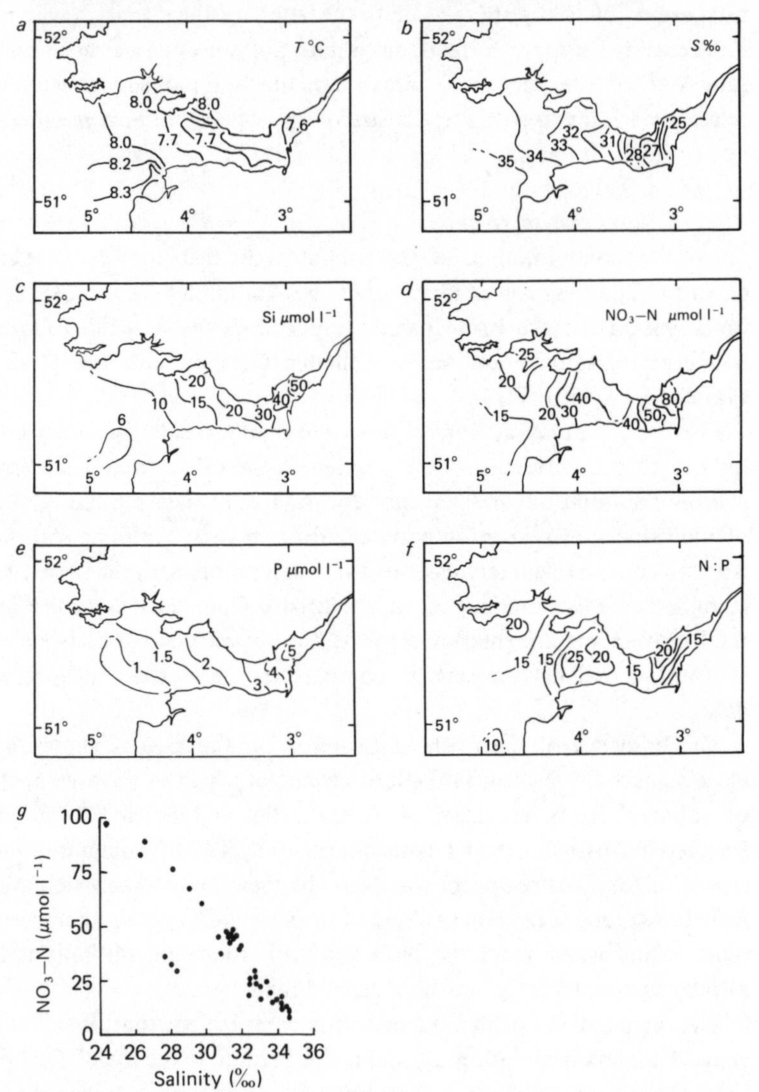

relationship Fig. 7.12*g* (as is discussed in detail on p. 318) this high nitrate water is difficult to detect but the existence of an input to the inner channel which is low in nitrate is suggested by the three data points with salinities of between 28 and 29‰ and nitrate concentrations of between 30 and 40 μmol l^{-1}. These three points lie well away from the rest of the set and show up as a band of low nitrate water bounded by the 40 μmol l^{-1} contour both upstream and downstream.

Longitudinal profiles

Longitudinal profiles or longitudinal sections as they are sometimes called are probably the most common format for the initial examination of data from estuarine surveys. As longitudinal and vertical variations in the chemical composition of estuarine waters are usually greater than lateral ones, the longitudinal profile gives a very good indication of the distribution of the different water masses in the estuary. If the form of the profiles derived from data on different constituents differ considerably, much can often be inferred about the reasons for the differences from a knowledge of the geography of the estuary and its inputs.

Where data from more than one depth are available, the individual data points are plotted on a diagram of the estuary which represents variations in depth with distance along the estuary. As such diagrams are not so readily available as maps or charts showing the general features of the estuary, it is usual to construct schematic diagrams from the depth data available. Some examples of longitudinal profiles of chemical data from estuaries are shown in Fig. 7.13 for the Tyne Estuary and Fig. 7.14 for the Clyde Estuary. The distributions for the Tyne, which were obtained before the interceptor sewers were constructed, indicate that even for this grossly polluted estuary the winter distribution of nitrate nitrogen and silicate closely resemble that of salinity, although there appears to be an input of water with a high nitrate and silicate concentration near the head of the estuary but downstream of the main fresh-water inflow. However, the distributions of ammonia and phosphate, which are present in untreated sewage in relatively greater amounts than nitrate and silicate, bear little resemblance to that of salinity. This latter situation probably results from the then discharge of untreated domestic and industrial wastes from about 180 separate outfalls along the banks of the estuary. The salinity profile suggests considerable stratification in the middle reaches of the estuary with low salinity water flowing seawards over the more saline coastal water.

The longitudinal profiles of the Clyde Estuary demonstrate the different types of distributions obtained for the trace metals copper and zinc during

Fig. 7.13. Longitudinal profiles of salinity, nitrate, silicate, ammonium and phosphate for the River Tyne Estuary at high water, March 1971 (Head, 1972).

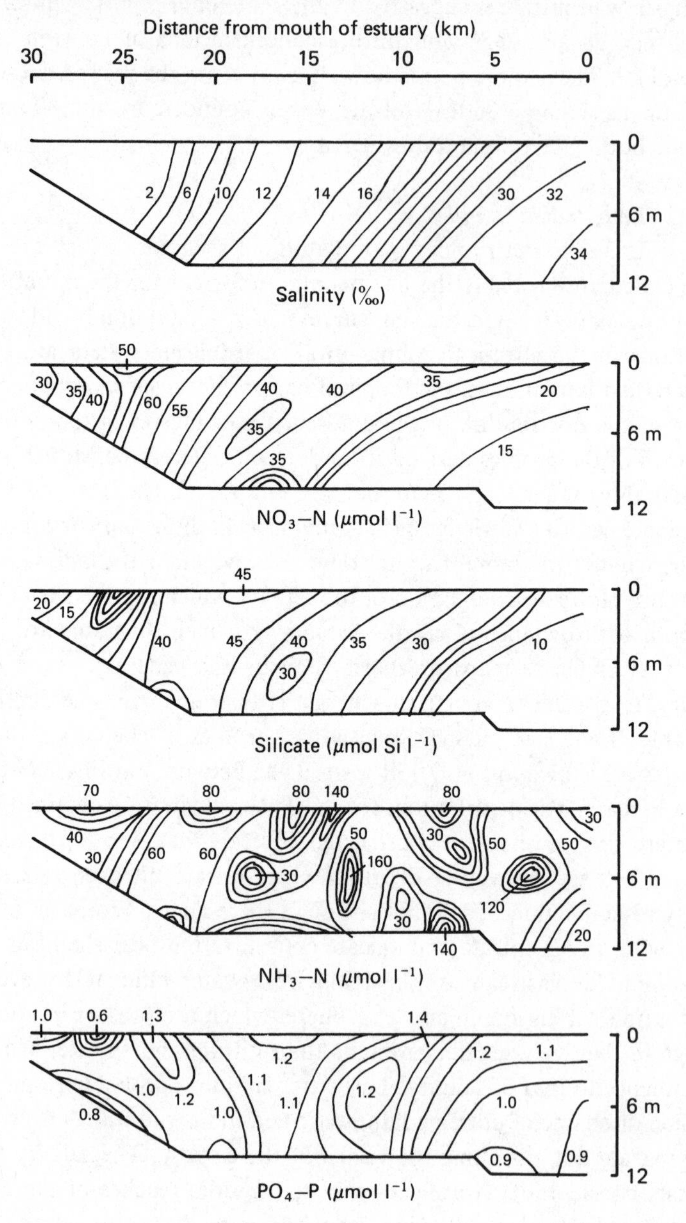

one survey of the estuary. Both show that the distributions are similar to that of salinity in the outer estuary, but that for copper differs markedly in the upper reaches as a result of an input to that part of the estuary. Although the input is to the surface waters, its effect is noticeable over the whole of the water column. Examination of the zinc data in the form of a plot of zinc concentration against salinity (see pp. 312–317) indicates that although the zinc concentration broadly follows that of salinity, concen-

Fig. 7.14. Longitudinal profiles of the concentrations of *a* zinc and *b* copper in the Clyde Estuary, October 1973 (Mackay & Leatherland, 1976).

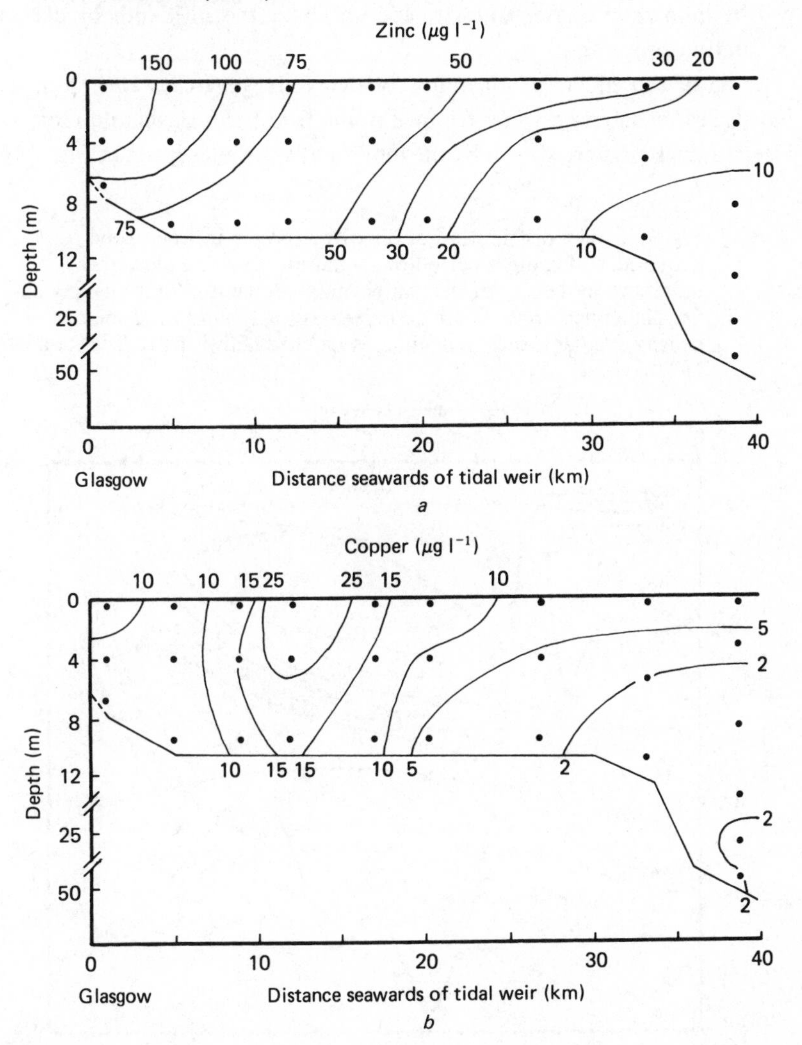

trations in the middle reaches are lower than would be expected from simple dilution.

Longitudinal profiles are particularly useful in studying deep estuaries with one or more basins cut off from each other or from the sea by shallow sills. Under these conditions, as occur in most fjords, vertical stratification of the water column can result in water becoming trapped in the deep basins for considerable periods during which its chemical composition can change radically, usually as a result of a reduction in concentration, or complete removal, of dissolved oxygen. Longitudinal profiles of such estuaries as is shown in Fig. 7.15 illustrates the extent of the trapped water masses and can give an insight into the probability of their being flushed out by inflows of denser water raised up above the sill depth by coastal circulation processes.

Some of the effects of an inflow of dense oxygenated water from the North Sea on stagnant water trapped in the Bornholm Basin and Gotland Deep in the Baltic are shown by the longitudinal profiles given in Fig. 7.16.

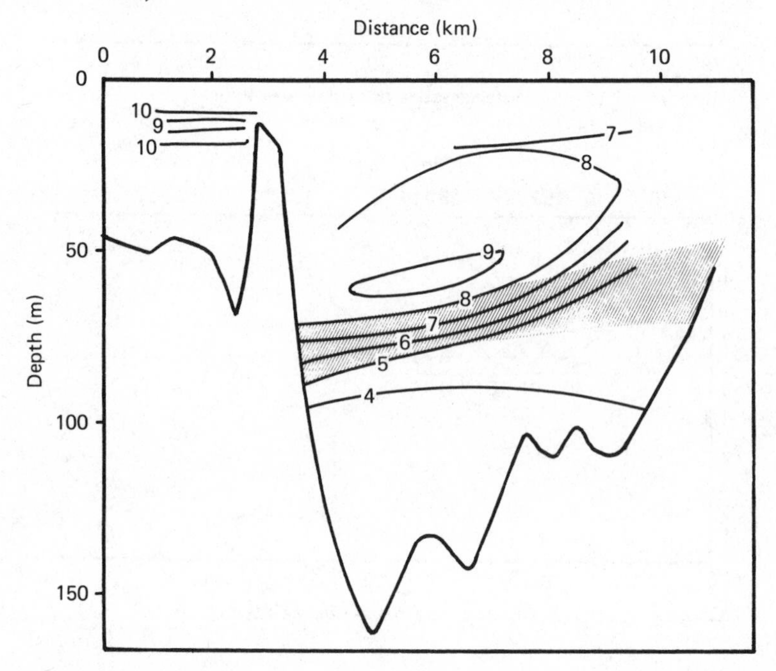

Fig. 7.15. The distribution of dissolved oxygen in Loch Etive, Scotland, following a period of stagnation resulting from the difference in density of two water masses separated by a shallow sill. Oxygen concentrations are expressed as mg l^{-1} and the zone of density gradient, the pycnocline, is hatched (Edwards & Edelsten, 1977).

Although not perhaps strictly an estuary, the Baltic exhibits many features which enable it to be considered as one. The first flow of North Sea water is not dense enough to displace the H_2S-containing water of the Bornholm Basin but flows over this into the Gotland Deep where it peels off the uppermost layers of water containing H_2S and pushes them further towards the upper reaches of the Baltic. Subsequent inflows of water displace all the H_2S-containing waters upwards to mix with oxygenated water where the H_2S is oxidized to SO_4^{2-} or elemental sulphur.

Such dramatic changes do not always occur in estuaries, particularly the more fully mixed ones, but longitudinal profiles can still be of use in interpreting data collected from them. In particular, where mixing is sufficiently strong to allow vertical and lateral variations to be neglected, diagrams of the concentration of dissolved constituents against distance

Fig. 7.16. The flushing of stagnant bottom water from the Bornholm Basin and Gotland Deep in the Baltic during the period September 1968 to January 1970. Oxygen concentrations are expressed as ml l^{-1} (Fonselius, 1970).

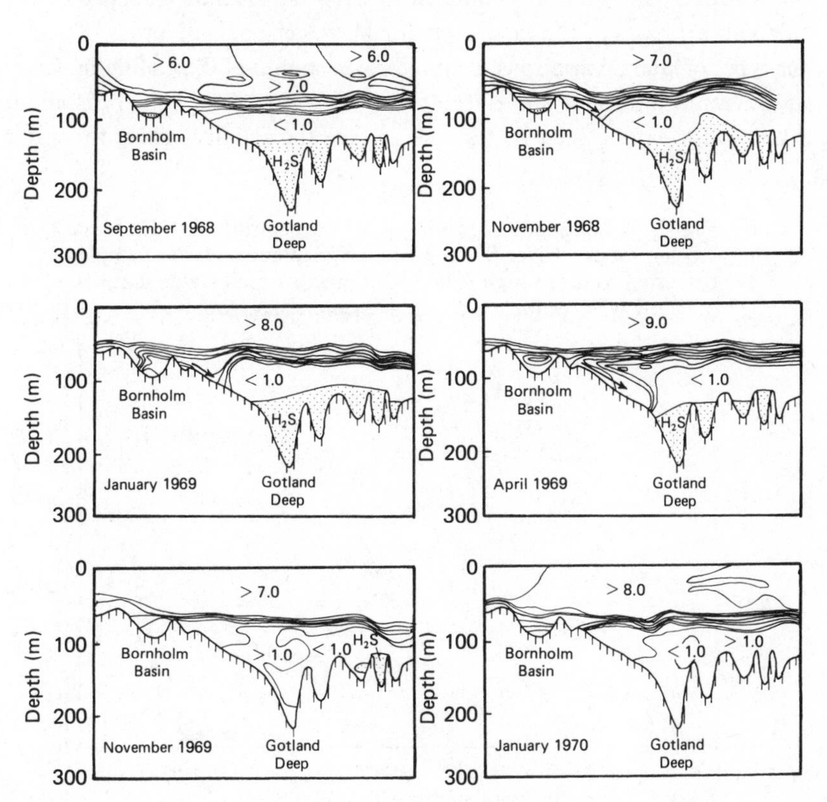

along the estuary can be very useful. The classic example of this technique is provided by the distribution of dissolved oxygen in polluted estuaries. As shown in Fig. 7.17, under conditions where there is a considerable input of material with a high oxygen demand to an estuary, the distribution of dissolved oxygen is reduced to a minimum value in the body of water which is subject to the largest pollutional load for the longest time. For the oxygen distribution, it is convenient to express the concentration in terms of percentage saturation so as to easily compare data from water masses with widely differing temperatures and salinities. If data are available for tidal corrections to be calculated, it is possible to estimate the distance the water with the minimum oxygen content travels up and down the estuary, or the time for which particular points are subject to the low oxygen concentrations.

Transverse profiles

Although lateral variations in the distributions of chemical constituents in estuaries are usually less significant than longitudinal or vertical ones they can be significant in wide estuaries at all states of the tide and for parts of the tidal cycle for narrower ones. Figure 7.18 shows the types of lateral variations recorded at the mouth of the Columbia River at different tidal states and river flows. The width of the estuary is about 4 km and it can be seen that at the surface the salinity values for water

Fig. 7.17. Longitudinal profile of the concentration of dissolved oxygen for the River Mersey Estuary, September 1976. Samples corrected to mean high water position. ○ – high water samples, × – low water samples, ● – intermediate samples.

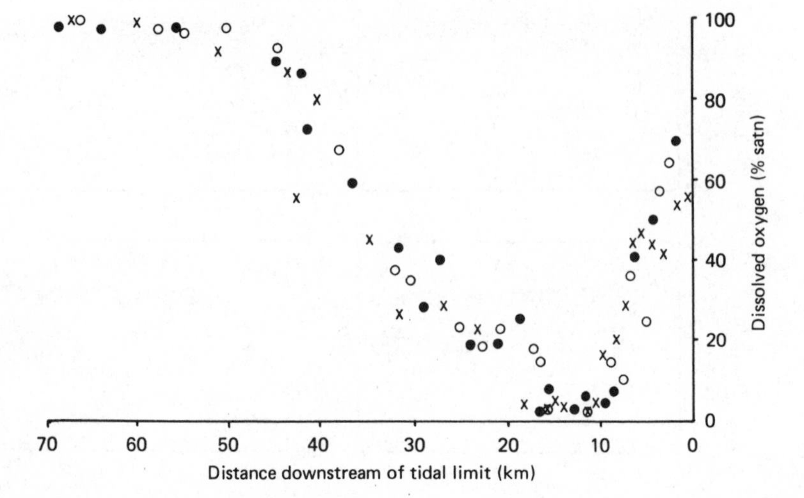

from either side of the estuary can vary by more than 7‰ (about 4000 mg l^{-1} Cl^{-}) or about 20 per cent of the salinity range found in the estuary. It would be expected that such salinity differences would also result in considerable differences in the concentrations of other chemical constituents of the water, although actual data are not available. It is interesting to note that under high discharge conditions the fresher river water is diverted towards the right bank of the estuary, as would be expected for

Fig. 7.18. Lateral variations in the distribution of *a* current velocity and *b* salinity at the mouth of the Columbia River Estuary under different fresh-water flow conditions. Current speed in knots (positive non-tidal current indicates seaward flow) salinity in ‰ (Hansen, 1965). Reproduced with permission of the Marine Technology Society.

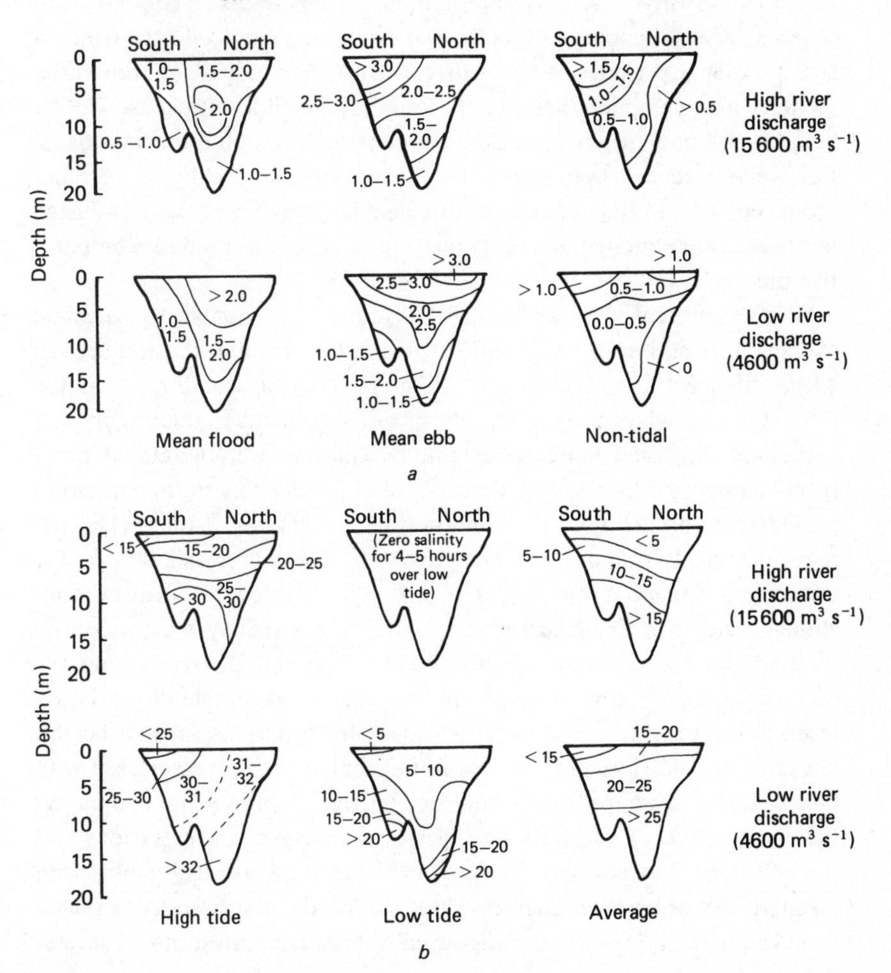

the northern hemisphere, by the Coriolis effect associated with the rotation of the earth. Under low discharge conditions it is suggested that tidal currents associated with the channel shape overcome the Coriolis Force and divert the least saline water along the left bank.

Temporal distributions

The presentation and interpretation of data collected from estuaries over an extended period of time is complicated by the problem of trying to detect and represent any long-term temporal variation in data which usually vary considerably in the short term and also with position in the estuary. It is difficult to obtain an estimate of any single statistic which will usefully summarize the data from an estuary as a whole, although the ratio of a dissolved constituent to a conservative mixing index such as chlorosity or salinity can be useful in this respect (see p. 316). In general, the most convenient method of examining data collected from an estuary over a period of time is to construct longitudinal or transverse. sections and compare these in the light of possible effects caused by variations in tidal height, river flow and general meteorological conditions. For well-mixed estuaries where depth variations are negligible, profiles produced by plotting concentration against distance, tidally adjusted where necessary, along the estuary (see Fig. 7.17) can be used for comparative purposes.

Where sufficient data are available, it may be possible to calculate average distributions for different seasons of the year and thus reduce the effects of short-term fluctuations in meteorological conditions on the distribution of water masses and chemical constituents in the estuary.

Perhaps the best example of examining and interpreting data collected from an estuary over a period of years is that provided by the investigation into changes occurring in the condition of the Thames Estuary (Department of Scientific and Industrial Research, 1964). For this study of a well-mixed estuary it was found that the distribution of chemical constituents, notably dissolved oxygen, could be adequately represented by plotting the concentration against tidally corrected distance along the estuary. Enough historical data were available to investigate the variation in dissolved oxygen concentrations caused by variations in tidal height, temperature and river flow. It was found that the effects associated with tidal height and temperature could be effectively removed by comparing the distributions of quarterly averages of the oxygen concentrations and the effects of flow allowed for by comparing these average values with standard curves for each quarter. These standard curves were constructed from the relationship between dissolved oxygen concentrations at several stations along the estuary and river flow. The appropriate standard curve was selected according to the average river flow for the quarter concerned.

Where variations occur both with depth and season at a particular station in an estuary, these variations can be effectively presented by means of the type of diagram shown in Fig. 7.19 where the individual values are plotted against season and depth and lines of equal concentration constructed. Figure 7.19 illustrates the annual cycle of nitrate for the whole water column at a station near the head of Southampton Water. Over the whole of the period there is considerable variation in the nitrate concentration with depth, probably reflecting the influence of river water with a high nitrate concentration flowing over the more saline estuarine water with a lower nitrate concentration. During the summer period, biological activity removes a considerable amount of nitrate from solution over the whole of the water column.

The comparison of chemical data collected from an estuary over a period of time is difficult without knowing the magnitude of effects attributable to the physical processes which determine the overall mixing and distribution of the water masses within the estuary. Mathematical models which quantify these physical processes may well be very useful in enabling data collected at different times to be compared by providing calculated distributions for a standard set of physical conditions.

Fig. 7.19. Seasonal variations in the concentration of nitrate with depth for a station at the landward end of Southampton Water, 1969–70. Nitrate concentrations expressed as μmol l^{-1} (Phillips, 1972).

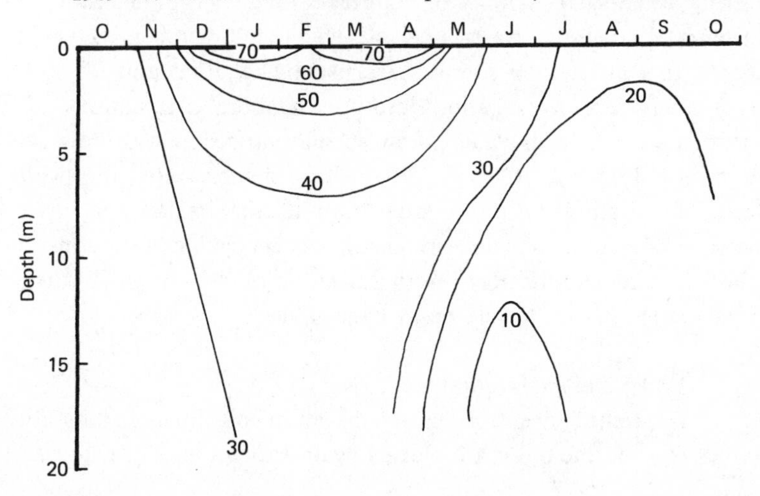

Simple descriptive models

Conservative behaviour and deviations therefrom

As discussed in Chapter 1, many chemical constituents entering an estuary dissolved in the fresh-water or coastal sea-water inflows may be expected to behave conservatively, i.e. their concentration will be directly related to the degree of mixing between the fresh and salt water. One of the most useful techniques available for interpreting data on the chemical composition of estuarine waters is to examine whether the data are consistent with conservative behaviour. This is done by testing the linearity of the relationship between the concentration of the constituent in question and an index of conservative mixing. Deviations from linearity indicate enrichment or depletion of a particular water mass in excess of that to be expected from simple mixing of a two-component system. In applying this technique, salinity or chlorosity are usually used as indices of conservative mixing but other parameters may also be used. When using salinity or chlorosity it must be borne in mind that although, in the region of the estuary where the fresh-water inflow is dominant, neither is a true index of the amount of fresh water present, they may for all practical purposes be used as such. Chlorosity is fundamentally a better index during the early stages of mixing but, when analytical and sampling problems are taken into consideration, the choice of which to use is probably best made on the equipment and personnel available. By examining estuarine data for conservative behaviour, it is usually possible to gain information about possible chemical interaction, biological activity or the existence of additional inputs. The latter information is of great use in pollution studies.

The main problems involved in applying the technique, apart from choosing a conservative mixing index, involve obtaining sufficient samples covering an adequate salinity or chlorosity range and in determining the concentrations of the fresh- and salt-water end-members. Liss (1976) suggests that at least 50 samples are usually required and Boyle *et al.* (1974) advise an investigation into the temporal fluctuations of the constituent in the fresh-water inflow so that variations with flow can be determined. Loder & Reichard (1981) have demonstrated the probable effects of variations in the composition of the fresh-water input on estuarine distributions. In determining the concentration of the constituent in the salt-water end-member it is important to obtain samples of the local coastal water and not to rely on average values.

Testing and interpretation

When the concentrations of a dissolved constituent in a number of samples of estuarine water are plotted against the salinity or chlorosity of the same samples the plot will probably approximate to one of the two distributions illustrated in Fig. 7.20. For conservative mixing, the data points will lie close to the straight line joining the end-members of the mixing series, often called the theoretical or ideal dilution line. Where a constituent behaves non-conservatively, or there is some other perturbation to the system, the data points will not lie on the theoretical dilution line, deviation above or below the line indicating, respectively, addition to or removal from the water. Liss (1976) has reviewed investigations into the behaviour of various dissolved constituents during estuarine mixing.

Examples of the type of results obtained from some of these investigations are given in Fig. 7.21. Where the results lie on or close to the theoretical dilution line and are not biased towards one side or the other it may be assumed that the constituent in question is behaving conservatively or that any addition or removal is being exactly balanced by other inputs or sinks. The latter may be the case in grossly polluted estuaries where there are many individual inputs to the estuary which in addition to contributing quantities of the constituent being studied may alter the chemistry of the water or the sediments to increase or decrease its concentration in the water. Figures 7.21*a* and *c* illustrate situations where conservative mixing appears to occur and Figs. 7.21*b* and *d* situations where processes of removal and addition are occurring within the estuary. Where non-conservative behaviour appears to occur, it is important to investigate whether the system can be adequately represented by the two

Fig. 7.20. Idealized representation of the relationship between concentration of a dissolved component and a conservative index of mixing for an estuary in which there are single sources of river and sea water. *a* For a component (A) whose concentration is greater in sea water than in river water and *b* for a component (B) whose concentration is greater in river water than in sea water (Liss, 1976).

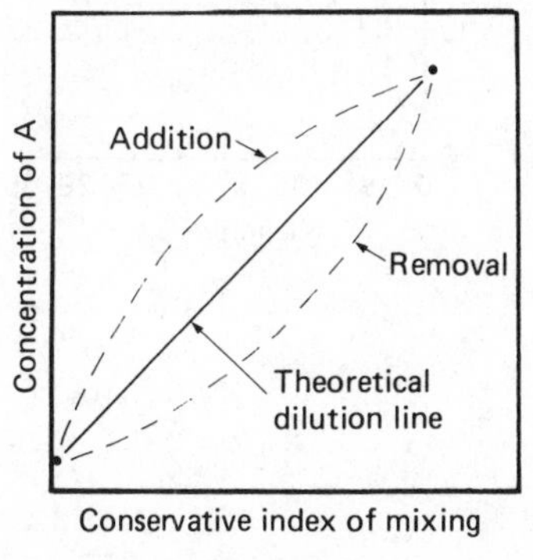

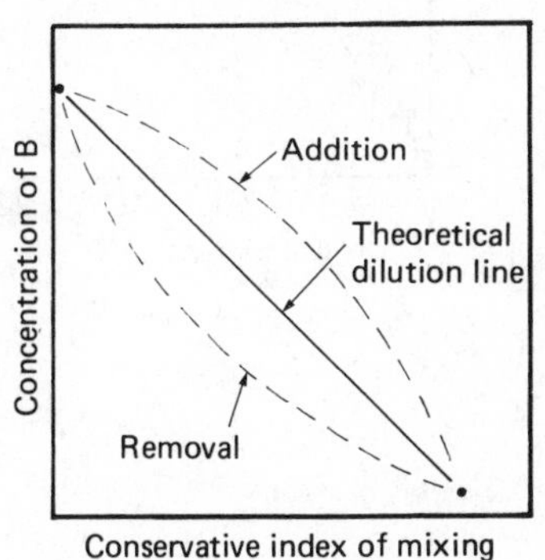

Fig. 7.21. Examples of the types of relationships found between dissolved chemical constituents and a conservative mixing index for estuarine waters: *a* and *c* indicate conservative behaviour, *b* and *d* indicate additions or removals within the estuary (*a* Burton *et al.*, 1970; *b* Holliday & Liss, 1976; *c* Windom, 1971; *d* Liss & Pointon, 1973).

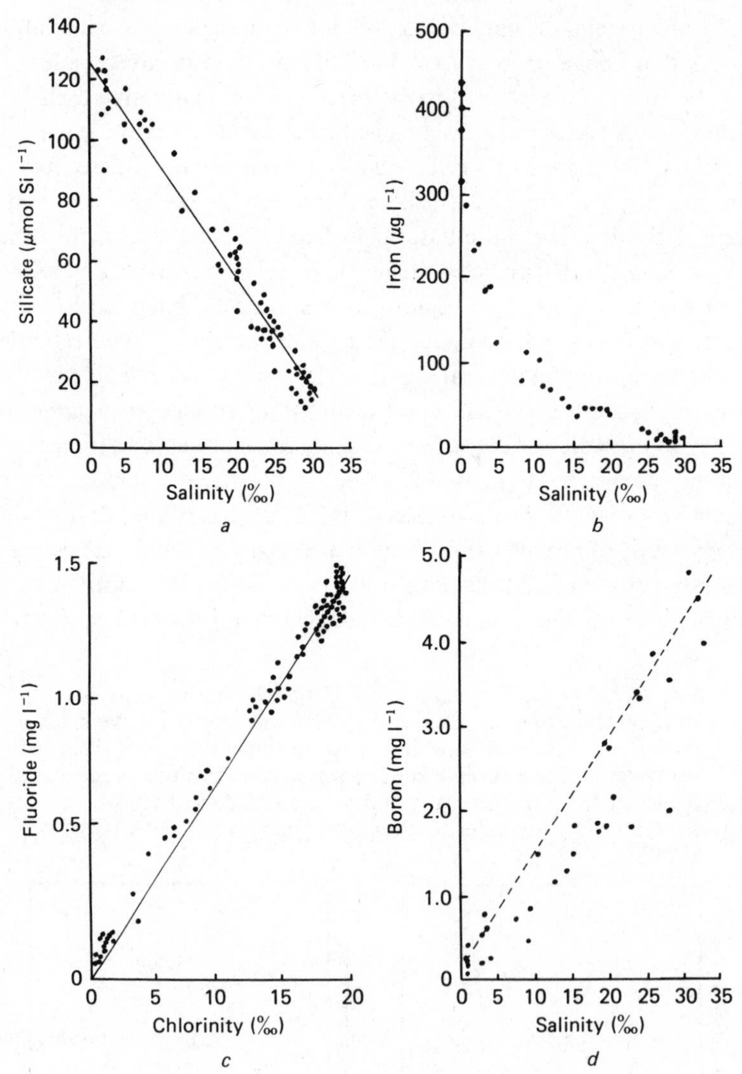

component mixing system, as in most estuaries there are inputs other than the main fresh-water and salt-water ones which may contain significantly different concentrations of the constituent in question. For a possible mathematical approach to quantifying the amounts of a constituent which need to be added or removed to establish non-conservative behaviour, the reader is referred to Boyle *et al.* (1974).

Where information about the variation of both dissolved and particulate species with salinity is available, it is possible to investigate the possibility of conversion from one state to the other. Figure 7.22 illustrates data on the concentrations of dissolved and particulate manganese, and suspended matter in the estuary of the Newport River (Evans *et al.*, 1977). The data for dissolved Mn are above the theoretical dilution line, particularly at low salinities, whereas those for particulate Mn show a maximum at higher salinities. This maximum is particularly apparent when the effect

Fig. 7.22. Variations with salinity of suspended matter and manganese species in the estuary of the Newport River (Postma, 1980). For explanation see text.

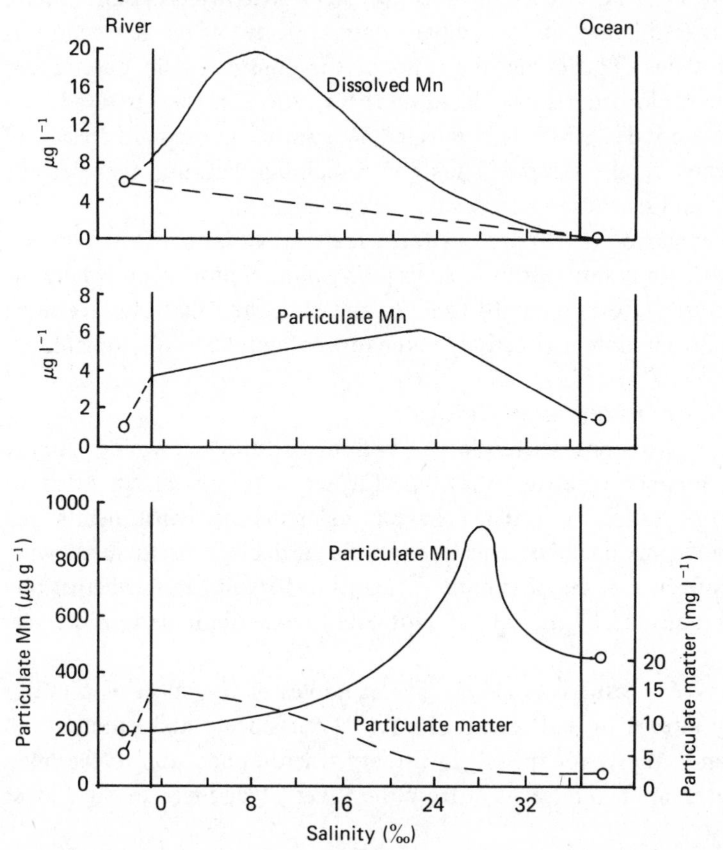

of higher suspended matter concentrations in the inner estuary is allowed for by examining the concentration of Mn per unit weight of the suspended matter. Additional data on the concentration of Mn in the deposited sediments indicated a net deposition of 1–4 mm per year; a concentration of Mn of about 150 $\mu g\ g^{-1}$, which is lower than in suspended matter; and a concentration of dissolved Mn of 25 $\mu g\ l^{-1}$, which is higher than the overlying water.

Taken together, the data suggest the cycling of Mn within the estuary with an input of dissolved Mn^{2+}, probably from the sediments, in the inner estuary and the precipitation of this in the outer estuary either by oxidation to MnO_2 or co-precipitation with iron. Some of the particulate Mn is carried upstream and deposited in sediments where reduction to Mn^{2+} can again occur.

The concept of conservative mixing may also be used as an aid to interpreting estuarine data on occasions when there is insufficient information available to construct reliable salinity distance plots from which to derive tidal corrections. Also it can be used to compare data from a preliminary investigation of an estuary in order to reveal areas which might require further investigation or to compare data collected from an estuary at different times. To do this the ratio of the constituent in question to salinity or chlorosity is calculated and this ratio can then be used as a comparative index which is unaffected by relative amounts of fresh and saline water in the sample. Thus any remaining variation will become apparent and its cause or causes can be investigated.

Head & Burton (1970) demonstrated seasonal variations in the molybdenum to salinity ratio for three stations in Southampton Water where the salinity varied from about 14 to 32‰ and suggested that these resulted primarily from biological activity in the form of uptake by phytoplankton.

Other mixing investigations

In situations where plots of various constituents against salinity indicate non-conservative behaviour, further information can often be obtained by testing for similar behaviour using another constituent as the conservative mixing index. This technique is particularly useful in allowing for the influence of major tributary streams or for indicating whether any removal indicated by the salinity plot could be attributable to biological activity.

Mackay & Leatherland (1976) give examples of the effect of a major tributary stream on the relationships between silicate and salinity, and nitrate and salinity for the Clyde Estuary where in addition to the main fresh-water input to the estuary from the River Clyde there are significant discharges from other rivers. From the silicate salinity plot shown in Fig. 7.23*a*, it can be seen that the data can be best represented by two straight lines which intersect near the point where the River Leven, which has a low silicate concentration, enters the estuary. A similar distribution is found for the nitrate data plotted against salinity but when the concentration of nitrate is plotted against that of silicate (Fig. 7.23*b*) the data points lie very close to the theoretical dilution line for the estuary. Investigations on other occasions indicated that the ratio of nitrate to silicate for the main inputs to the estuary were similar on most occasions, even though they were both taken up by algae during the spring and summer, and that silicate could be used as a conservative mixing index for interpreting the nitrate distributions. On occasions when data points were well below the theoretical dilution line they were associated with water masses or low dissolved oxygen concentrations where bacterial denitrification was occurring.

Data on the Clyde Estuary can also be used to illustrate how data on the distribution of three dissolved constituents, nitrate, phosphate and silicate

Fig. 7.23. Relationships between *a* silicate and salinity and *b* silicate and nitrate for the Clyde Estuary, January 1973. The influence of the River Leven is responsible for the non-linearity of the silicate salinity relationship. The ratios of nitrate to silicate in the fresh-water inputs from the Clyde and the Leven are similar and thus the plot of nitrate against silicate for the whole estuary is linear (Mackay & Leatherland, 1976).

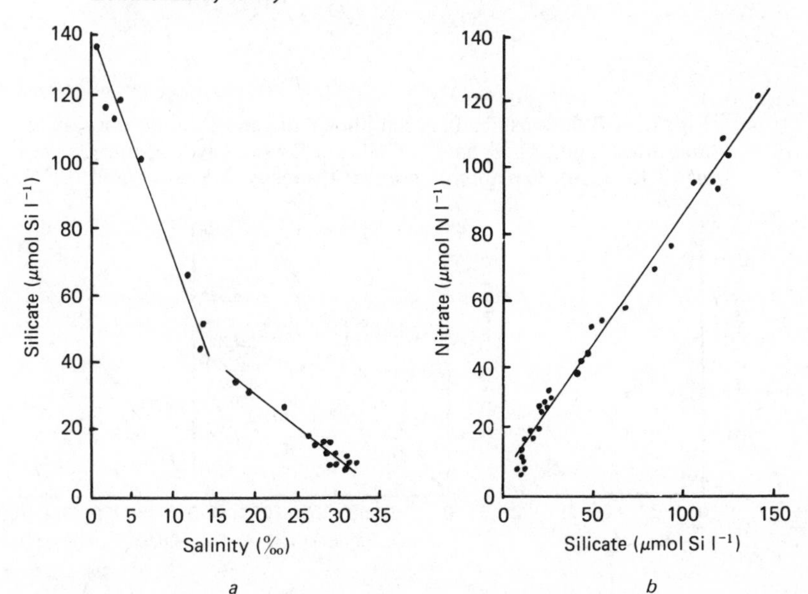

in addition to salinity can be used to elucidate possible causes for the observed distribution of one of them, in this case phosphate.

A survey of the estuary in early spring revealed the phosphate salinity relationship shown in Fig. 7.24*a*, indicating considerable phosphate removal within the estuary. To help to decide whether this removal resulted from physico–chemical processes, biological activity or the inflow of water of markedly low phosphate concentration from a tributary, the data on phosphate concentration were plotted against those for nitrate and silicate (Fig. 7.23*b* and *c*). If the phosphate removal was a biological phenomenon, it would be probable that nitrate would also be taken up and a linear relationship between phosphate and nitrate would result. Figure 7.24*b* shows that, for much of the estuary, the phosphate concentrations were lower than would be expected for biological removal in the ratio indicated by the theoretical dilution line. Figure 7.23*c* indicates that the inflow of water from the River Leven was not responsible for the low phosphate concentrations in the middle salinity range because this inflow has both low phosphate and silicate concentrations and yet the non-linear relationship, between phosphate and silicate again indicates that the phosphate concentrations in the upper and middle parts of the estuary were lower than would be expected from the silicate concentration. Thus on this occasion it is probable that physico–chemical processes were more important than biological processes or tributary inputs in determining the distribution of phosphate within the estuary.

Fig. 7.24. Relationships between phosphate and salinity, phosphate and nitrate, and phosphate and silicate for the Clyde Estuary, April 1973. For explanation see text (Mackay & Leatherland, 1976).

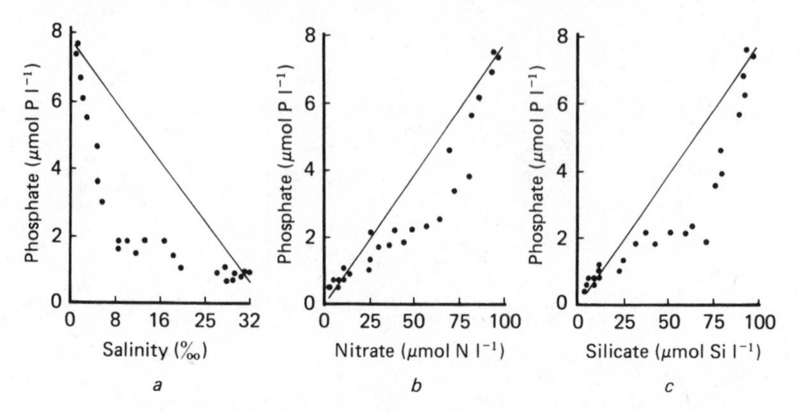

Simple flux calculations

In many estuarine investigations it is useful to be able to obtain an estimate of the flux of a particular constituent through the estuary or part of it. Such an estimate can be obtained from a plot of the constituent against an index of fresh water such as salinity or chlorosity and a knowledge of the fresh-water inflows to the system. It was shown by Stommel (1953) that for a well-mixed estuary the net seaward flux F of a dissolved constituent through any section x is:

$$F_{(x)} = \underset{\text{Advective flux}}{Qc} + \underset{\text{Diffusive flux}}{\left(-K_x A \frac{\mathrm{d}c}{\mathrm{d}x}\right)} \tag{7.1}$$

where Q = fresh water flow
c = concentration of constituent
K_x = longitudinal dispersion coefficient
A = cross-sectional area normal to the flow

The diffusive flux term is negative because material will diffuse from high concentrations to low concentrations.

Officer (1979) showed that it was possible to combine this equation with a similar one describing the salt flux to eliminate the term $K_x A$ and give $F_{(x)}$ in terms of Q, c and s. This can be done because under steady state conditions the net flux of salt $F_{(s)}$ through any section of an estuary is zero. Thus:

$$F_{(s)} = Qs - K_x A \frac{\mathrm{d}s}{\mathrm{d}x} = 0 \tag{7.2}$$

where s = salinity.

By rearranging, equation 7.2 becomes

$$K_x A \frac{\mathrm{d}s}{\mathrm{d}x} = Qs$$

or $$K_x A = Qs \frac{\mathrm{d}x}{\mathrm{d}s}$$

which can be substituted in equation (7.1).

$$F_{(x)} = Qc - Qs \frac{\mathrm{d}x}{\mathrm{d}s} \frac{\mathrm{d}c}{\mathrm{d}x}$$

or $$F_{(x)} = Q\left(c - s\frac{\mathrm{d}c}{\mathrm{d}s}\right)$$

or $$F_{(x)} = Qc_o{}^* \tag{7.3}$$

where $c_o{}^* = (c - s(\mathrm{d}c/\mathrm{d}s))$.

Thus given the fresh-water flow, the salinity gradient and the concentration of the particular constituent at section x, it is possible to calculate the net seaward flux of the constituent. If there are no gains or losses of the constituent during its passage through the estuary, $c_o{}^*$ will correspond to the concentration of the constituent in the fresh-water inflow but if gains or losses occur, a higher or lower value will be obtained. The extent of any gains or losses can be calculated if concentration and salinity data are available for stations near the seaward and landward ends of the estuary. The gain or loss Y will be the difference between the flux in at the landward end and the flux out at the seaward end. Thus:

$$Y = Q(c_o{}^*)_r - Q(c_o{}^*)_s$$

or

$$Y = Q[(c_o{}^*)_r - (c_o{}^*)_s]. \tag{7.4}$$

Where the subscript r indicates measurements at the landward station and the subscript s measurements at the seaward station.

If no measurements of flow are available, an estimate of the fractional gain or loss Z in the estuary may be obtained, as in dividing the gain or loss by the river input the river flow is eliminated. Thus:

$$Z = Q[(c_o{}^*)_r - (c_o{}^*)_s]/Qc_o$$

or

$$Z = [(c_o{}^*)_r - (c_o{}^*)_s]/c_o \tag{7.5}$$

where c_o is the measured concentration of c in the river inflow. If the landward station is situated sufficiently close to the river input, then $s = 0$ and $(c_o{}^*)_r$ becomes c_o and equation 7.5 becomes

$$Z = (c_o - c_o{}^*)/c_o \tag{7.6}$$

In Fig. 7.25, data for the Ribble Estuary are used to assess the flux of ammonium nitrogen through the estuary in relation to the known inputs. The data for chlorosities greater than about 500 mg l^{-1} approximate to a straight line, suggesting that ammonium nitrogen is behaving conservatively and that the flux is thus constant and equivalent to the fresh-water flow multiplied by an apparent fresh-water concentration ($c_o{}^*$) of 138 μmol l^{-1}. This flux of 3100 kg d^{-1} (221×10^3 mol d^{-1}) is considerably greater than the known input of about 1025 kg d^{-1} derived from measurements of the flows and concentrations of ammonium nitrogen in the three main tributaries to the estuary. Additional inputs of about 2075 kg d^{-1} to the estuary are required to explain the flux estimated from the estuarine data. Measurements of the ammonium nitrogen load discharged to the estuary on the day of the survey from sewage works and industrial sources give a total input of 2260 kg d^{-1} which is in good agreement with the estimated deficit of 2075 kg d^{-1}. Thus it would appear that the observed ammonium nitrogen distribution is consistent with the known inputs.

As with any of the techniques involving the assumption of conservative behaviour it is important to bear in mind that changes in magnitude and composition of the inputs can influence the type of distributions found within the estuary and further investigations with different input conditions are necessary to establish the validity of the fluxes calculated.

Regression techniques

Simple regression

The probable geographical distribution of a conservatively behaving substance in an estuary can be estimated from a knowledge of the salinity or chlorosity distribution. Although, in general, the salinity of water in an estuary does not vary linearly with distance for the whole length of the estuary, it is often the case that a linear relationship does exist for a particular state of tide over a considerable portion of the middle reaches. Thus a linear regression equation relating salinity or chlorosity to distance along the estuary can be used to calculate the probable value at any point. From a knowledge of the relationship of any other substance to salinity an estimate of its probable value at any point in the estuary can be made.

Dyer (1973) gives examples of estuaries where the depth mean salinity

Fig. 7.25. Calculation of ammonium nitrogen flux for the Ribble Estuary. For explanation see text.

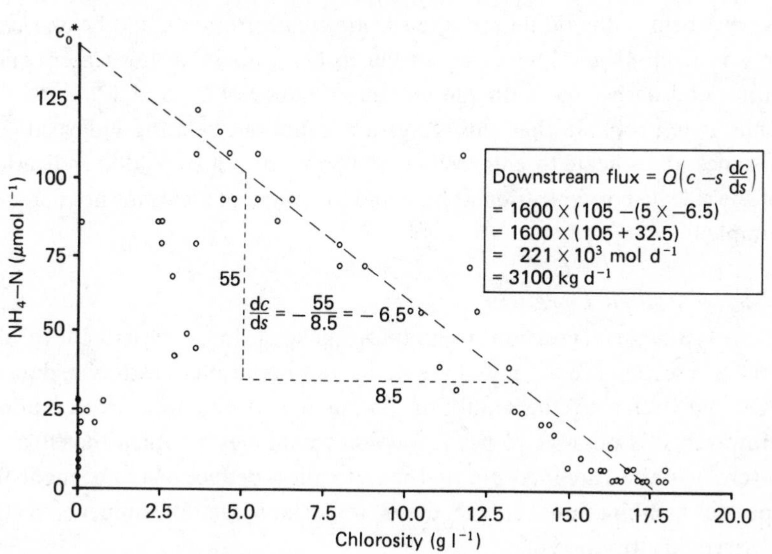

varies linearly with distance for a considerable proportion of their length. In the Mersey Estuary there is usually a good linear relationship between the surface salinity and distance along the estuary for water with a chlorosity of between about 3 and 16 $g\,l^{-1}$. Thus in February 1977 the chlorosity distribution in the estuary at high water for the chlorosity range 2 to 16 $g\,l^{-1}$ could be represented by the relationship:

$$Cl^{-}(g\,l^{-1}) = 0.41x-2.7, \quad r = 0.980, \text{ 13 degrees of freedom} \qquad (7.7)$$

where x = the distance downstream of the tidal limit (km).

At low water a relationship of the following form was found for the chlorosity range 2 to 14 $g\,l^{-1}$:

$$Cl^{-}(g\,l^{-1}) = 0.336x-6.2, \quad r = 0.971, \text{ 13 degrees of freedom} \qquad (7.8)$$

where x = the distance downstream of the tidal limit (km).

Using these relationships, it is possible to calculate the chlorosity at high and low water at any point along the estuary with good precision providing it is in the range 2 to 14 $g\,l^{-1}$. The calculated values for a point 30 km downstream of the tidal limit are 9.6 and 3.8 $g\,l^{-1}$ and may be compared with the observed values of 10.2 and 2.5 $g\,l^{-1}$.

For this survey, a good linear relationship between silicate and chlorosity was found to exist which could be described by the following equation:

$$\text{silicate } (\mu mol\,l^{-1}) = 157.3-7.967\ Cl^{-1}\,(g\,l^{-1}), \quad r = 0.984, \text{ 49 degrees of freedom.} \qquad (7.9)$$

Using this relationship and the calculated chlorosities for a point 30 km downstream of the tidal limit, the silicate concentration could be expected to vary from 81 $\mu mol\,l^{-1}$ at high water to 127 $\mu mol\,l^{-1}$ at low water which is in good agreement with the measured values of 79 and 128 $\mu mol\,l^{-1}$. Thus it is probable that the regression equations relating chlorosity to distance and silicate to chlorosity could be used to give a good indication of the silicate concentration at high and low water at locations not actually sampled.

Multiple regression

Simple regression techniques such as those described in the previous section are of limited use in the interpretation of estuarine data as they only refer to the conditions occurring at the time of sampling. However, it is possible to use regression techniques in which the effect of several variables are incorporated so as to allow deductions to be made for specific combinations of the more important factors influencing the observed distributions.

An exercise of this type has been carried out for the Clyde Estuary with the view to using the regression equations to calculate the effect of various changes to the system. The project was carried out by the Clyde River Purification Board and is described in detail in papers by Mackay (1973), Mackay & Fleming (1969) and Mackay & Gilligan (1972). In this polluted estuary it was important to be able to determine the effects of various sewerage and sewage treatment schemes, and industrial developments on the concentration of dissolved oxygen at various locations along the length of the esuary. It was found that the dissolved oxygen concentration at a number of fixed stations could be well represented by multiple regression equations relating the depth averaged dissolved oxygen concentrations at each station to the fresh-water flow, temperature and tidal range. Mackay & Fleming (1969) demonstrated that the dissolved oxygen values varied with the $\log_{10}$ flow of the upstream rivers entering the system, although the scatter around the regression line was large. Some of this scatter could be removed by using flow data averaged for different periods prior to sampling for different sets of stations in calculating the regressions. It was found that the best agreement occurred when the average flow for the three preceding days was used for the upstream stations and the flow averaged over up to six preceding days for the more seaward stations. However, considerable

Table 7.5. *Predictive equations for dissolved oxygen in Clyde Estuary*

Station	Dissolved oxygen		Constant	Log_{10} flow	Temp	Tidal range	Standard error of estimate
0	y	=	-18.67	$+57.89\,x_1$	$-2.46\,x_2$	$+11.25\,x_3$	11.8
3	y	=	-19.20	$+59.83\,x_1$	$-2.26\,x_2$	$-3.08\,x_3$	11.1
6	y	=	-6.24	$+51.40\,x_1$	$-2.47\,x_2$	$-4.07\,x_3$	11.5
10	y	=	26.64	$+35.77\,x_1$	$-2.72\,x_2$	$-10.11\,x_3$	11.3
13	y	=	18.75	$+38.59\,x_1$	$-2.61\,x_2$	$-9.09\,x_3$	10.5
17	y	=	16.66	$+36.86\,x_1$	$-2.30\,x_2$	$-8.66\,x_3$	10.9
19	y	=	22.55	$+31.31\,x_1$	$-2.45\,x_2$	$-8.79\,x_3$	11.3
22	y	=	42.11	$+25.31\,x_1$	$-1.72\,x_2$	$-12.53\,x_3$	10.5
26	y	=	68.89	$+9.45\,x_1$	$-1.65\,x_2$	$-8.60\,x_3$	11.2
29	y	=	59.45	$+6.36\,x_1$	$-0.72\,x_2$	$-3.12\,x_3$	10.6
32	y	=	106.30	$-10.22\,x_1$	$-0.90\,x_2$	$-4.66\,x_3$	10.8
35	y	=	141.05	$-26.50\,x_1$	$-0.26\,x_2$	$-3.54\,x_3$	13.6
38	y	=	119.99	$-22.45\,x_1$	$+1.17\,x_2$	$-1.05\,x_3$	17.5

Station number refers to distance from tidal weir (km).
y = Dissolved oxygen expressed as percentage saturation.
x_1 = Log_{10} of upstream fresh-water input to estuary ($m^3\,s^{-1}$).
x_2 = Temperature – average of readings at five depths (°C).
x_3 = Tidal range (m).

unexplained variance remained and it was decided to investigate the effect of other factors. From an examination of the data it was apparent that, in addition to fresh-water flow, temperature and tidal range were important in controlling the dissolved oxygen concentrations in the estuary. Thus multiple regression equations were calculated for each of 13 stations in the estuary giving the relationship between the dissolved oxygen concentration and the fresh-water flow, tidal range and temperature. From these equations, given in Table 7.5, it is possible to estimate the concentration of dissolved oxygen to be expected at the fixed stations for any set of conditions within the limits measured during the data gathering exercise, providing there are no significant changes to the major polluting discharges to the estuary.

It was found that this set of equations were extremely useful in investigating the effects of variations in the fresh-water flow and increased heated discharges to the estuary on the dissolved oxygen regime. Also the effects of an improvement in the quality of a major polluting input to the estuary could be quantified by comparing the observed dissolved oxygen concentrations with those calculated for similar flow, temperature and tidal range conditions prior to the improvement. With this information it was possible to estimate the probable effects of other changes to the systems.

Conclusions

In attempting to understand processes causing the observed distributions of chemical constituents in estuaries, much can be achieved using the relatively simple techniques described in this chapter. Most of the techniques are designed to yield representations of the field situation which can be fairly easily compared with what might be expected to happen in an idealized estuary. Where the observed distributions do not agree well with those predicted by simple dilution models, the deviations highlight areas where further investigations are required to determine whether additional sources and sinks are present. Much has been achieved using these techniques in determining the main factors affecting the distribution of various substances in large estuaries much altered by man's activities as well as in small estuaries with limited inputs where the situation might be expected to be similar.

In particular, the concept of conservative behaviour and any deviations from it has formed the basis of attempts to quantify the importance of many processes on the distribution of constituents. This has been illustrated with reference to the Clyde Estuary (see pp. 316–318) where the influence of a tributary stream with a low nitrate concentration must be taken into consideration when interpreting the distribution of nitrate in the estuary. Simpson *et al.* (1975) were able to demonstrate the magnitude of the contribution of phosphorus and silicon to the estuary of the Hudson River from New York City and the surrounding urban areas by comparison with the simple dilution of Hudson River water by coastal sea water. Greisman & Ingram (1977) investigated the St Lawrence Estuary and from salinity measurements were able to calculate the proportion of fresh water at various positions along the length of the estuary (see Chapter 1, p. 26) and by assuming conservative behaviour the distribution of nitrate and nitrite derived from the fresh-water and sea-water sources. In this large estuary slightly less than one-quarter of the nutrients in the surface layer some 400 km downstream of the tidal limit could be attributed to the river input. Comparison of the calculated values with field observations indicated a removal of nitrate and nitrite for most of the estuary probably as a result of phytoplankton production. Yentsch (1975) has used the same concept to estimate the amount of phosphate taken up by biological processes but in this case has extended it to the coastal waters off New England. Tilley & Dawson (1971) have applied the technique to the stratified Duwamish River Estuary and demonstrated that the concentrations of ammonium nitrogen and phosphate were greater in the stratified bottom water than would be expected from the entrainment of surface water of known ammonium and phosphate concentration. In this case they suggest that this is the result of the sinking of phytoplankton from the upper layers and their subsequent decomposition in the bottom water. Mackay & Leatherland (1976) in their study of the Clyde Estuary investigated the amounts of various metals derived from fresh-water and sea-water sources. By calculating the concentration of the fresh-water component at various points in the estuary they were able to extrapolate to the concentration expected in a river input with zero salinity. This was then compared with the measured concentration in the river input and the cause of any differences investigated. In the case of zinc they were able to conclude that about 53 per cent of the river input was lost to the estuarine sediments.

Foster & Foster (1977) combined the concept of conservative behaviour and regression analysis to investigate the distribution of organic material in the Milford Haven Estuary. They found that a plot of ultraviolet (UV) absorbing material in the estuary against salinity indicated a source of this material within the estuary. As there is a large oil terminal and oil refineries situated on the estuary they were interested to try to discover whether the excess organic material might be derived from the oil installations. By examining the relationship between the concentration of UV absorbing material at each station over a tidal cycle and salinity they were able to demonstrate good linear relationship between the two for stations in the middle and inner estuary but only a poor correlation for stations in the outer estuary. This they suggest indicates the dominance of physical mixing processes in determining the distribution of the UV absorbing material in the inner estuary and probably biological or industrial factors in the outer estuary. From previous work (Foster, 1973) it was probable that any organic material originating from the oil installations would exhibit a different UV absorbance spectrum to that of dissolved organic matter produced naturally in the estuary. By examining the absorbance of the material in the estuary at two different portions of the UV spectrum, it was apparent that for all the samples collected from the estuary a linear relationship existed between the absorbances at the two different wavelengths. Thus it could be inferred that any material introduced into the estuary in the region near the oil installations coincidentally had a similar UV absorption spectrum to the naturally occurring organic matter or more likely that contributions from such sources are insignificant relative to the concentrations of organic material present derived from natural sources.

The importance of river flow in determining the residence time of material within an estuary and thus the type of distribution found has been demonstrated by Callaway & Specht (1982) for silicon in the Yaquina Estuary. They point out that residence time must be fully taken into consideration in any discussion of chemical processes in estuaries.

In general, the procedures described in this chapter should enable the reader to process the data gathered from field investigations in a manner which will allow the major sources and sinks for any particular constituent to be identified.

Once this information is available it should be possible to further investigate the processes which result in additions, removals and transformations within the estuarine system.

Acknowledgement

I wish to thank the North West Water Authority for permission to use data collected by various of its officers as part of the Authority's programme for monitoring the effect of discharges to the aquatic environment.

References

Abdullah, M. I., Dunlop, H. M. & Gardner, D. (1973). Chemical and hydrographic observations in Bristol Channel during April and June 1971. *Journal of the Marine Biological Association of the United Kingdom*, **53**, 299–319.

Almgren, T., Dyrssen, D., Elgquist, B. & Johansson, O. (1976). Dissociation of hydrogen sulphide in sea water and comparison of pH scales. *Marine Chemistry*, **4**, 289–97.

Bowden, K. F. (1967). Circulation and diffusion. In *Estuaries*, ed. G. H. Lauff, pp. 15–36. Washington, D.C.: American Association for the Advancement of Science.

Bowden, K. F. (1980). Physical factors: salinity, temperature, circulation, and mixing processes. In *Chemistry and Biogeochemistry of Estuaries*, ed. E. Olausson & I. Cato, pp. 37–70. Chichester: John Wiley & Sons.

Boyle, E., Collier, R., Dengler, A. T., Edmond, J. M., Ng, A. C. & Stallard, R. F. (1974). On the chemical mass-balance of estuaries. *Geochimica et Cosmochimica Acta*, **38**, 1719–28.

Burton, J. D. (1976). Basic properties and processes in estuarine chemistry. In *Estuarine Chemistry*, ed. J. D. Burton & P. S. Liss, pp. 1–36. London: Academic Press.

Burton, J. D. & Liss, P. S. (eds) (1976). *Estuarine Chemistry*. London: Academic Press.

Burton, J. D., Liss, P. S. & Venugopalan, V. K. (1970). The behaviour of dissolved silicon during estuarine mixing. 1. Investigations in Southampton Water. *Journal du Conseil Permanent International pour l'Exploration de la Mer*, **33**, 134–40.

Callaway, R. J. & Specht, D. T. (1982). Dissolved silicon in the Yaquina Estuary, Oregon. *Estuarine, Coastal and Shelf Science*, **15**, 561–7.

Cooper, L. H. N. (1933). A system of rational units for reporting nutrient salts in sea water. *Journal du Conseil Permanent International pour l'Exploration de la Mer*, **8**, 331–4.

Cox, R. A. (1965). The physical properties of sea water. In *Chemical Oceanography*, ed. J. P. Riley & G. Skirrow, vol. 1, pp. 73–120. London: Academic Press.

Department of Scientific and Industrial Research. (1964). Effects of polluting discharges on the Thames Estuary. *Water Pollution Research Technical Paper*, No. 11. London: Her Majesty's Stationery Office.

Dyer, K. R. (1973). *Estuaries: A Physical Introduction*. London: John Wiley & Sons.

Dyer, K. R. (1981). The measurement of fluxes and flushing times in estuaries. In *River Inputs to Ocean Systems*, ed. J.-M. Martin, J. D. Burton & D. Eisma, pp. 67–76. Paris: UNEP/UNESCO.

Edwards, A. & Edelsten, D. J. (1977). Deep water renewal of Loch Etive: a three basin Scottish fjord. *Estuarine and Coastal Marine Science*, **5**, 575–95.

Evans, D. W., Cutshall, N. H., Cross, F. A. & Wolfe, D. A. (1977). Manganese cycling in the Newport River estuary, North Carolina. *Estuarine and Coastal Marine Science*, **5**, 71–80.

Foster, P. (1973). Ultraviolet absorption/salinity correlation as an index of pollution in inshore sea waters. *New Zealand Journal of Marine and Freshwater Research*, **7**, 369–79.

Foster, P. & Foster, C. M. (1977). Ultraviolet absorption characteristics of water in an industrialized estuary. *Water Research*, **11**, 351–4.

Fonselius, S. H. (1970). On the stagnation and recent turnover of the waters of the Baltic. *Tellus*, **22**, 533–44.

Gieskes, J. M. (1982). The practical salinity scale 1978: a reply to comments by T. R. Parsons. *Limnology and Oceanography*, **27**, 387–8.

Greisman, P. & Ingram, G. (1977). Nutrient distribution in the St Lawrence Estuary. *Journal of the Fisheries Research Board of Canada*, **34**, 2117–23.

Grasshoff, K., Ehrhardt, M. & Kremling, K. (eds) (1983). *Methods of Seawater Analysis*, 2nd edn. Weinheim: Verlag Chemie.

Hansen, D. V. (1965). Currents and mixing in the Columbia River Estuary. In *Ocean Science and Ocean Engineering*. Transactions of a Joint Conference. Washington, D.C.: Marine Technology Society and American Society of Limnology and Oceanography.

Head, P. C. (1972). Nutrient studies. In *Pollution of the River Tyne Estuary*, ed. A. James, pp. 13–17. Bulletin of the Civil Engineering Department, University of Newcastle upon Tyne, No. 42.

Head, P. C. & Burton, J. D. (1970). Molybdenum in some ocean and estuarine waters. *Journal of the Marine Biological Association of the United Kingdom* **50**, 439–48.

Holliday, L. M. & Liss, P. S. (1976). The behaviour of dissolved iron, manganese and zinc in the Beaulieu Estuary. *Estuarine and Coastal Marine Science*, **4**, 349–53.

Ketchum, B. H. (1952). Circulation in estuaries. *Proceedings of* 3rd *Conference on Coastal Engineering*, pp. 65–76.

Kester, D. R. (1975). Dissolved gases other than CO_2. In *Chemical Oceanography*, ed. J. P. Riley & G. Skirrow, 2nd edn., vol. 1, pp. 497–556. London: Academic Press.

Liss, P. S. (1976). Conservative and non-conservative behaviour of dissolved constituents during estuarine mixing. In *Estuarine Chemistry*, ed. J. D. Burton & P. S. Liss, pp. 93–130. London: Academic Press.

Liss, P. S. & Pointon, M. J. (1973). Removal of dissolved boron and silicon during estuarine mixing of sea water and river waters. *Geochimica et Cosmochimica Acta*, **37**, 1493–8.

Loder, T. C. & Reichard, R. P. (1981). The dynamics of conservative mixing in estuaries. *Estuaries*, **4**, 64–9.

Mackay, D. W. (1973). A model for levels of dissolved oxygen in the Clyde Estuary. In *Mathematical and Hydraulic Modelling of Estuarine Pollution*, ed. A. L. H. Gameson, Water Pollution Research Technical Paper, No. 13, pp. 85–94. London: Her Majesty's Stationery Office.

Mackay, D. W. & Fleming, G. (1969). Correlation of dissolved oxygen levels, freshwater flows and temperatures in a polluted estuary. *Water Research*, **3**, 121–8.

Mackay, D. W. & Gilligan, J. (1972). The relative importance of freshwater input, temperature and tidal range in determining levels of dissolved oxygen in a polluted estuary. *Water Research*, **6**, 183–90.

Mackay, D. W. & Leatherland, T. M. (1976). Chemical processes in an estuary receiving major inputs of industrial and domestic wastes. In *Estuarine Chemistry*, ed. J. D. Burton & P. S. Liss, pp. 185–218. London: Academic Press.

Morris, A. W. (ed.) (1983). *Practical Procedures for Estuarine Studies.* Swindon: Natural Environment Research Council.

Officer, C. B. (1979). Discussion of the behaviour of nonconservative constituents in estuaries. *Estuarine and Coastal Marine Science*, **9**, 91–4.

Olausson, E. & I. Cato. (eds) (1980). *Chemistry and Biogeochemistry of Estuaries.* Chichester: John Wiley & Sons.

Parsons, T. R. (1982). The new physical definition of salinity: biologists beware. *Limnology and Oceanography*, **27**, 384–5.

Phillips, A. J. (1972). Chemical processes in estuaries. In *The Estuarine Environment*, ed. R. S. K. Barnes & J. Green, pp. 33–50. London: Applied Science.

Postma, H. (1980). Sediment transport and sedimentation. In *Chemistry and Biogeochemistry of Estuaries*, ed. E. Olausson & I. Cato, pp. 153–86. Chichester: John Wiley & Sons.

Riley, J. P. & Chester, R. (1971). *Introduction to Marine Chemistry.* London: Academic Press.

Sharp, J. H. & Culberson, C. H. (1982). The physical definition of salinity: a chemical evaluation. *Limnology and Oceanography*, **27**, 385–7.

Simpson, H. J., Hammond, D. E., Deck, B. L. & Williams, S. C. (1975). Nutrient budgets in the Hudson River Estuary. In *Marine Chemistry in the Coastal Environment*, ed. T. M. Church, ACS Symposium Series, No. 18, pp. 618–35. Washington, D.C.: American Chemical Society.

Stommel, H. (1953). Computation of pollution in a vertically mixed estuary. *Sewage and Industrial Wastes*, **25**, 1065–71.

Strickland, J. D. H. & Parsons, T. R. (1972). A practical handbook of seawater analysis. *Bulletin of the Fisheries Research Board of Canada*, No. 167, 2nd edn.

Tilley, L. H. & Dawson, W. A. (1971). Plant nutrients and the estuary mechanism in the Duwamish River Estuary, Seattle, Washington. *United States Geological Survey Professional Paper* 750C, C185–91.

UNESCO (1981). Background papers and supporting data on the practical salinity scale 1978. *UNESCO Technical Papers in Marine Science*, No. 37. Paris: UNESCO.

Wallace, W. J. (1974). *The Development of the Chlorinity/Salinity Concept in Oceanography.* Elsevier Oceanography Series, No. 7. Amsterdam: Elsevier.

Whitfield, M. (1982). The salt sea – accident or design? *New Scientist*, **94**, 14–17.

Wilson, T. R. S. (1975). Salinity and the major elements of sea water. In

Chemical Oceanography, ed. J. P. Riley & G. Skirrow, 2nd edn., vol. 1, pp. 365–413. London: Academic Press.

Windom, H. L. (1971). Fluoride concentration in coastal and estuarine waters of Georgia. *Limnology and Oceanography*, **16**, 806–10.

Yentsch, C. S. (1975). New England coastal waters – an infinite estuary. In *Marine Chemistry in the Coastal Environment*, ed. T. M. Church. ACS Symposium Series, No. 18, pp. 608–17. Washington, D.C.: American Chemical Society.

Index

The numbers in *italics* refer to the figure on the page indicated.
The suffix 't' refers to the table on the page indicated.